T0275868

Key Concepts in Environmental Chemistry

Grady Hanrahan

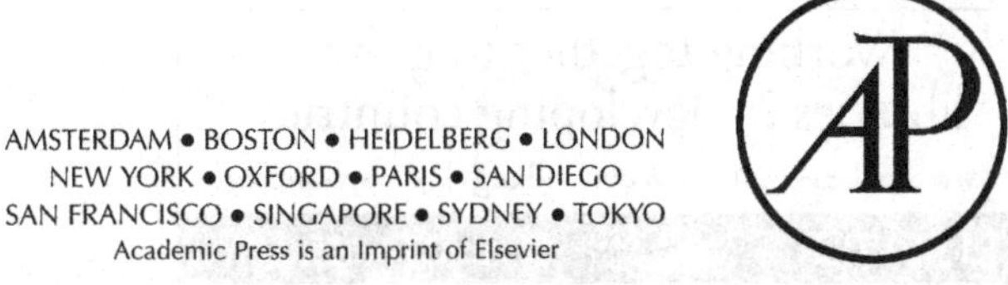

Academic Press is an imprint of Elsevier
225 Wyman Street, Waltham, MA 02451, USA
The Boulevard, Langford Lane, Kidlington, Oxford, OX5 1GB, UK

Notices

Knowledge and best practice in this field are constantly changing. As new research and experience broaden our understanding, changes in research methods, professional practices, or medical treatment may become necessary.

Practitioners and researchers must always rely on their own experience and knowledge in evaluating and using any information, methods, compounds, or experiments described herein. In using such information or methods they should be mindful of their own safety and the safety of others, including parties for whom they have a professional responsibility.

To the fullest extent of the law, neither the Publisher nor the authors, contributors, or editors assume any liability for any injury and/or damage to persons or property as a matter of product liability, negligence or otherwise, or from any use or operation of any methods, products, instructions, or ideas contained in the material herein.

Library of Congress Cataloging-in-Publication Data

Hanrahan, Grady.

Key concepts in environmental chemistry / Grady Hanrahan.

p. cm.

ISBN 978-0-12-810350-0

1. Environmental chemistry. 2. Environmental toxicology. I. Title.

TD193.H36 2012

628–dc23

2011019245

British Library Cataloguing-in-Publication Data

A catalogue record for this book is available from the British Library.

For information on all Academic Press publications visit our website at www.elsevierdirect.com

Printed in the United States of America

11 12 13 14 9 8 7 6 5 4 3 2 1

Contents

PHILOSOPHICAL APPROACH

New instrumental and data analysis capabilities have led to fresh insights into the field of environmental chemistry and questions regarding the future of this rapidly evolving area of scientific study. As teachers and scholars, we have the ability to influence the curious minds of students, simultaneously guiding them through the path of knowledge and truth. And when considering the dynamic nature of environmental systems, varying both spatially and temporally across a wide range of matrices, there is an increasing need for training on tools that can account for complexity, variability, scale, and purpose.

Key Concepts in Environmental Chemistry provides a modern and concise introduction to environmental chemistry principles and the dynamic nature of environmental systems. It offers an intense, one-semester examination of fundamental chemical concepts encountered in this field of study and provides integrated data analysis and research tools to help elucidate complex chemical problems of environmental significance. Principles typically covered in more comprehensive textbooks are well integrated into general chapter topics and application areas. Unique to this book is its integrated approach—a distinctive blend of theoretical and practical material from all aspects of environmental chemistry including statistical analysis and modeling concepts not covered (or limited coverage) in current textbooks. By presenting basic principles together with research-based applications and methods in the field of environmental chemistry, students will fully comprehend the power of such tools in explaining complex chemical problems of environmental importance.

PURPOSE AND ORGANIZATION

The purpose of this textbook is to provide students with an indispensable resource for learning the basic concepts of environmental chemistry from an easy to follow, condensed, application and inquiry-based perspective. Additional statistical, sampling, modeling, and data analysis concepts and exercises are introduced for greater understanding of the underlying processes of complex environmental systems and fundamental chemical principles. Each chapter houses problem-oriented exercises that stress the central concepts covered. Research applications and case studies from experts in the field are directly tied to theoretical background offered in each chapter. This textbook opens with a broad definition and overview of the field of environmental chemistry. It provides detailed understanding of chemical, physical, and biological processes that will shape the foundation of material covered in subsequent chapters. In addition, in-depth examination of the hydrologic cycle will be beneficial in understanding surface and groundwater concepts and the interactions that occur within such systems. Chapter 2 deals with theoretical and applied statistical concepts useful in solving real-world environmental problems. The basic methodology presented—such as sample collection, data distribution, hypothesis testing, analysis of variance (ANOVA), and multivariate analysis—provides necessary background to study and characterize complex chemical data sets and the myriad of physical measurements accompanying extensive environmental analyses.

Chapters 3 and 4 examine the chemical processes affecting the distribution of chemical compounds in natural waters. Chapter 3 begins by reviewing the chemical composition and physical properties of water, including the role of hydrogen bonding, density, salinity, and temperature in governing aqueous systems. The focus of the chapter then shifts to the study of the carbonate system, oxidation/reduction reactions, acid/base chemistry, adsorption processes, and chemical precipitation. In Chapter 4, detailed understanding of the hydrologic cycle presented in Chapter 1 is beneficial in grasping surface and groundwater concepts and the chemical/physical interactions that occur within such systems. A review of typical concentration units encountered in aqueous chemistry is presented, followed by information on drinking water standards, physico-chemical indicators, and selected environmental regulations relative to chemical, biological, and physical contaminants.

Chapter 5 presents a solid overview of wastewater treatment processes and related technologies including activated carbon, reverse osmosis, and ion exchange processes as applied to real-world situations. The next two chapters examine atmospheric chemical and physical processes and the impact of natural and anthropogenic pollutant releases. Chapter 6 provides an introduction to atmospheric layers, a review of basic photochemistry, and information on transport, transformation, and exchange processes between the biosphere and the atmosphere. Chapter 7 provides a more detailed investigation of atmospheric pollutants including the identification of primary sources and secondary precursor compounds. Particular importance is given to the mechanism of ozone formation and information on heterogeneous atmospheric reactions.

Chapter 8 presents a condensed view of soil chemistry including soil formation, composition, and structural concepts. The role of pH and organic matter is critically examined to help explain the adsorption and transportation of selected contaminants. The final chapters survey environmental toxicology concepts and green chemistry and sustainable chemical processes. Chapter 9 provides an overview of environmental toxicology and hazardous waste characterization including toxicity testing and examination of dose-response relationships. Finally, Chapter 10 presents material related to the burgeoning area of green chemistry. It starts with coverage of the 12 Principles of Green Chemistry and their relation to chemical and industrial processes. Lastly, information on example sustainable chemical and related processes is presented, including emerging environmental technologies and sustainability metrics.

Preface to Students

You are enrolled in a course that has considerable relevance to today's global society—this is true not only for human health and well-being, but also for the fragile ecosystems in which we live, work, and play. When it comes to safe-guarding global natural resources; air, soil, and water quality; human health; and social and political systems, we as educated citizens have the ability to make important decisions that affect these vital considerations. *Key Concepts in Environmental Chemistry* has been written to provide a condensed examination of key environmental concepts and the development of analytical skills. Numerous tools are provided to help you succeed in both. It is essential to recognize that those concepts and skills will help foster learning and provide you with the ability to examine and interpret environmental data as they relate to real-world problems.

I encourage you to master both theoretical and application-based content presented within this textbook. Provided are worked example problems, case studies, and research applications to complement the learning process. As you work through the end of chapter problems, go back and reflect upon information gathered and learned in relevant chapters. Admittedly, this text will not provide you with information on all aspects of environmental chemistry. However, it does present a concise and detailed look at key concepts of environmental chemistry in which you will likely encounter along your academic and work-related career path. You are encouraged to explore the external sources of information recommended in this textbook. Finally, I encourage you to further develop your analytical skills through hands-on, practical-based laboratory exercises that accompany the lecture course. With practice and patience, you will be sure to contribute significantly to furthering the knowledge base of environmental topics for years to come.

Acknowledgments

The author is grateful to many of those who supported this effort. First and foremost, I wish to thank Piyawan and my family for their continued patience and support of my academic pursuits. To Kanjana Patcharaprasertsook for the meticulous illustrations housed in this book. To Professor Greg Santillan for stimulating conversations and helpful comments regarding the Lake Nyos information presented in Chapter 1. I thank all of the reviewers for their valuable comments and constructive criticism during the early stages of this book, and all of the publishers who have granted permission for use of external material housed in this book. I thank Elsevier and Academic Press for believing in the concept of this book. In particular, I would like to acknowledge Linda Versteeg, Anita Koch, Lisa Lamenzo, Emily McCloskey, and the supporting editorial staff. To Vicki Wright and Jennifer Arceo for their help with scanning and formatting of chapter material. Finally, to my colleagues Drs. Krishna Foster, Haco Hoang, Mehdi Jalali-Heravi, Nate Carlson, and Crist Khachikian for their continued research collaboration in the environmental field.

About the Author

Grady Hanrahan received his Ph.D. in Environmental Analytical Chemistry from the University of Plymouth, UK, in 2001. With experience in directing undergraduate and graduate research, he has taught in the fields of Analytical Chemistry and Environmental Chemistry at California State University, Los Angeles (CSULA), and California Lutheran University (CLU). He is currently the John Stauffer Endowed Professor of Analytical Chemistry and Assistant Dean of the College of Arts and Sciences at CLU. He has considerable experience in environmental analysis and modeling, particularly in areas of low-income and underrepresented populations. He was a major participant in a National Institutes of Health (NIH) grant award studying the correlation between the chemical composition of air pollutants in the East Los Angeles region and the potential impact of long-term exposure to these chemicals on human health. Dr. Hanrahan has recently completed work on a U.S. EPA-funded project where he developed an integrated modeling approach to assess the health of aquatic systems in Southern California watersheds. He continues to be active in the field and is the author of five books and 45 peer-reviewed, scientific papers.

Chapter 1

Abstracts and Keywords

Abstract

This chapter introduces key chemical and physical concepts to help provide a broad definition and overview of the field of environmental chemistry for the study of chemical, biological, and integrated processes. It commences by examining biotic and abiotic factors that affect living organisms. Subsequently, a thorough review of energy, thermodynamics, kinetics, and chemical equilibrium concepts is presented. Lastly, critical information on the global water and elemental cycles is presented. Understanding these processes will help build a foundation for the remaining chapters, and help infer relationships between pollutant behavior, transport, and fate of chemical substances within and between environmental compartments, and perturbation by human-induced activities.

Keywords

Environmental chemistry, Ecosystems, Thermodynamics, Chemical kinetics and equilibrium, Water and elemental cycles

Chapter 1

Introduction to Environmental Chemistry

Environmental chemists are interdisciplinary in nature and have a solid foundation in chemistry and related training in environmental science, biological sciences, and engineering disciplines. They are routinely involved in analytical testing, new product design and development, field analysis, remediation, modeling efforts, consultation with users of chemicals, and safety, management, and regulatory issues.

FIGURE 1.1 *Photo by Brian Stethem.*

Over the last decade there has also been a movement toward **green-collar jobs**, blue- and white-collar workforce type opportunities created by firms and organizations whose mission is to improve environmental quality and promote sustainability. Such positions may include renewable energy and water conservation, residential solar energy, whole home performance appraisal and construction, local procurement, recycling and reuse, weatherization, open space, zero waste, and earth-friendly agricultural practices. Students are encouraged to explore these dynamic and rapidly growing fields of employment for a rewarding and enriching career.

1.1. OVERVIEW AND IMPORTANCE OF ENVIRONMENTAL CHEMISTRY

Environmental chemistry is a multidisciplinary study of chemical, biological, and integrated processes of the environment at large. The natural environment includes components of air, water, soil and land, as well as biota found in those compartments. The man-made environment is represented by human settlements consisting of physical elements. The complex interactions that occur between the natural and physical environments are key to defining the broad definition of environmental chemistry, and vital for understanding the larger context of environmental science. Its scope is considerable, ultimately encompassing targeted areas of study including atmospheric chemistry, biogeochemistry, chemical toxicology, marine chemistry, soil and sediment chemistry, water chemistry, astrochemistry, fire chemistry, chemistry of energy, Earth and geochemistry, and climate change (Figure 1.2).

In order to gain a true understanding of the importance of environmental chemistry, and to foster knowledge on how living organisms interact with each other and their surroundings, an integrated approach to facilitating links between the targeted areas above must be adopted. Environmental chemistry draws on a myriad of concepts from chemistry, biology, statistics, geology,

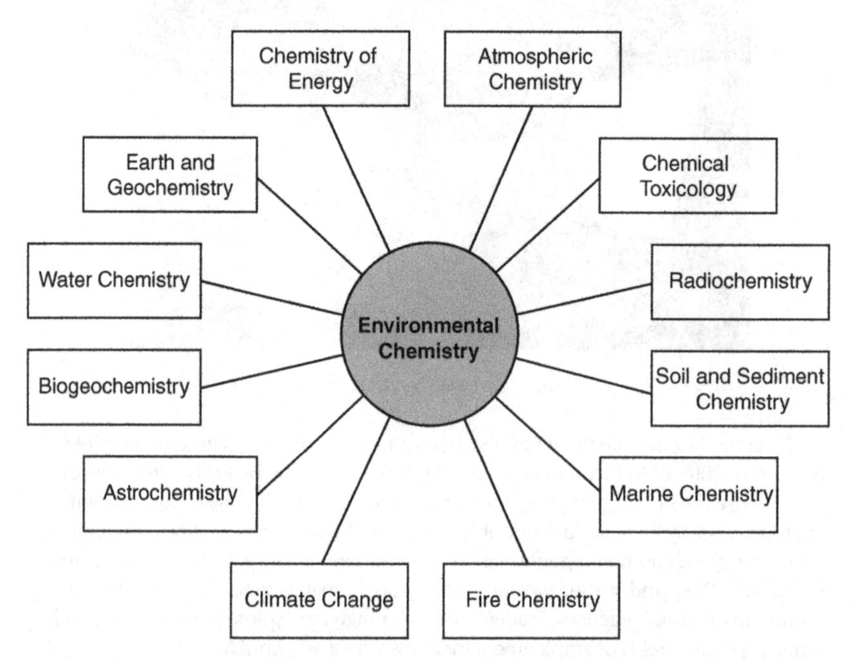

FIGURE 1.2 Environmental chemistry encompasses many targeted areas of specialization with each requiring an integrated approach to study.

and environmental science to assist in defining the sources, reactions, transport, effects, and fates of chemical species in various environmental **matrices.**

TEXTBOX 1.1

Environmental **matrices** are surrounding substances in which elements or chemical compounds of interest originate, or are contained.

For example, phosphorus is a vital macronutrient, and the accurate determination of phosphorus species in environmental matrices such as natural waters and soils is essential for understanding the biogeochemical cycling of the element, studying its role in ecosystem health and monitoring compliance and legislation (Worsfold, et al., 2005). At the watershed scale, for example, phosphorus export from both **point** and **non-point sources** can result in increased primary production and **eutrophication**—the overenrichment of aquatic ecosystems leading to algal blooms and anoxic events. For accurate assessment, knowledge of the source, transport, and fate of phosphorus species, including dissolved colloidal and particulate fractions as inorganic and organic compounds, is paramount. More detailed description of phosphorus dynamics is provided in section 1.5.4.

TEXTBOX 1.2

Point source pollution is discharged from an identifiable point, including pipes, outfalls, and municipal and industrial treatment processes.

1.2. ECOSYSTEM CONSIDERATIONS

A closer examination of the term *environment* reveals both biological (**biotic**) and physical (**abiotic**) factors that affect living organisms on earth. The environment can be subdivided into functional units termed **ecosystems**, where dynamic interactions between plants, animals, microorganisms, and abiotic factors (e.g., soil, water, nutrients, and light) exist. The global sum of all ecosystems is referred to as the **biosphere**, one of four separate components of the **geosphere**, the other three being **lithosphere**, **hydrosphere,** and **atmosphere.** A related term, **biome**, is an extensive area comprising many similar ecosystems and characterized by dominant biological communities. Biomes can be grouped into eight major types: freshwater, marine, desert, savannah, coniferous forests, deciduous forests, tropical rainforests, and tundra.

All ecosystems are characterized by flow of energy through **food webs**, production and degradation of organic matter, and transformation and cycling of nutrients. Food webs are networks of interrelated **food chains** in a given area. It is the production of organic matter that serves as an energy platform for biological communities within ecosystems. At the base of the platform, producers (or autotrophs) actively convert solar energy into stored chemical energy through a process termed **photosynthesis**:

$$6H_2O + 6CO_2 \xrightarrow{\text{Light and Chlorophyll}} C_6H_{12}O_6 + 6O_2.$$

Here, light energy is absorbed by chlorophyll with subsequent conversion of H_2O and carbon dioxide (CO_2) from the air to the simple carbohydrate glucose ($C_6H_{12}O_6$). Further reactions between glucose and other sugar molecules combine to form cellulose, the structural component of plants. **Respiration** is the process by which most living organisms (including animals, plants, fungi, and microorganisms) use to obtain energy from glucose and oxygen with CO_2 and H_2O forming as wastes:

$$C_6H_{12}O_6 + 6O_2 \rightarrow CO_2 + H_2O + \text{Energy}.$$

Consumers, including herbivores, carnivores, omnivores, and decomposers must consume plants or other creatures to obtain nutrients and energy needed to sustain life. The general model of energy flow through ecosystems shown in Figure 1.3 ultimately depicts decomposers feeding on **detritus**, the decomposed remains of producers, herbivores, and carnivores. This process allows for essential nutrients to be recycling through a given ecosystem.

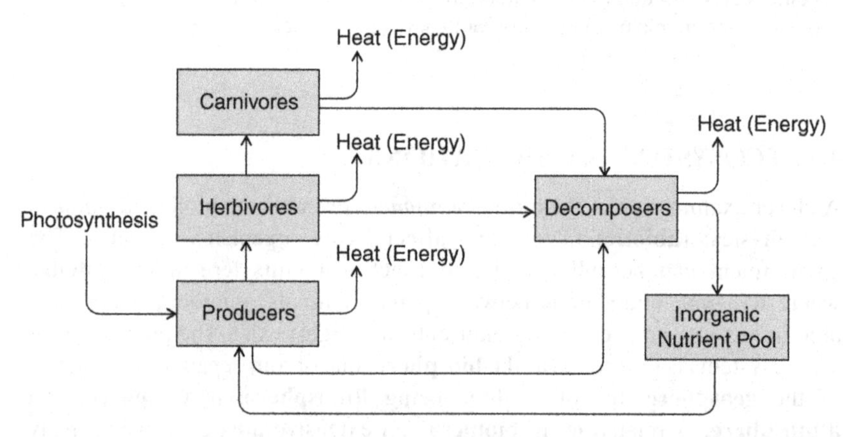

FIGURE 1.3 A general diagram showing how both energy and inorganic nutrients flow through a given ecosystem. The ultimate source of energy is the sun. Energy and nutrients are passed from producers to herbivores to carnivores through the food chain. Decomposers remove the last energy from the remains of organisms, and inorganic nutrients are cycled back to the producers. Ultimately, the energy originally captured is transformed and lost as heat. Thus, energy is not recycled.

1.3. REVIEW OF ENERGY AND THERMODYNAMIC CONCEPTS

Let's consider the term *energy* in greater detail. A technical definition would be the ability to do work. All forms of energy can be classified into two categories: **kinetic** and **potential**. Kinetic energy is the energy that an object possesses by virtue of its motion. For example, flowing water can cause a turbine to turn and thus produce electricity. Potential energy is that which is stored as a result of its composition or its position with respect to another object. For example, energy is stored as chemical compounds in fuel. When combusted kinetic energy is released, the object (e.g., motor vehicle) is set in motion.

Energy transformations are inherent in almost every system in the universe—from heat production in simple Earth and physical systems—to those involved in the growth, dying, and decay of living systems. The principal energy laws that govern every system are derived from two famous laws of thermodynamics. The **first law of thermodynamics**, also termed the law of conservation of energy, states that energy can be neither created nor destroyed—it can only be transformed from one form to another (the total energy remains constant). If energy is transferred between the system and surroundings, then the total amount of energy contained in the system has changed:

$$\Delta E_{\text{System}} = q + w, \tag{1.1}$$

where q is the heat transferred into/out of the system and w is the work done by/on the system.

Modern industrial processes, although efficient and productive, are inherently extractive and dependent upon finite reserves of non-renewable resources. These processes will ultimately deteriorate and deplete the resources on which they depend, and thus, are not ideally sustainable. How does this relate to the first law of thermodynamics? The production of quality goods from industrial processes results in waste generation—matter that does not disappear, and ultimately must go somewhere. For living organisms this means they must collect and convert existing energy into a form readily suitable for biological processes (consider our discussion on photosynthesis above). The first law of thermodynamics has also been used unsuccessfully to argue against the Theory of Evolution and self-organization because of the assumption that the Earth is an isolated system and the tendency for some systems to become disorganized when energy inputs cease. The Earth is a "closed" system and receives continual energy inputs from the sun in the form of radiation, so it is possible for systems on Earth to become more organized and complex at the expense of its surroundings (sun).

The **second law of thermodynamics** describes the nature of processes and chemical reactions as follows: processes occur spontaneously if and only if by their process, the entropy change in the universe, is greater than or equal to zero. **Entropy (*S*)** is a quantity associated with the number of states (in terms of either energy, mass distribution, or information) possible in a system. The greater the allowable states, the larger the S. The greater the S, the greater is the **disorder** of the system. In other words, spontaneous processes lead to greater entropy in general, even if the entropy may decrease in a system as long as the surroundings, i.e., the rest of the universe, increase in entropy enough to make the **entropy change,** ΔS, less than or equal to the entire process positive. Because the system and surroundings may each experience an entropy change, the second law can be written as follows:

$$\Delta S_{universe} = \Delta S_{system} + \Delta S_{surroundings}, \quad (1.2)$$

where $\Delta S_{universe} > 0$ for a spontaneous process. In a physical sense, thermodynamic entropy applies to the universe as a whole and cannot be disregarded. It measures the progression of relatively **closed systems** toward equilibrium (distribution of energy from a concentrated form to a diluted form). Such systems can maintain organization by importing sufficient energy from outside influences. For example, a plant is able to organize randomized CO_2 and water into cellulose (system) only at the expense of the energy loss of the sun (surroundings). Conversely, dead plant tissue will break down into less organized and less energetically concentrated chemicals (system) because there is no input of energy (surroundings). Students needing a more comprehensive review of basic thermodynamic principles can refer to the plethora of general chemistry textbooks in circulation.

TEXTBOX 1.3

Entropy refers to the inevitable and steady deterioration or degradation of a system. Closed systems do not exchange matter with their surroundings but can exchange heat and work (energy). In contrast, open systems continually exchange matter with their surroundings. An isolated system can exchange neither heat nor matter with its surroundings.

1.4. REVIEW OF CHEMICAL KINETICS AND CHEMICAL EQUILIBRIUM

We learned that thermodynamics describes the changes in the form of energy when a reaction occurs, for example, converting chemical energy to heat. However, it does not tell us the rate at which a reaction occurs or the various factors that affect it. The area concerned with rates of reaction is termed **chemical kinetics**. Although thoroughly covered in general chemistry courses, a brief review of the main concepts is important, especially due to its broad significance in environmental chemistry involving chemical equilibrium, and for topics covered in subsequent chapters. For example, knowledge of chemical kinetics is vital when determining whether the formation and depletion of ozone in the upper atmosphere are balanced. The relative rates of decomposition and depletion determine how extensively ozone-destroying molecules are involved in destruction of the ozone layer. For example, scientists have proposed that the presence of chlorine (Cl) atoms accelerates the destruction of ozone (O_3) by acting as a catalyst. The catalyzed reaction is written as follows:

$$O_{3(g)} \xrightarrow{Cl} 2O_{2(g)}.$$

In addition to catalysts, key factors that influence reaction rates include: physical state of the reactants, concentration of the reactants, and the temperature at which the reaction occurs. One can gain valuable information about the rate of a reaction by focusing on how the rate changes with changes in concentration. The change in rate as the reaction proceeds is often best represented in graphical form. The quantitative relationship between the two is expressed by a **rate law**. Considering the following general reaction:

$$aA + bB \rightarrow cC + dD,$$

the rate law can take on the following form:

$$R = k[A]^m[B]^n, \tag{1.3}$$

where k = the **rate constant** and the exponents m and n are termed the **reaction orders**. For a **first-order reaction**, the rate of reaction is directly proportional to the concentration of one of the reactants, with $R = k[A]$. Since the concentration of reactants decreases, a negative sign must be added to equate the loss of reactant A with the positive rate of the reaction. Upon integration, this relationship can be transformed to

$$\ln[A]_t = -kt + \ln[A]_0. \tag{1.4}$$

This equation relates the concentration of A at time 0 to its concentration at any other time, t, $[A]_t$, during the reaction. Natural nuclear decay is one example of a first-order reaction process. For a **second-order reaction**, the rate of reaction is directly proportional to the square of the concentration of one of the reactants, with $R = k[A]^2$. Integrating both sides from zero to t results in

$$\frac{1}{[A]_t} = kt + \frac{1}{[A]_0} \tag{1.5}$$

Gas phase reactions involving atmospheric radicals that convert hydrocarbons from auto exhaust to CO_2 would be an example of a second-order reaction. For a **zero-order reaction**, the rate of reaction is a constant, with $R = k$. For example, zero-order kinetics is followed when a substrate has saturated a given enzyme (rate is independent of substrate concentration). The **overall reaction order** is equal to the sum of the orders of reactants.

The time it takes for the concentration of a reactant to reach one-half of its initial value is termed the **half-life** of the reaction, $t_{1/2}$. The half-life of a first-order reaction is determined by (by substituting $[A]_t = 1/2[A]_0$ and replacing t with $t_{1/2}$):

$$t_{1/2} = -\frac{\ln(1/2)}{k} = \frac{0.693}{k}, \tag{1.6}$$

which does not depend on the starting concentration. However, the half-life of a second-order reaction depends on both the rate constant and initial concentration of A:

$$t_{1/2} = \frac{1}{k}[A]_0. \tag{1.7}$$

Intuition should tell you that the faster the reaction, the shorter the half-life. In addition, the rate of the reaction is proportional to the rate constant. Thus, the larger the rate constant, the shorter the half-life. For practical purposes, knowledge of the half-life is essential in environmental remediation strategies. For example, Sr-90, a radioactive byproduct of uranium fission, has a half-life of 29 years and can contaminate a food supply, incorporate into bones, and potentially cause cancer. Given the relatively long half-life, it is worth the effort to remove the isotope from the area of contamination, such as through topsoil removal, since the danger from radiation will linger for several generations.

The term **reaction mechanism** refers to detailed steps that occur in the course of a given reaction. For example, one can study potential reaction mechanisms for photochemical smog, strongly emphasizing the relation between the concentration of oxides of nitrogen (NO_x) and organic reactants in the atmosphere and the evolution of the tropospheric ozone. Defined rate laws can be determined for each step involved in the multistep mechanism. This complex process will be covered in detail in Chapter 7.

Chemical equilibrium provides us with a basic framework for environmental chemistry, with the equilibrium state being the stationary state in a closed system as previously discussed. It is interesting that many chemical reactions can proceed simultaneously in opposite directions. Reactions are always in search of a situation where the ratio of concentrations of reactants and products is constant. Consider the reaction below. If at least one of the components in the reaction is a constituent of a homogeneous mixture, that is, a solution (aqueous, gaseous, or solid), the reaction exhibits equilibrium with a defined equilibrium constant. One can deduce the equilibrium constant by the ratio of their activities.

$$a\mathrm{A} + b\mathrm{B} \overset{K}{\leftrightarrow} c\mathrm{C} + d\mathrm{D}$$

Because activity is directly proportional to the molar concentration (except at high concentrations), it is customary to express the equilibrium constant in terms of the molarities at equilibrium:

$$K_c = \frac{[\mathrm{C}]^c[\mathrm{D}]^d}{[\mathrm{A}]^a[\mathrm{B}]^b}.$$

Example Problem 1.1

Given the following dissociation of substance BC, prove that [C] = [B] using the equilibrium constant expression.

Answer: The equilibrium concentration of B, C, or BC can be calculated from

$$K_C = \frac{[B][C]}{[BC]}$$

To find the concentration of B, [C] and [BC] are represented in terms of the unknown. Note that if the dissociation of BC is the only source of B and C, their equilibrium concentrations must be equal.

When gases are involved in the equation, the **partial pressure,** ***P***, of the gases may be inserted in the *K* expression instead of the concentration. The partial pressure (pressure exerted by a particular component of a mixture of gases) is expressed in units of bar (B, equal to 100,000 pascals, the S.I. unit of pressure). The commonly used unit of pressure, the atmosphere, is approximately equal to 1.01 bar). The constant, *K*, is written as K_p, where the subscript *p* indicates "pressure":

$$K_p = \frac{P_C^c P_D^d}{P_A^a P_B^b}.$$

If we can assume the gases to behave close to ideality, we can use the **ideal gas equation,**

$$PV = nRT, \tag{1.8}$$

where $R = 0.082057$ atm L mol^{-1} K^{-1}, *T* is the absolute temperature in K, and *n* is the quantity of gas (normally expressed in moles). The partial pressures can be expressed as

$$P = (n/V)RT = MRT, \tag{1.9}$$

where *M* is the molarity of the chemical component.

The activities for reactants or products that are pure solids or pure liquids are assigned a value of one or "unity." In reactions in dilute solutions, which involve water as a reactant or product, the activity of the water is also expressed as unity because the change in the concentration of water from that of pure water is negligible, due to its great amount of activity compared to the other chemical species. Another type of equilibrium involving the phase of gases (i.e., involving the dissolved and the gaseous phases) is given by **Henry's Law**, which states that the solubility (the equilibrium concentration) of a gas in a liquid solution is directly proportional to the partial pressure of the gas above the solution. When the dissolved gas exceeds this solubility due, for example, to a drop in the partial pressure, a supersaturated solution ensues leading to bubbling as observed after a soda bottle if opened or, worse, in limnic explosions (see Textbox 1.4).

TEXTBOX 1.4

Unstable equilibrium in Lake Nyos, Cameroon

Lake Nyos is a crater lake in Cameroon, Africa, the remains of a seemingly extinct volcano. In August 21, 1986, it suddenly released an estimated up to 1 km^3 of carbon dioxide gas, forming a lethal heavier-than-air blanket that killed 1,700 people and most living things within a 15-mile radius of the crater. This natural explosion is referred to as a limnic eruption or lake overturn, caused by the sudden release of CO_2 gas from the depths of the lake, which are saturated with CO_2. This is an example of an **unstable equilibrium** that may be triggered by any number of mechanical factors—such as the sudden exposure to higher temperatures or partial pressures—which may be likened to the effervescence of a bottle of soda upon being opened or disturbed. To prevent future disasters, a group of French scientists have installed long pipes to siphon the CO_2 from the depths of the lake and release it on the surface. The formation and expansion of the CO_2 bubbles as they rise in the pipe allow the water to rise as its density decreases, thus creating a self-sustaining flow of CO_2 and water to the surface. The team's website is provided here as a link for further inquiries: http://pagesperso-orange.fr/mhalb/nyos.

Henry Le Châtelier in 1884 enunciated a principle (Le Châtelier's Principle) bearing his name relating to how systems in equilibrium respond to factors that disturb the equilibrium. His principle states that a system initially at equilibrium, upon being disturbed from equilibrium, would readjust itself by minimizing the factors that shifted it from the equilibrium. These factors are normally factors such as concentration, volume, pressure, or temperature. This principle is useful for manipulating a reversible equilibrium to affect a desirable result. For example, it may be preferable to decrease the solubility of a product that needs to be isolated from a solution (e.g., recovering heavy metals in wastewater treatment processes).

Environmental chemists are increasingly called upon to analyze, for example, the rate of pollutant conversion in both chemical and biological systems, and in determining relevant rates of photoreactions in aquatic systems. Chemical kinetics and chemical equilibrium concepts are also crucial in understanding the various biogeochemical cycles presented below. Elements are often governed by kinetics and may involve the input of energy, so that chemical equilibrium states are not attained. Students of environmental chemistry must therefore be aware of the significance and applications of chemical equilibrium and kinetics. Theoretical concepts and application-based discussion will be examined in detail in subsequent chapters to ensure complete and formalized learning.

1.5. GLOBAL WATER AND ELEMENT CYCLES

Global water and element cycles are controlled by long-term, cyclical processes, with many depicted schematically in Figure 1.4. Understanding such processes

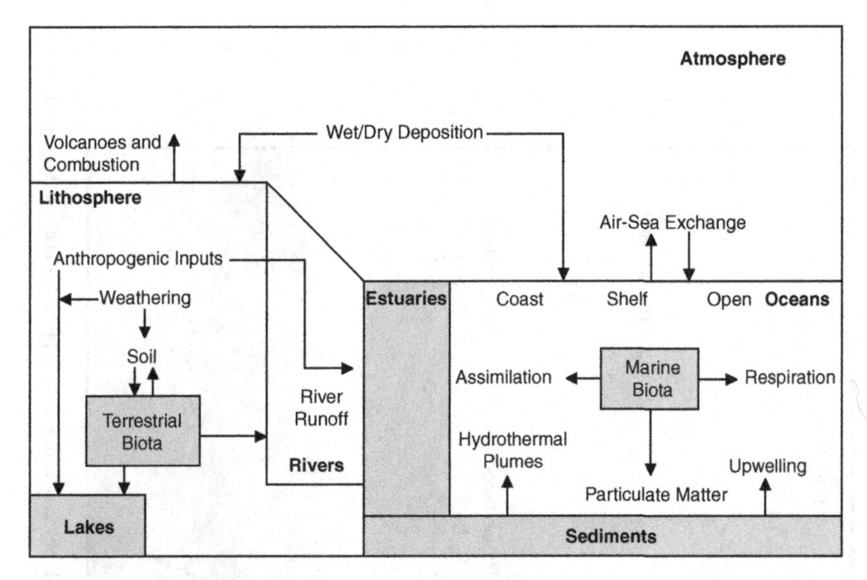

FIGURE 1.4 Illustrated diagram of the major global environmental processes. This depiction considers chemical, physical, geological, and biological processes, together with their interactions, which determine the characteristics of environmental compartments.

is key in interpreting the environmental behavior, transport, and fate of chemical substances within and between environmental compartments, environmental equilibria, transformations of such chemicals, and assessing the influence of and perturbation by **anthropogenic** activities. When covering elemental cycles in this opening chapter, we have kept our discussions focused on carbon, nitrogen, and phosphorus. It should be noted, however, that a number of other elements (e.g., silicon and iron) are essential for biogeochemical reactions and the growth and maintenance of biomass.

TEXTBOX 1.5

Anthropogenic activities are those within ecosystems caused by sustained direct human influence, including agriculture, forestry, and other land uses, urbanization, and industrially-related processes.

1.5.1. The Global Water Cycle

The ubiquitous nature of water on Earth plays a crucial role in climate variability and change, as well as aiding in the integration of chemical, physical, and biological processes that sustain the overall health of ecosystems. Detailed understanding of the **global water cycle** (conceptual landscape diagram shown in Figure 1.5) is thus necessary and provides a solid foundation

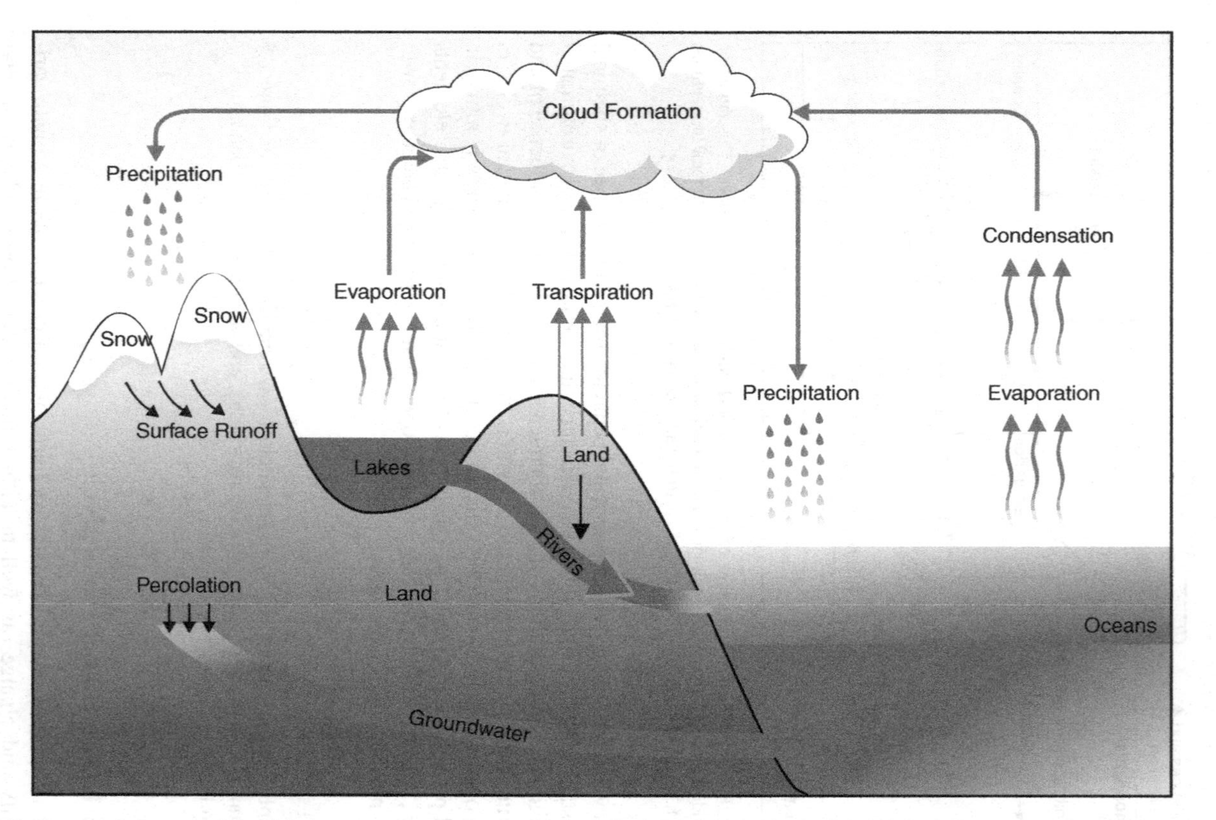

FIGURE 1.5 A general diagram of the global water cycle. It plays a major role in the chemical, physical, and biological processes that sustain ecosystems and influence climate.

for environmental and social decision making and policy development and implementation.

An estimation of global water distribution is provided in Table 1.1. As expected, a vast bulk of it, 96.5%, is in the global oceans. Hence, any examination of the global water cycle should appropriately start here. Radiation from the sun drives the water cycle by heating the world's oceans, causing water to evaporate as vapor into the air. The water vapor rises where cooler temperatures cause condensation and cloud formation. Air currents direct the movement of clouds, and as the atmosphere becomes unstable the **convection** intensifies and clouds develop. Clouds are formed as air parcels are forced to rise, cool, and eventually condense. As warm cloud droplets grow they form **precipitation** by collision. In contrast, cold clouds form precipitation by deposition and aggregation. If precipitation falls as a snowy-ice mixture it can accumulate as ice

TABLE 1.1 Estimated Global Water Distribution

	Volume (1000 km^3)	% of Total Water	% of Fresh Water
Oceans, Seas, and Bays	1,338,000	96.5	–
Ice Caps, Glaciers, and Permanent Snow	24,064	1.74	68.7
Groundwater	23,400	1.7	–
Fresh	(10,530)	(0.76)	30.1
Saline	(12,870)	(0.94)	–
Soil Moisture	16.5	0.001	0.05
Ground Ice and Permafrost	300	0.022	0.86
Lakes	176.4	0.013	–
Fresh	(91.0)	(0.007)	0.26
Saline	(85.4)	(0.006)	–
Atmosphere	12.9	0.001	0.04
Swamp Water	11.47	0.0008	0.03
Rivers	2.12	0.0002	0.006
Biological Water	1.12	0.0001	0.003
Total	1,385,984	100.0	100.0

Data source (Gleick, 1996).

caps, glaciers, and permanent snow. Similarly, **evaporation** and **transpiration** return water to the atmosphere nearly everywhere, with rates varying considerably according to climatic conditions. Transpiration is the evaporation of water from plants occurring at the leaves while their stomata are open for the passage of CO_2 and O_2 during photosynthesis. Factors that affect the rate of transpiration include light, temperature, humidity, wind, and soil water content.

Surface runoff is a major component of the global water cycle and occurs when water from precipitation, snowmelt, or other sources flows over the land surface. Ideally, portions of runoff are directed through rivers and streams with movement toward the world's oceans. Much of the runoff is stored as freshwater in lakes or soaks into the ground as infiltration to replenish saturated subsurface rock, termed *aquifers*. Water in the saturated zone is referred to as groundwater. The upper surface of this zone is referred to as the water table. As shown in Table 1.1, the percent of total water housed in groundwater is small (0.76%), but does represent over 30% of the global freshwater. Due to urbanization, however, the increased production of pavement, buildings, and other impervious surfaces is evident with surface runoff on the rise. This ultimately lessens the percolation of water through the soil to the subsurface and increases the rate of runoff into rivers and streams where erosion and siltation issues are created.

When water first infiltrates the land surface, numerous chemical reactions that affect the biological and geochemical characteristics occur. For example, upon infiltration, microorganisms in the soil have a significant effect on water chemistry by degrading organic matter and thus producing high concentrations of dissolved carbon dioxide. Such a process lowers the pH by increasing the carbonic acid (H_2CO_3) concentration in the soil water, which initiates a number of mineral-weathering reactions. Contact times between water and minerals are important. Where contact times are short, the dissolved-solids concentration in the water is generally at the lower range, thus limiting chemical changes that may take place before groundwater is discharged to **surface water**. This movement provides a major pathway for chemical transfer between terrestrial and aquatic systems, ultimately affecting the biological and chemical characteristics of such systems downstream (including oceans).

TEXTBOX 1.6

An area of land where surface runoff is directed to a common point is termed a **watershed**.

1.5.2. The Global Carbon Cycle

Carbon is an essential element of life and is therefore of fundamental importance when studying the biosphere and its most basic mechanisms. In order to fully comprehend the importance of carbon, and in determining the defining characteristics of the **carbon cycle**, consideration of chemical, physical, biological, and geological processes on varying time scales is paramount. The global carbon cycle (Figure 1.6) is one of the Earth's major biogeochemical cycles, where considerable amounts of carbon continuously cycle between the Earth's atmosphere, oceans, land biota, marina biota, and mineral reservoirs in both short- and long-term cycles. The largest **fluxes** occur between the atmosphere and terrestrial biota and between the atmosphere and surface water of the oceans. The exchange in oceans takes place on a very large scale, but it is often thought of to be a very rapid process, absorbing and releasing CO_2 in short-term cycles. Greater discussion on the carbonate system in aqueous solutions is provided in Chapter 3.

The oceans are natural carbon reservoirs or **sinks**, with the process by which oceans remove CO_2 from the atmosphere termed **ocean carbon sequestration**. It is estimated that oceans currently take up a third (roughly two billion metric tons per year) of the carbon emitted by anthropogenic influences. Terrestrial processes also play crucial roles in global carbon cycling. Plant life removes CO_2 from the atmosphere and stores it in biomass. Over a given time period, plants remove and release CO_2, and net sequestration results if the rate of removal is higher than the rate of release. Soil itself is reported to hold two to three times the amount in biomass, much in the form of dead organic matter. A more significant amount is stored in fossil fuel reserves. A complete list of estimated global carbon reserves is presented in Table 1.2. The predominant source of anthropogenic CO_2 emissions is the combustion of fossil fuels through electricity generation, industrial processes, and transportation. Forest clearing, other biomass burning, and some non-energy-production processes (e.g., cement production) also emit significant quantities of CO_2. The physical and economic impacts of increased CO_2 emissions, along with further explanation of sequestering processes, will be covered in detail in Chapter 3.

Research Application I – Priming the Biological Carbon Pump

One major way in which oceans sequester carbon is termed the *biological pump*—a process by which phytoplankton in the surface waters take up dissolved CO_2 during photosynthesis and convert the carbon to cellular material. As phytoplankton die off or are eaten, they sink as whole cells and aggregates. A portion of this sinking carbon is re-dissolved back into the ocean, with the remaining ending up as sediment on the sea floor. It is this flurry of dead cells and fecal matter (termed *marine snow*) that "pumps" carbon from the surface of the ocean to the sea floor.

What role does iron play in this process? The availability of iron limits primary productivity and the associated uptake of carbon over large areas of the ocean. Since the early 1990s scientists have used this knowledge to design and deploy artificial iron fertilization experiments. Artificial iron fertilization is carried out by releasing a solution of ferrous sulfate into small controlled areas of the ocean to

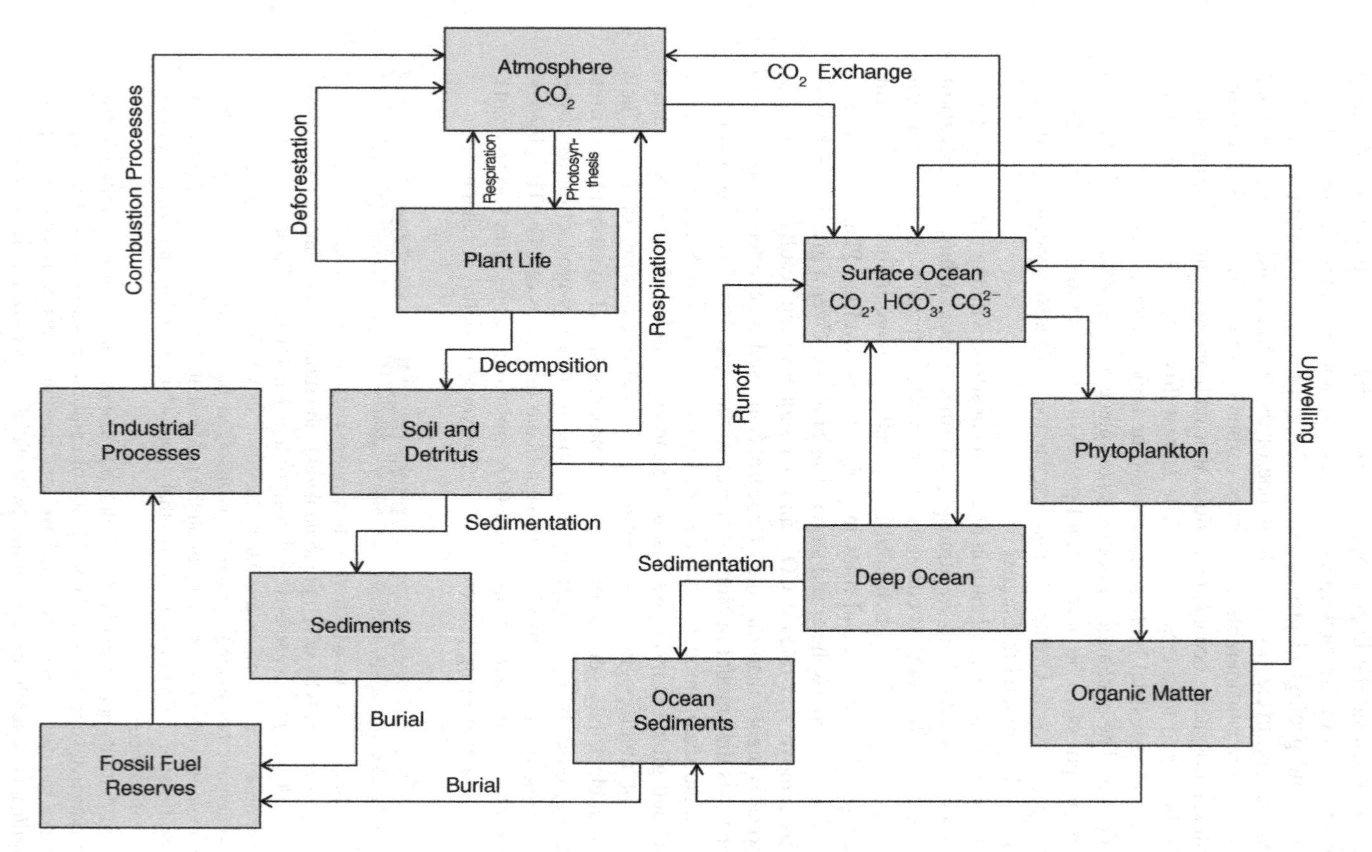

FIGURE 1.6 A generalized global carbon cycle showing the cycling of carbon between the Earth's atmosphere, oceans, land biota, marine biota, and mineral reserves.

increase phytoplankton production. This in turn increases the ratio of $C_{organic}/C_{carbonate}$, and thus the efficiency of the biological pump. Iron thus plays an important role in the carbon cycle, and changes in its supply to the surface ocean may have had a significant effect on atmospheric CO_2 concentrations over glacial–interglacial cycles (Blain, et al., 2007).

Just over a dozen iron fertilization experiments have been performed since 1993 by scientists from a number of countries. They have proven this technique to be a powerful method to study and quantify ecological and biogeochemical processes in the ocean. A recent study performed over the Kerguelen plateau in the Southern Ocean, for example, showed that the efficiency of fertilization (ratio of the carbon export to the amount of iron supplied) was at least ten times higher than previous estimates from short-term blooms induced by iron-addition experiments (Blain, et al., 2007). Such findings show that changes in iron supply may have a more significant effect on atmospheric CO_2 concentrations than previously thought.

The ocean iron fertilization concept, however, is not without its critics. In 2007, natural and social scientists along with environmental advocates, policymakers, economists, and legal experts from around the globe met at Woods Hole Oceanographic Institution (WHOI) in Massachusetts to discuss the benefits and consequences of ocean iron fertilization as a means to mediate global warming. They argued that the efficacy and ecological impacts of iron fertilization remain uncertain, particularly with larger-scale experiments. Clearly, more work is needed to draw solid conclusions.

TABLE 1.2 Estimated Global Carbon Reserves

Reservoir	Amount of Carbon (in Gigatons (GT) of C)
Atmosphere	750
Soils	1,580
Forests	610
Oceans (surface)	1,020
Oceans (deep)	38,100
Fossil Fuels:	
Coal	4,000
Oil	500
Natural Gas	500

Data source: U.S. Global Change Research Information Office (http://www.gcrio.org/).

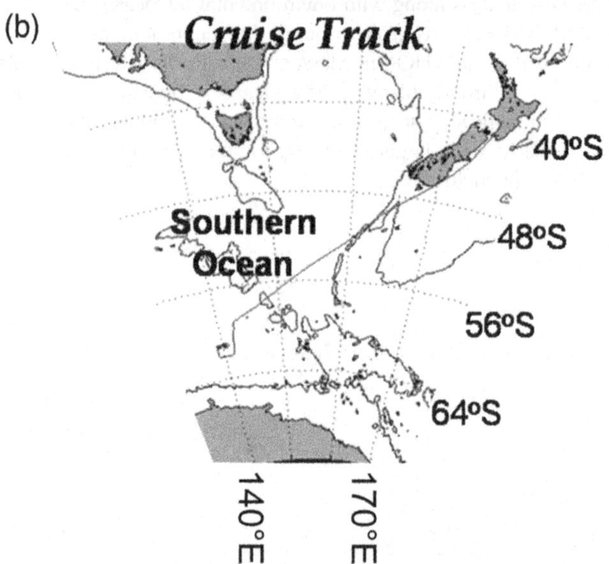

FIGURE 1.7 a) Research vessel *Tangaroa* utilized during the Southern Ocean Iron Enrichment Experiment (SOIREE). b) The *Tangaroa*'s cruise tract from 31 January to 1 March 1999 in the polar Southern Ocean during the SOIREE *(Photo source: http://www.niwavessels.co.nz/tangaroa/; ship cruise tract diagram provided by Dr. Andrew Bowie).*

1.5.3. The Global Nitrogen Cycle

A schematic of the global nitrogen cycle is shown in Figure 1.8. The elemental gas dinitrogen (N_2) is the most abundant but least available form of nitrogen in the global environment. Fortunately, many biochemical transformations can convert N_2 into dissolved inorganic species, including nitrate (NO_3^-), nitrite

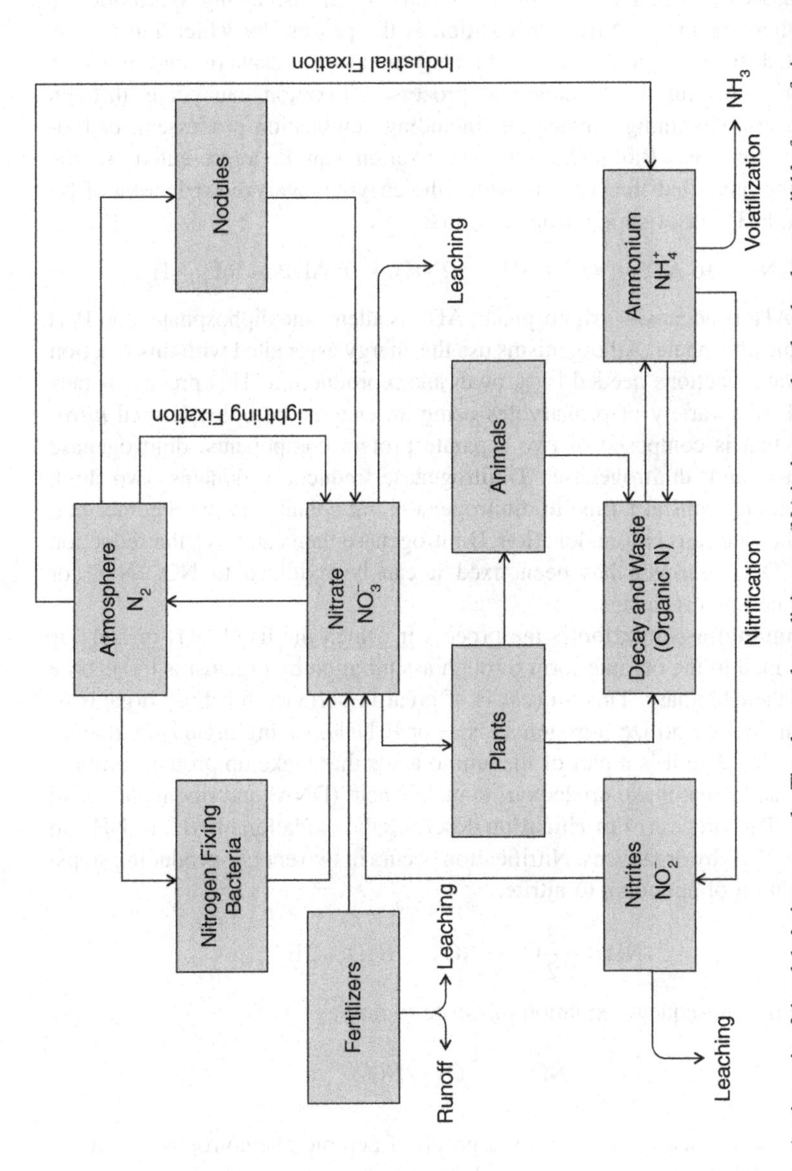

FIGURE 1.8 A schematic of the global nitrogen cycle. The elemental gas dinitrogen (N_2) is the most abundant but least available form of nitrogen in the global environment. Biochemical transformations convert N_2 into dissolved inorganic species, including nitrate (NO_3^-), nitrite (NO_2^-), ammonium (NH_4^+), and organic nitrogen compounds in both dissolved and particulate forms.

(NO_2^-), ammonium (NH_4^+), and organic nitrogen compounds in both dissolved and particulate forms.

These processes are mediated by various types of microorganisms with some processes being energy producing and others occurring symbiotically with other organisms. **Nitrogen fixation** is the process by which nitrogen is converted from its molecular form (N_2) in the atmosphere into nitrogen compounds useful for biochemical processes. Fixation can occur through atmospheric (lightning), industrial (including combustion processes), or biological processes. Biological nitrogen fixation can be represented by the following redox half-reaction, in which the enzyme-catalyzed reduction of N_2 to NH_3, NH_4^+, or organic nitrogen occurs:

$$N_2 + 16\ ATP + 8e^- + 8H^+ \rightarrow 2NH_3 + 16\ ADP + 16P_i + H_2,$$

where ATP is adenosine triphosphate, ADP is adenosine diphosphate, and P_i is inorganic phosphate. All organisms use the energy associated with this reaction to activate reactions needed for growth and reproduction. This process is performed by a variety of prokaryotes using an enzyme complex termed *nitrogenase* that is composed of two separate protein components: dinitrogenase reductase and dinitrogenase. Dinitrogenase reductase donates two high potential electrons at a time to dinitrogenase and contains an Fe-S center that holds the electrons before donation. Dinitrogenase then catalyzes the reduction of N_2. Once nitrogen has been fixed it can be oxidized to NO_2^-/NO_3^- or assimilated by organisms.

Ammonia assimilation is the process in which the fixed NH_3 or NH_4^+ is transformed to the organic form through assimilation by organisms to become part of their biomass. This process is of great importance for those organisms that can directly utilize nitrogen as NH_3 or R-NH_2. Living organisms require nitrogen because it is a part of the amino acids that make up proteins and the nucleic acids that make up deoxyribonucleic acid (DNA) and ribonucleic acid (RNA). This process of **nitrification** describes the oxidation of NH_3 or NH_4^+ to NO_2^- or NO_3^- by organisms. Nitrification occurs in two energy-producing steps: 1) oxidation of ammonia to nitrite,

$$NH_4^+ + \frac{3}{2}O_2 \rightarrow NO_2^- + H_2O + 2H^+$$

and, 2) the subsequent oxidation of nitrite to nitrate,

$$NO_2^- + \frac{1}{2}O_2 \rightarrow NO_3^-.$$

This process is used primarily by a group of aerobic chemotrophs called the nitrifiers. While some low levels of heterotrophic nitrification do occur, rates are low, and quantities of nitrate produced are relatively small compared to those by the chemotrophs. The oxidation of ammonia into nitrite is done by *Nitrosomonas* species, with the second step performed by *Nitrobacter* species.

In the global environment, nitrogen is assimilated (immobilized) into biomass in the form of NH_3. This reduction is catalyzed by two assimilatory enzymes (nitrite reductase and nitrate reductase) and can be carried out by plants, fungi, and prokaryotes. This process is likely to dominate when reduced nitrogen is in low supply (e.g., during aerobic conditions).

Denitrification (also termed *dissmilatory nitrate reduction*) is the microbial reduction of NO_3^- to N_2. The reaction steps require an organic carbon as electron donor (CH_2O generically used for simple carbohydrates, such as glucose) and are shown below:

$$2NO_3^- + CH_2O \rightarrow 2NO_2^- + CO_2 + H_2O$$

$$4NO_2^- + 3CH_2O + 4H^+ \rightarrow 2N_2 + 3CO_2 + 5H_2O.$$

Many denitrifying bacteria consume organic matter while utilizing NO_3^- as their electron acceptor. Thus, denitrification can readily occur in the presence of organic matter, in soils devoid of oxygen, and with the help of denitrifying bacteria. These denitrifiers are represented by many diverse species: both the Gram-positive and Gram-negative bacteria, as well as archaea; some are thermophilic and others are halophilic.

Finally, the **ammonification** process involves breakdown of organic nitrogen compounds (from soil or aquatic organic compounds) into NH_3 or NH_4^+. Heterotrophic bacteria are principally responsible for the process of ammonification, utilizing organic nitrogen compounds and leaving behind NH_3 and NH_4^+. These nitrogen species can then be recycled back into the biosphere.

The global distribution of nitrogen is presented in Table 1.3. As shown, the atmosphere is the principal nitrogen reservoir, with over 99% of the total in the N_2 form. In terrestrial systems, nitrogen occurs mainly as soil organic matter. In the world's oceans, N_2 in the dissolved form is most prevalent. Nitrogen also occurs in various inorganic forms (e.g., nitrate, nitrite, ammonia, hydrazine, nitrous oxide, and nitrogen dioxide) and organic forms (e.g., amino acids, amines, and amides). More detailed information on the most prevalent nitrogen species (and techniques for their analytical determination) that exist in the global environment by oxidation state is presented in Chapters 3 and 4. It is important to note that nitrogen is also tightly coupled with other elements (carbon, phosphorus, sulfur, and trace metals), and understanding these relationships will help in determining the role of living matter in biogeochemical cycles. The effect of human activity on the global nitrogen cycle is also of interest. The realization that nitrogen is commonly a limiting nutrient in plant growth has led to the invention and large-scale usage of nitrogen fertilizers, which in turn account for more than half of the human perturbation to the global nitrogen cycle. In addition, the widespread use of fossil fuels has led to the increased production of nitrogen oxides, which ultimately contribute to photochemical smog and acid precipitation. The latter two topics will be explored in subsequent chapters.

TABLE 1.3 Global Nitrogen Reservoirs and Fluxes

Reservoir	10^{15} grams N
Lithosphere	190,000,000
Atmosphere	3,900,000
Oceanic	23,000
Terrestrial	500
Flux	10^{15} grams N year^{-1}
Dry and Wet Deposition (terrestrial and oceanic)	0.160–0.450
Denitrification (terrestrial)	0.043–0.390
Denitrification (oceanic)	0.10–0.330
Biological Nitrogen Fixation (terrestrial)	0.044–0.200
Biological Nitrogen Fixation (oceanic)	0.001–0.130
River Runoff	0.13–0.040

Data source: (Hanrahan, et al., 2002).

Example Problem 1.2

How might feedback processes operate and control the interactions between the global water cycle and related systems (e.g., energy flow and the carbon cycle)?

Answer: As discussed in the text, numerous physical processes are responsible for transporting water vapor, cloud formation, and precipitation. Such processes play important feedback roles, e.g., in altering surface and atmospheric heating/cooling rates. For example, an increase in water vapor (a dominant greenhouse gas) could lead to positive temperature feedback. This may in turn lead to variable atmospheric circulation and thus result in altered precipitation patterns.

1.5.4. The Global Phosphorus Cycle

The distribution and transformation of phosphorus in environmental systems is shown in Figure 1.9. Unlike nitrogen, the phosphorus cycle does not have a significant atmospheric component. A chemical distribution of phosphorus between aquatic and particulate components occurs via, e.g., adsorption and precipitation processes. Major reservoirs and fluxes of phosphorus are shown in Table 1.4. Other sources of phosphorus include marine sediments and crystal rocks and soil. Major point sources include sewage and industrial discharge. Land use, in particular agriculture, is a significant non-point source. The majority

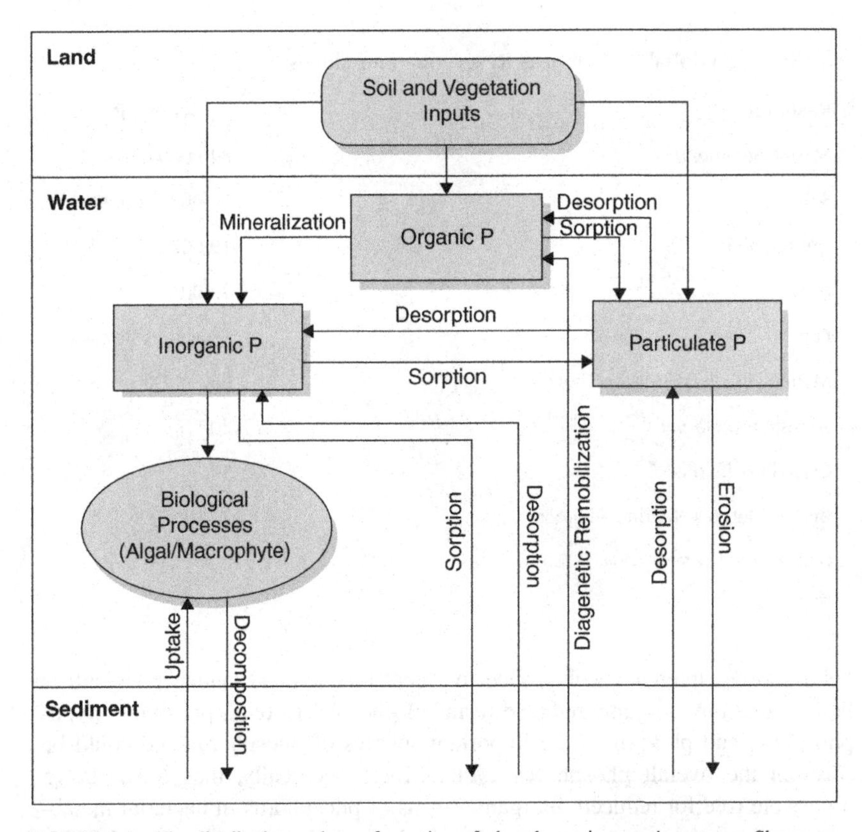

FIGURE 1.9 The distribution and transformation of phosphorus in aquatic systems. Shown are examples of processes that can distribute and transform a species between matrices (soil, water, and sediments). Unlike nitrogen, phosphorus does not have a significant atmospheric component. Chemical "speciation" of phosphorus between dissolved and particulate components does occur via, e.g., adsorption and precipitation processes. In addition, biological processes (e.g., algal) are involved in uptake and decomposition.

of phosphorus transport to watersheds, from both point and non-point sources, occurs during periods of increased discharge (e.g., storm events due to rainfall), which demonstrate the importance of the water cycle in studying such transport. When considering aquatic systems, phosphorus species are found in "dissolved," "colloidal," and "particulate" fractions, as inorganic and organic compounds, and in biotic and abiotic particles. Common **operationally-defined** aquatic forms of phosphorus include total phosphorus (TP), total reactive phosphorus (TRP), filterable reactive phosphorus (FRP), and total filterable phosphorus (TFP).

Overall, much research has been conducted to provide an understanding of how phosphorus is cycled through the environment. However, most of the current literature on the global biogeochemical cycles uses "phosphate" and

TABLE 1.4 Global Phosphorus Reservoirs and Fluxes

Reservoir	Teragrams P
Marine Sediments	840,000,000
Soil	96,000–160,000
Crystal Rock	19,000
Biota	2,600
Flux	Teragrams P $year^{-1}$
Marine Dissolved–Marine Biota	600–1,000
Terrestrial Biota-Soils	200
Crystal Rock-Soils	14
Marine Detritus–Marine Sediment	2–13

Data source: (Hanrahan, et al., 2002).

"phosphorus" interchangeably. Due to recent biochemical evidence, scientists believe that the inorganic reduced forms of phosphorus (e.g., phosphite, hypophosphite, and phosphine) are important sources of phosphorus and could be focal in the overall phosphorus cycle. More specifically, they have shown a possible role for reduced, inorganic forms of phosphorus in bacterial metabolism and as evolutionary precursors of biological phosphate compounds. (See section 3.3.3 for greater discussion on the equilibria between phosphorus species.) Regardless, scientists believe that human activity has roughly doubled the phosphorus flux to oceans, mostly through increasing the dissolved inorganic flux and flux adsorbed to colloidal particles. This recent evidence points to the need for better use and control of global phosphorus resources, including the inefficient use of phosphorus-containing fertilizers and inadequate treatment of wastewater.

TEXTBOX 1.7

Operationally-defined species are defined by the methods used to separate them from other forms of the same element, which may be present. Nutrient and metal species, for example, can be separated between "soluble" and "insoluble" forms by passing the aqueous sample through a 0.20 or 0.45 μM membrane filter. Students will also see the term **speciation** used throughout this book. Speciation refers to the identification and quantitation of specific forms of elements traditionally given by operationally-defined results to identify such forms as "bioavailable," those readily used by biological organisms.

Example Problem 1.3

What are the major anthropogenic sources of nitrogen and phosphorus in aquatic environments?

Answer: Anthropogenic sources of nutrients such as nitrogen and phosphorus can arise via point and non-point sources. Nutrients can originate from runoff of fertilizers, agricultural and residential lands, livestock, and human wastes. Point sources of nutrients include industrial discharges, municipal sewage-treatment facilities, and agricultural animal production facilities.

1.6. AN INTRODUCTION TO ENVIRONMENTAL DATA ANALYSIS

As discussed above, environmental systems are dynamic in nature, varying both spatially and temporally across a wide range of matrices. The development, application, and validation of data analysis tools used to aid in analytical measurements, and in investigating these complex systems, have thus been well received and widely utilized in university laboratories, research centers, and public and federal agencies. With increasing application in environmental systems come a number of key considerations: complexity, variability, scale, purpose, robustness, and sensitivity analysis. Such demands have led to the development of sophisticated statistical techniques and models utilizing today's computational capabilities and assessable resources. Such techniques will be covered in detail in subsequent chapters.

This section describes the initial stages of evaluating environmental data sets, in particular data processing and the use of Microsoft Excel spreadsheets and related functions to assist in collecting, organizing, manipulating, and visualizing the large amount of information gathered during the environmental analysis process. Such discussion will be valuable in your study of subsequent chapter material, and aid in understanding the numerous applications housed throughout this book. Raw data generated from environmental analyses, for example, often contain measurement data from multiple variables, which are recorded in varying units. Consider the hypothetical water quality data provided in Table 1.5. As shown, multiple variables were determined for ten different samples at a given location. Before engaging in data analysis, it is often useful to **transform** such data into unitless values to allow comparison, for example, the distribution among two or more of the monitored variables. Furthermore, environmental data sets often contain variables that are measured out of convenience rather than selecting the "best" variable from a matrix or system point of view. Such "noise" can obscure trends both in time (e.g., monthly occurrences) and in other variables, thus some form of **data smoothing** is recommended prior to data analysis. These concepts will be evident in our discussion of multivariate data analysis in Chapter 2.

TABLE 1.5 Hypothetical Water Quality Data Set Showing Multiple Variables with Varying Units

Sample	Phosphate (mg/L)	Nitrate (mg/L)	pH	Conductivity (mS/cm)
1	1.05	3.78	6.81	22.44
2	1.13	4.01	6.75	19.05
3	0.90	2.08	6.80	21.44
4	0.96	5.07	7.01	21.76
5	0.88	4.99	7.05	35.75
6	1.01	3.85	6.95	31.05
7	0.88	5.10	6.8	26.55
8	0.64	4.88	7.05	25.41
9	0.85	4.07	6.55	21.78
10	0.67	3.65	6.77	20.95

Spreadsheet tools are a valuable resource in data analysis and introductory-level statistics. Excel is one of today's most widely used spreadsheet tools, with its basic nature and ease of application making it popular with beginning learners, although higher-level computing tools are readily available. Everything is built in a spreadsheet program with rows and columns, formatting controls, and mathematical functions ready to use. (See Appendix I for common Excel shortcut and combination keys utilized in typical applications.) A useful addition to Excel is the Analysis ToolPak add-in that allows custom commands and procedures to aid in statistical analyses. Excel spreadsheets also have a number of two-dimensional graphing functions, including linear, scatterplot, bar plots, and histograms. Linear and scatterplots are particularly useful in environmental analyses, for example, when plotting two or more variables against each other for relationship determination. For more advanced computing, modeling framework can be integrated with Excel to perform such tasks as water quality simulation, modeling bioaccumulation of organic chemicals in food webs, prediction of sediment nutrient fluxes, and studying the CO_2 system in oceans. Regardless of the application, Excel is a powerful program available to students for enhanced learning. We will utilize many of its features throughout this book.

1.7. END OF CHAPTER PROBLEMS

1.1 Describe the characteristics of point and non-point sources of pollution. Provide specific examples of each.

1.2 What are meant by the terms abiotic and biotic factors? Provide specific examples of each.
1.3 Describe the differences between kinetic and potential energy.
1.4 Water in a sealed container would be an example of:
  a An open system
  b A closed system
  c An isolated system
  d None of the above
1.5 Name two ways in which energy can be transferred.
1.6 Using the concepts presented in Equations 1.8 and 1.9, derive a relationship where $K_P = K_c$.
1.7 Consider the Earth's system as a heat engine operating between thermal reservoirs with different temperatures (equator and poles). Describe how both kinetic and potential energy and the global water cycle are involved in this process.
1.8 Explain why chemical kinetic concepts are important when studying complex environmental systems.
1.9 The water cycle ends with what process?
  a Evaporation
  b Condensation
  c Percolation
  d None of the above
1.10 Describe how increased surface runoff could possibly lead to drought conditions.
1.11 Explain how increased emissions of airborne nitrogen might affect the global carbon cycle.
1.12 Nitrogen gas (N_2) is the most abundant gas in the atmosphere, comprising nearly 78% of the air we breathe. However, N_2 is not readily available to most living organisms. Explain why and what is needed to convert to more biologically relevant forms.
1.13 Which of the following is indicative of the process that describes the oxidation of NH_3 or NH_4^+ to NO_2^- or NO_3^- by organisms?
  a Ammonia assimilation
  b Denitrification
  c Nitrification
  d Assimilatory nitrate reduction
1.14 What organisms are primarily responsible for the process of ammonification?
1.15 The growth-limiting nutrient in biological processes is the one in low proportion or limited supply to the others that will be exhausted first. In freshwater systems, phosphorus is considered the limited nutrient. What would you expect from even a modest increase in phosphorus concentration to such a system?
1.16 Discuss why the interactions between the elemental cycles and climate are of significant importance.

1.17 Which of the following does not have a significant atmospheric component to its cycle?
  a Nitrogen
  b Carbon
  c Phosphorus
  d Water
1.18 What are the major fluxes and reservoirs of phosphorus in the global environment?

1.8 REFERENCES

Blain, S., Quéguiner, B., Armand, L., Belviso, S., Bombled, B., Bopp, L., Bowie, A., et al. (2007). Effect of natural iron fertilization on carbon sequestration in the Southern Ocean. *Nature, 446*, 1070–1074.

Gleick, P. H. (1996). Water resources. In S. H. Schneider (Ed.), *Encyclopedia of climate and water, vol. 2* (pp. 817–823). New York: Oxford University Press.

Hanrahan, G., Gardolinski, P., Gledhill, M., & Worsfold, P. J. (2002). Environmental monitoring of nutrients. In F. Burden (Ed.), *Environmental monitoring handbook* (pp. 8.2–8.3). New York: McGraw-Hill.

Worsfold, P. J., Gimbert, L., Mankasingh, U., Ndukaku, O., Hanrahan, G., Gardolinski, P., Haygarth, P., Turner, B., Keith-Roach, M., & McKelvie, I. (2005). Sampling, sample treatment and quality assurance issues for the determination of phosphorus species in environmental matrices. *Talanta, 66*(2), 273–293.

Chapter 2

Abstracts and Keywords

Abstract

The purpose of this chapter is to provide students with a solid background in theoretical and applied statistical concepts useful to real environmental problems. Coverage begins with sample collection, preservation, and storage concepts to complement the chemical measurement process. Next, discussion on the distribution of data is presented to aid in making the sampling process more efficient and to allow precise estimates in the face of uncertainty. This is followed by a comprehensive description of inferential statistics, including significance tests and confidence intervals. Presentation of outlying results and analysis of variance (ANOVA) follows. Lastly, a brief introduction to multivariate data analysis is presented. Such discussion is vital to understanding and characterizing complex chemical data sets and the myriad of physical measurements accompanying widespread environmental analyses.

Keywords

Sampling, Inferential statistics, Uncertainty, Analysis of variance (ANOVA), Hypothesis testing, Multivariate data analysis, Precision, Data outliers

Chapter 2

Environmental Statistical Analysis and Sampling Considerations

FIGURE 2.1 For sampling aquatic systems at shallow depths, sampling scoops may be used. Generally, scoops should be constructed of high-density polyethylene or Teflon for the determination of aquatic inorganic species.

Our knowledge of the natural environment is based to a very large extent on **samples** taken from a larger **population**—an entire collection of samples of air, soil, water, organisms, or other material that could potentially be sampled. As will be discussed in this chapter, samples should be collected in properly preserved and prepared sample containers and handled in accordance with the method of analysis. Documentation, data validation, and production of a final report that ensures that the laboratory data produced will be **representative** of the conditions sampled and eligible for submission to clients or regulatory agencies is then provided. Representativeness is a measure of the degree to which data accurately and precisely represent a population parameter at a sampling point or for a process condition. Regardless of sampling requirements, a sample's life can be detailed in the seven steps provided by Popek (2003) as follows:

1. The sample is planned or "conceived";
2. The sampling point(s) is/are identified;

3. The sample is collected or "born";
4. The sample is transferred to the laboratory;
5. The sample is analyzed;
6. The sample expires and is discarded;
7. The sample is reincarnated as statistically validated chemical data.

2.1. INTRODUCTION TO ENVIRONMENTAL STATISTICS

The classical probabilistic and statistical methods outlined in Figure 2.2 are wide ranging and provide environmental scientists with basic tools for quantitative methods of evaluation. They are especially important in environmental chemistry due to the explosion in automated instrumental methods and data collection systems, advanced computing power, interactive software, public and ecological health concerns, and the continuing need for analysis to support environmental assessment, policymaking, and regulation efforts.

TEXTBOX 2.1

Accuracy measures the closeness of a measurement to an accepted reference or true value.
Precision is the agreement among individual measurements of the same property under a prescribed set of conditions.

Environmental statistics has rapidly emerged as a solid branch of statistics, with increased emphasis on applied mathematical statistics, statistical methodology, and data interpretation for addressing contemporary environmental analysis and health concerns. The aim of this chapter is to improve student understanding of the quality, relevance, and effectiveness of research on environmental problems through the use of theoretical and applied statistical concepts. It is impossible to cover every topic in detail given the space limitations of this chapter. However, coverage will include many of the areas highlighted in Figure 2.2, backed by related case studies and applications in the natural environment. We will explore data collection, analysis, uncertainty, significance, presentation, and interpretation of environmental data for research, policy, and regulation considerations.

2.2. SAMPLE COLLECTION

The path to reliable analysis of environmental variables in complex matrices starts with properly planned sample collection. If the sample collection effort

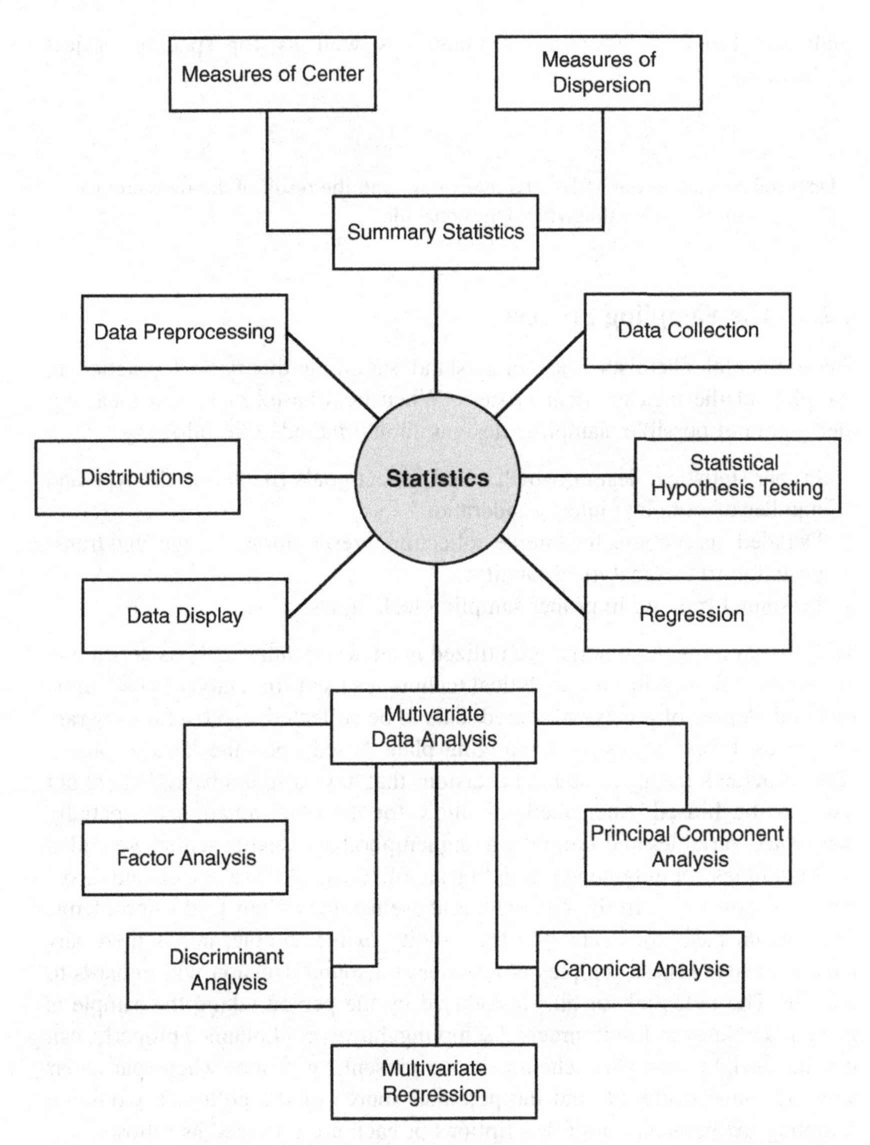

FIGURE 2.2 A schematic of general statistical methods and techniques commonly employed to qualitatively and quantitatively study applications in the natural environment, the biota within environmental media, and human settlements.

is not properly planned and statistically designed, then the resulting data set will likely have a number of limitations that lend to increased **uncertainty**. A true understanding of the underlying theories and principles would thus allow environmental professionals to select and adapt proper sampling techniques, effective preservation and storage protocols, and subsequent analytical tools for specific pollutants, as well as for specific project applications.

> **TEXTBOX 2.2**
>
> **Uncertainty** can be defined as the interval around the result of the measurement that contains the true value with high probability.

2.2.1. The Sampling Process

Environmental chemists must understand sampling theory and practice to compliment the measurement process. When field-based or *in situ* measurements are not possible, sampling designs should include the following:

1. Proper statistical design, which takes project goals (including the type and number of samples) into consideration;
2. Detailed instructions for sample collection, preservation, storage, and transportation to the analytical facility;
3. Personnel training in proper sampling techniques.

Multiple sampling techniques are utilized in environmental analysis depending on site-specific conditions, analytical techniques used for analysis, costs to be incurred, degree of precision desired, data to be collected, and overall program objectives. Properly designed sampling plans based upon the laws of probability provide a means of making decisions that have a sound basis and are not likely to be **biased**. Such methods allow for the determination of spatially distributed variables and sampling along temporal and spatial scales. So-called "grab samples" or judgmental sampling are often used to do a "quick and dirty" evaluation of a site. In the legal arena, these samples often lead to problems. There is no basis for evaluating the validity of the sample, nor is there any means for using these samples in arriving at a sound decision with regards to the site. The potential for bias introduced by the person taking the sample is great and unknown. Environmental sampling, however, if planned properly, can provide insight into what chemicals are present on a site, where particular activities have occurred, and the potential source of the pollutant. Common sampling methods and brief descriptions of each are provided as follows.

> **TEXTBOX 2.3**
>
> **Bias** is systematic distortion of a measurement process that causes error in one direction.

Simple random sampling gives each sample unit an equal probability of being selected, and hence is unbiased in its approach. However, the application of simple random sampling in environmental analyses has been questioned owing to its tendency to be affected by the distribution of the data (see section 2.3) and non-uniformity. This is of obvious concern since most environmental applications deal with data from natural distributions, which can be highly skewed. **Stratified random sampling**, also termed proportional or quota random sampling, involves dividing the population into homogeneous subgroups and then taking a simple random sample from each. The most important issue in this type of sampling is how to stratify the data. Ostensibly, the population is divided into non-overlapping groups (i.e., strata) N_1, N_2, N_3, ... N_i, such that $N_1 + N_2 + N_3 + ... + N_i = N$. A simple random sample of $f = n/N$ in each strata is then performed. When one uses the same sampling fraction within strata, proportionate stratified random sampling is performed. When one uses different sampling fractions in the strata, we call this disproportionate stratified random sampling. As long as the samples are sufficiently homogeneous, a representative sample can be drawn from the population. Overall, stratified random sampling will generally have more statistical precision than simple random sampling.

Systematic random sampling involves preselected sampling points based on the number of samples needed and one randomly selected point as the first sample to draw. If the samples are relatively unordered, then systematic sampling should yield a fairly random sample. Systematic sampling involves sampling the target population at regular intervals (space and time). Here, periodicity of the sampling cycle corresponds to natural cycles or trends. Like stratified sampling, a grid pattern can be constructed to divide the area into distinct units. However, samples are collected from specified locations within the unit (e.g., center of each grid, Figure 2.4). If a periodic trend is observed, samples should be taken frequently enough to give an accurate estimate of the true signal.

Students will also likely encounter a variety of other named sampling techniques as they progress in this field, including **cluster sampling**, **double sampling**, and **composite sampling**. Cluster sampling is useful when population groups are found in groups or clusters. This technique is considered cheaper and more investigator friendly when compared to simple random sampling. Double sampling comes in three forms: ranked set sampling, weighted double sampling, and double sampling with ratio estimation. All three methods have been shown to increase precision and lower sampling costs when compared to simple random sampling. Composite sampling involves taking several physical samples from an area, mixing them, and taking a sub-sample of the mixture. Note that this is physical averaging rather than mathematical averaging and not considered the best sampling technique when information on high values or variability is desirable.

Example Problem 2.1

Suppose a chemical release occurred from a leaking barrel on an unpaved industrial lot with a total area of 200 m^2. Environmental professionals need to assess the extent of this release and design an appropriate sampling design for subsequent analytical testing. What approach should they take?

Answer: Here, the proportion of samples taken in the "suspected" contaminated area in relation to the overall area of the lot is considered. Stratified random sampling is a logical choice given the limited time and possible observer bias during this crucial period. It is suspected that 50 m^2 (25% of area) is contaminated as shown by Figure 2.3. A total of 12 grid units are sectioned. Three samples (25%) are taken within the suspected contamination area, with an additional nine (75%) located in other unit areas. This design will likely allow for any spatially distributed movement of the chemical of interest. Samples from each unit or strata are analyzed separately.

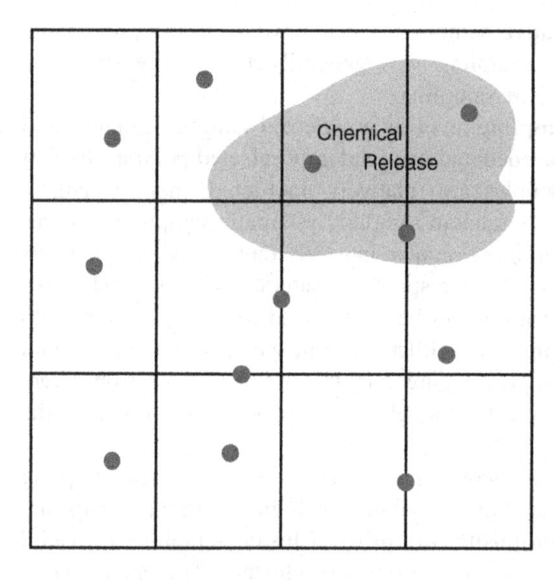

FIGURE 2.3 A simulated chemical release and suspected contaminated area of an unpaved industrial plot.

2.3. THE DISTRIBUTION AND TRANSFORMATION OF DATA

The ultimate goal is to make the sampling process more efficient with the intent of developing methods of sample selection that provide estimates that are precise enough for a given purpose at the lowest possible cost. Regarding precision, one cannot exactly quantify the extent of error present in an estimate in any specific sampling situation, for this would require knowledge of the true value for the entire population of samples. As a result, the

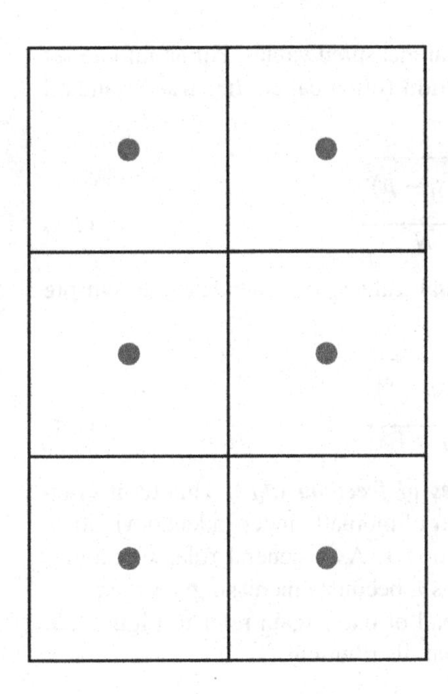

FIGURE 2.4 Visual representation of sampling target populations at regular space intervals. Grid pattern shown.

precision of a sampling campaign can be judged by examining relative distributions.

2.3.1. Normal Distribution and Log-Normal Distributions

Sampling size is of great consideration. In contrast to the classical theory of sampling, environmental sampling involves a population with a **finite** number of units. Even with a finite number as typically experienced in environmental analysis, the **normal (Gaussian) distribution** is often assumed to describe the random variation that occurs in generated data. The shape of the normal distribution resembles that of a bell (Figure 2.5) and can easily be characterized by two values: the **mean** μ and the **standard deviation** σ, so that the data sets generated from a given population can be described by the expression $\mu \pm \sigma$. Here, the population mean is calculated as

$$\mu = \frac{\sum x_i}{N}, \tag{2.1}$$

where N = number of data values in the population and x_i = individual data values. The **arithmetic mean** $\bar{x}$ is the "average" value of a sample, which is a finite series of measurements, and is calculated as follows:

$$\bar{x} = \frac{\sum x_i}{n}, \tag{2.2}$$

where n = sample size and x_i = individual measured values. For an infinite set of data, the **population standard deviation** (often called the "true" standard deviation) is represented by σ:

$$\sigma = \sqrt{\frac{\sum_i (x_i - \mu)^2}{N}}. \tag{2.3}$$

To measure the **spread** of the individual values, we calculate the **sample standard deviation (*s*)** as follows:

$$s = \sqrt{\frac{\sum_i (x_i - \bar{x})^2}{n - 1}}, \tag{2.4}$$

where the quantity $n - 1$ = the ***degrees of freedom*** **(*d.f.*)**. This term often confuses students and refers to the number of mutually independent deviations, represented by $(x - \bar{x})$, used in calculating s. As a general rule, s becomes a more reliable expression of precision as n becomes increasingly larger.

To further our discussion on the spread of data, again refer to Figure 2.5. The empirical rule states that for a normal distribution:

1. 68.2% of the data will fall within roughly 1 σ of the mean;
2. 95.4% of the data will fall within roughly 2 σ of the mean;
3. 99.7% of the data will fall within roughly 3 σ of the mean.

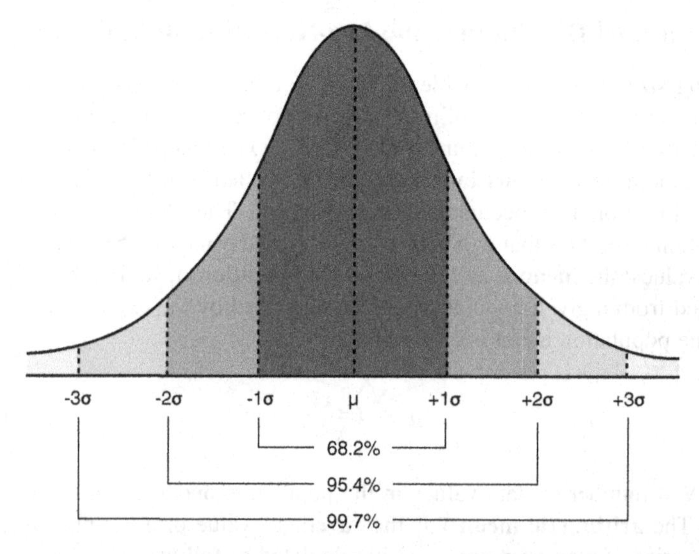

FIGURE 2.5 The bell-shaped normal (Gaussian) distribution. The most frequent value (greatest height of the curve) is the mean, with smaller or greater values decreasing in frequency the more they depart from the mean. This schematic shows the distribution or area of this curve measured in terms of departures from the mean (standard deviations).

In addition, we can find z-scores (standardized values) that correspond to areas under the curve (Figure 2.6). A z-score is found by converting a value to a standardized scale. In practice, a z-score is the number of standard deviations that a data value is from the mean and calculated for the population as

$$z = \frac{x - \mu}{\sigma} \tag{2.5}$$

and for the sample as

$$z = \frac{x - \overline{x}}{s}. \tag{2.6}$$

For extended discussion of the distribution of environmental data (and whether or not a normal distribution is expected and exhibited), consider the following two case studies:

1. **Case Study 1**: The sampling and analysis of 50 groundwater samples over a 24-hour period for nitrate (NO_3-N (mg L^{-1})) levels. Nitrate concentrations in pristine environments with limited human activities are fairly consistent with the natural processes in such environments.

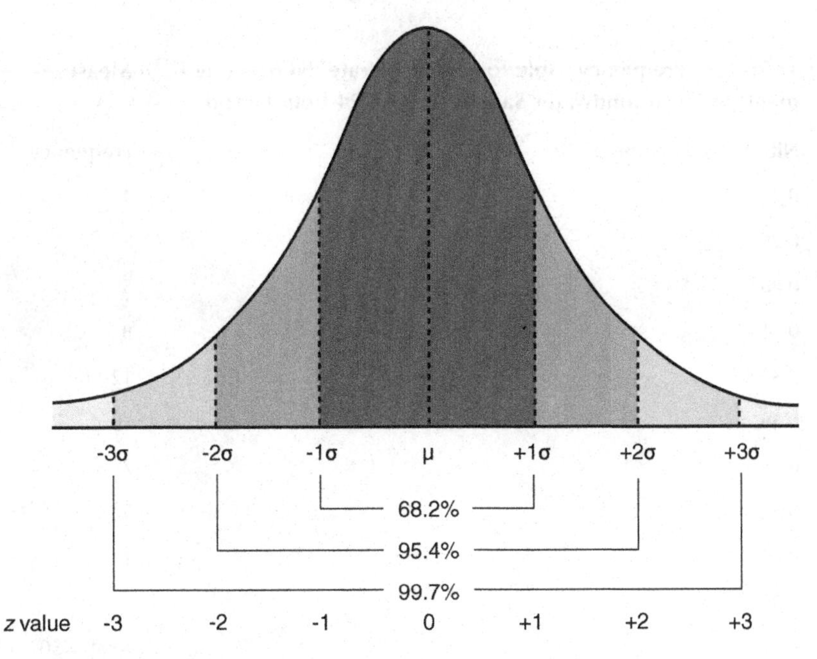

FIGURE 2.6 The bell-shaped normal (Gaussian) distribution showing z-scores (standardized values) that correspond to areas under the curve.

2. **Case Study 2**: The sampling and analysis of 52 samples over a 24-hour period for cadmium (Cd (μg L^{-1})) at a location just downstream from an industrial effluent pipe. Such an environment is largely influenced by human activities and is expected to exhibit higher variability in regards to pollutant concentrations.

For Case Study 1, consider the frequency table (Table 2.1) of all 50 samples and the frequency of each measured concentration in the samples obtained. The **frequency distribution** (Figure 2.7a) shows with slight exceptions that the distribution of the measurements is roughly symmetrical about the mean value (0.33 mg L^{-1}). This is expected since nitrate conditions remain fairly stable with moderate concentration levels in such systems. The smooth curve superimposed line on the distribution (Figure 2.7b) is a plot of the mathematical function for a nearly ideal normal distribution with a population mean μ of 0.33 mg L^{-1} and a population standard deviation σ of ± 0.014. If independent, repeated samples are taken from the same population with a **confidence interval** calculated for each, a certain percentage (**confidence level**) of the intervals will include the unknown population attribute. Here, the extreme values of the range are called **confidence limits**. Let's assume that the sample size is ≤ 30 and calculate according to

$$\mu = \bar{x} \pm \frac{ts}{\sqrt{n}}, \tag{2.7}$$

TABLE 2.1 Frequency Table for the 50 Nitrate (NO_3-N (mg L^{-1})) Measurements in 50 Groundwater Samples over a 24-hour Period

Nitrate (NO_3-N (mg L^{-1}))	Frequency
0.23	1
0.26	2
0.28	5
0.31	8
0.33	12
0.36	10
0.38	7
0.41	3
0.44	1
0.46	1
	Total = 50

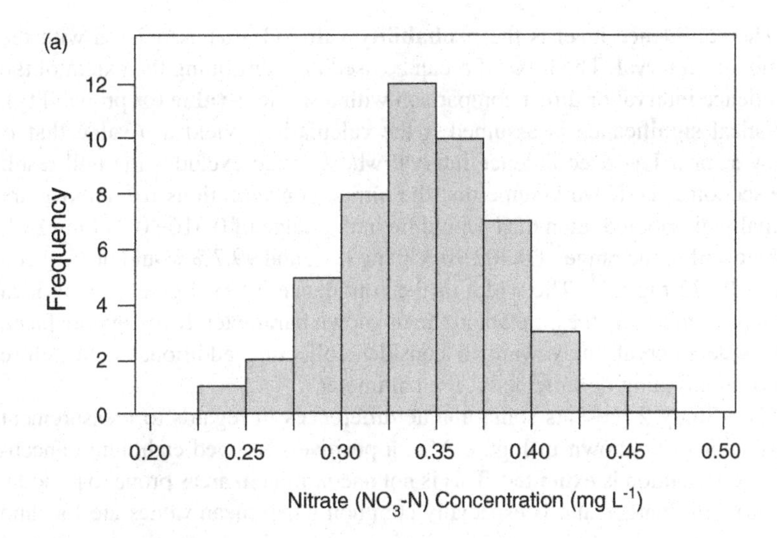

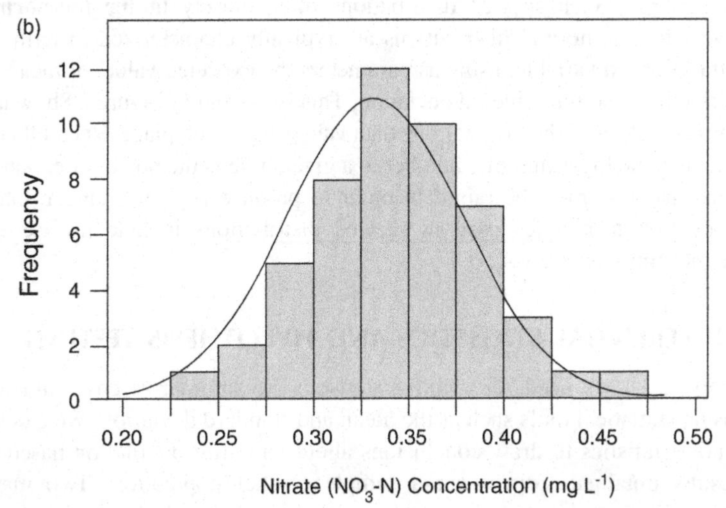

FIGURE 2.7 a) The frequency distribution showing with slight exceptions that the distribution of nitrate (NO_3-N (mg L^{-1})) measurements is roughly symmetrical about the mean value (0.33 mg L^{-1}). b) A smooth curve is superimposed on the distribution. Shown is a plot of the mathematical function for a nearly ideal normal distribution with a μ of 0.33 mg L^{-1} and a σ of $\pm$ 0.014.

where s = the measured standard deviation, n = number of measurements, and t = Student's t, obtained from Appendix II. The **Student's *t*** is a statistical tool used to express confidence intervals. Note that in Appendix II, the *d.f.* are equal to n-1. If there are seven data points, for example, the *d.f.* are six. If independent, repeated samples are taken from the same population with a confidence interval calculated for each, a certain percentage (confidence level) of the intervals will include the unknown population attribute.

The confidence level is the **probability value** (1 - α) associated with the confidence interval. The level of α can be used in determining the extent of the confidence interval or direct comparison with a single ***p*-value** (or probability). Statistical significance is assumed if the calculations yield a p-value that is below α, or a $1 - \alpha$ confidence interval whose range excludes the null result (see section 2.4). If we assume that the nitrate concentrations from above are normally distributed, then 68% would lie in the range of 0.316–0.344 mg L^{-1}, 95% lie within the range of 0.302–0.350 mg L^{-1}, and 99.7% would lie between 0.288–0.372 mg L^{-1}. The width of the confidence interval gives one an idea about how uncertain they are about the unknown parameter. Investigators faced with wide intervals may want to consider collecting additional data before reporting anything definite about the parameter.

Case Study 2 presents some unique differences in regards to measurement distributions. As shown in Figure 2.8a, a positively skewed cadmium concentration distribution is exhibited. This is not uncommon in areas prone to periodic industrial discharges and is especially common when mean values are low and variances large. Such skewed distributions often closely fit the **log-normal distribution**. Log-normal distributions are typically characterized in terms of a log-transformed variable, using as parameters the expected value, or mean, of its distribution, and the standard deviation. This is evident in Figure 2.8b, where log transformation of the original raw data values has taken place. Recall that in mathematics the logarithm of a number to a given base is the power or exponent to which the base must be raised in order to produce a given number. Other transformation techniques used in skewed distributions include inverse and square root transformations.

2.4. INFERENTIAL STATISTICS AND HYPOTHESIS TESTING

In section 2.3.1 we used "descriptive statistics" to summarize environmental data using statistical tools such as the mean and standard deviation. We can use inferential statistics to draw conclusions about an entire population based on the results obtained from a small sample of that population. Two major activities of inferential statistics are i) significance tests and ii) examination of confidence intervals.

The determination of **statistical significance** is the cornerstone of any analytical method and has obvious importance in quantitative environmental analysis. We covered this topic briefly in section 2.3.1, learning that significance is assumed if the calculations yield a p-value that is below α, or a $1 - \alpha$ confidence interval whose range excludes the null result. Here, a test of the truth of the **null hypothesis (H_0)** is performed. Ideally, we are asking how likely it is that the null hypothesis is true. This answer is usually in the form of a p-value. The error of rejecting H_0 when it is actually true is termed a **Type I error**. Conversely, the error made when H_0 is false (but accepting it as truth) is termed a **Type II error**. The latter is more prevalent when α is decreased.

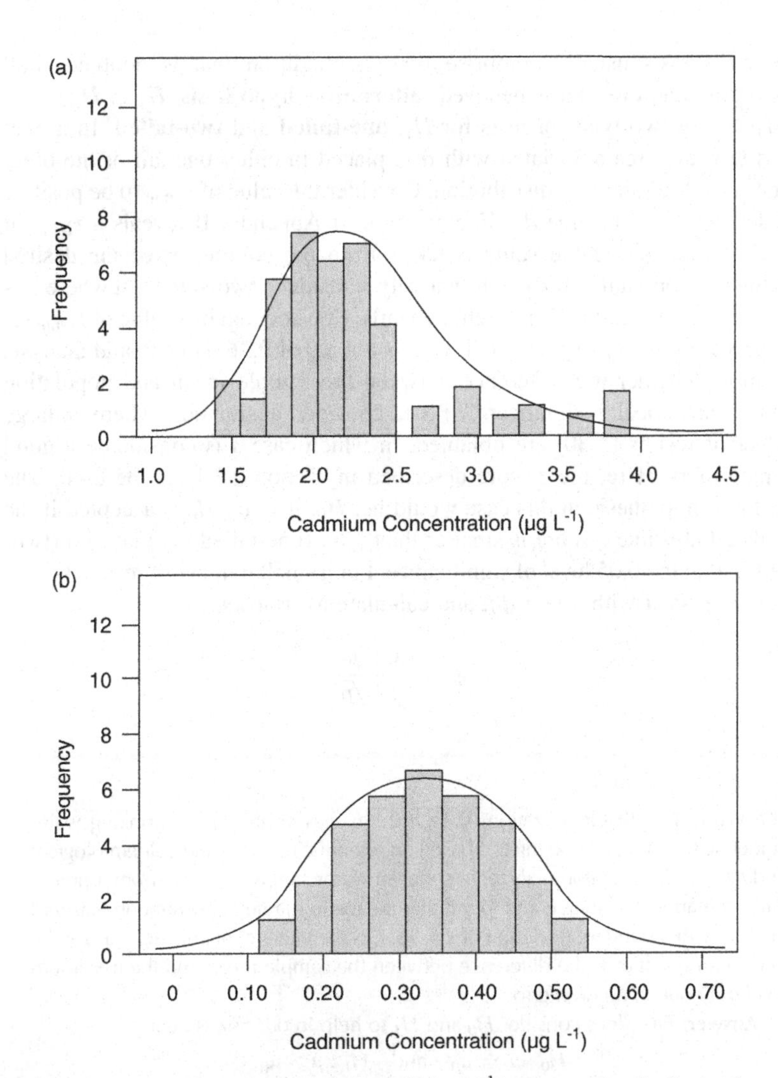

FIGURE 2.8 a) A positively skewed cadmium (Cd (μg L^{-1})) concentration distribution from the analysis of 52 samples over a 24-hour period is exhibited. b) Log transformation of the original raw data values showing a symmetrical distribution at the log level.

If one assumes that the null hypothesis is true, statistical analysis can be used to calculate the probability that the observed difference between the sample and population mean is due solely to random errors. In general, a *p*-value of 0.05 indicates insufficient evidence to say that they differ. A *p*-value of 0.05 indicates significant difference and thus ample evidence to say they do in fact differ. It is important to note that rejecting a hypothesis is to conclude that it is "false." Be critically aware that to accept a hypothesis does not mean that it is true, only that there is insufficient evidence to deem otherwise. Thus, hypothesis tests usually comprise both a condition that is doubted (null hypothesis) and one that is believed (**alternative hypothesis, H_1** or **H_A**).

There are two types of tests for H_1, **one-tailed** and **two-tailed**. In a one-tailed test, the area associated with α is placed in either one tail or the other (+ or −) of the sampling distribution. Consider the value of t_{Table} to be positive and set to 0.05 with nine *d.f.* Examination of Appendix II reveals a t_{Table} of +1.83 (remembering the value is taken from the column twice the desired *p*-value for a one-tailed test). Alternatively, consider a two-sided test where α is placed in both the left (−) and right (+) tails. Consider again a value of t_{Table} set to 0.05 with nine *d.f.* Appendix II reveals a t_{Table} of 2.26. How would analysts determine whether the difference between the sample mean and population mean is statistically significant? First, consider a situation where a large number of results ($\geq$30) are obtained, in which case s is considered a good estimate of σ. Here, the *z*-score described in section 2.3.1 can be used. The stated null hypothesis in this case would be: H_0: $\mu = \mu_0$. H_0 is accepted if the calculated absolute *z*-score is smaller than 1.64 (one-tailed test) or 1.96 (two-tailed test) at the 0.05 level of significance. For a small number of results ($\leq$30), we use the ***t-test*** with $n - 1$ *d.f.* and calculate as follows:

$$t_{Calc} = \frac{\bar{x} - \mu_0}{s/\sqrt{n}}. \tag{2.8}$$

Example Problem 2.2

The World Health Organization (WHO; 2006) has established a drinking water guideline for nitrate of 500 mg L^{-1} based on scientific research and epidemiological findings. An investigator analyzed a series of water samples ($n = 7$) from a potentially contaminated source and found that the mean nitrate concentration was 540 mg L^{-1} with a standard deviation of $s = 30.4$. Is the water contaminated (above the maximum level) or is the difference between the sample mean and the maximum level only due to random error?

Answer: First, let's consider H_0 and H_1 to help in our assessment:

$$H_0 : \mu = \mu_0 \quad \text{and} \quad H_1 : \mu > \mu_0,$$

where μ = the true mean and μ_0 = 500 mg L^{-1}, the maximum level. A one-tailed test can now be set up. Note that the hypothesis established is one-tailed, since we were looking at the probability that the sample mean was either > or ≤ 500 mg L^{-1}. Inserting the values into Equation 2.8 reveals

$$t_{Calc} = \frac{540 - 500}{30.4/\sqrt{7}} = 3.48.$$

Referring to Appendix II, we see that at the 95% confidence level (p=0.05) with 7−1 = 6 *d.f.*, the t_{Table} = 1.94. If the $t_{Calc} > t_{Table}$, we reject the null hypothesis. In the case of the investigation, the t_{Calc} = 3.11 > t_{Table} = 1.94, thus rejecting H_0 and concluding that the water is contaminated above the maximum suggested level.

You have just learned how to perform a *t*-test for the comparison of means, and a statistically sound way to detect systematic errors in your analyses. Knowing this, how would you effectively compare standard deviations (variances) of two sets of data? In other words, this is a comparison of precision and thus the amount of random error in two data sets. If we assume a normal distribution, an ***F*-test** that compares the difference between two sample variances can be used to test $H_0 = \sigma_1^2 = \sigma_2^2$ as follows:

$$F_{\text{Calc}} = s_1^2/s_2^2, \tag{2.9}$$

with the *d.f.* for the numerator and denominator being $n_1 - 1$ and $n_2 - 1$, respectively. Consider the one-tailed problem below.

Example Problem 2.3

Let's assume that the nitrate concentrations of the following water samples were measured by two different methods, each method tested a total of 10 times. The results were as follows:

Method	Mean (mg L^{-1})	Standard deviation (mg L^{-1})
1	570	23.4
2	570	12 1

Determine if s_1^2 (Method 1) exceeds s_2^2 (Method 2)—a one-tailed characteristic.

Answer: Our null hypothesis is stated: $H_0 : s_1^2 \leq s_2^2$ and $H_1 : s_1^2 > s_2^2$. Using Equation 2.9 we have

$$F_{\text{Calc}} = 23.4/12.1^2 = 3.74.$$

The table value (Appendix III) for *d.f.* in each case and a one-tailed, 95% confidence level is $F_{\text{Table}} = 3.179$. Here, $F_{\text{Calc}} > F_{\text{Table}}$, thus we reject H_0, and s_1^2 exceeds s_2^2 significantly.

2.5. OUTLYING RESULTS

Examine the data set below showing six conductivity measurements (mS cm^{-1}) from a given river monitoring station:

53.7 53.9 53.8 57.9 53.1 53.5.

Upon visual inspection it appears that 57.9 is an **outlier**, a result that is outside of the range of what random errors would typically produce. To effectively determine whether it is an outlier, one would need to employ a proper statistical test such as the **Dixon's *Q*-test**. This test compares the difference between the suspect value (in this case, 57.9) and the nearest value with the

TABLE 2.2 Critical Values of Q for a Two-Sided Test ($p = 0.05$)

Sample Size	Q_{Table}
3	0.970
4	0.829
5	0.710
6	0.625
7	0.568
8	0.526
9	0.493
10	0.466

difference between the highest and lowest values producing a ratio calculated from:

$$Q_{Calc} = \frac{|\text{suspect value} - \text{nearest value}|}{(\text{largest value} - \text{smallest value})}. \tag{2.10}$$

Q_{Calc} values are found in Table 2.2 for a two-tailed test ($p = 0.05$). Note that if $Q_{Calc} > Q_{Table}$, rejection of the suspect value is warranted. For the data set above, list in increasing value and calculate Q:

53.1 53.5 53.7 53.8 53.9 57.9

$$Q_{Calc} = \frac{|57.9 - 53.9|}{(57.9 - 53.1)} = 0.833.$$

Table 2.2 gives a $Q_{Table} = 0.568$ for seven samples. Since $0.833 > 0.568$, the suspect value is rejected. It is in fact an outlier. Remember to report the Q-test and removal of the suspect data point when preparing final results for evaluation purposes (e.g., when calculating the sample mean).

2.6. ANALYSIS OF VARIANCE

Using hypothesis testing one can determine whether the difference between the sample mean and population mean is significant. How does one go about determining differences in means between groups or factors for statistical significance? This can be accomplished by performing an **analysis of variance (ANOVA)**. In ANOVA, the total variance is portioned into components, those

TABLE 2.3 ANOVA Summary Table

Source of Variation	Sum of Squares (SS)	*d.f.*	Mean Square (MS)	*F*
Between-Groups (Treatment)	$SS_{treatment} = \sum_{i=1}^{p}\sum_{j=1}^{n_1}(\overline{x}_i - \overline{x})^2$	$p-1$	$MS_{treatment} = \frac{SS_{treatment}}{p-1}$	$F = \frac{MS_{treatment}}{MS_{error}}$
Within-Groups (Error)	$SS_{error} = \sum_{i=1}^{p}\sum_{j=1}^{n_1}(x_{ij} - \overline{x}_i)^2$	$n-p$	$MS_{error} = \frac{SS_{error}}{n-p}$	
Total	$SS_{total} = \sum_{i=1}^{p}\sum_{j=1}^{n_1}(x_{ij} - \overline{x})^2$	$n-1$		

due to random error (within-groups) and the components that are due to differences between means (between-groups). Subsequently, the latters components are tested for statistical significance using *F*. If significant, H_0 is rejected and we accept the H_1 that the means (population) are in fact different. A generalized ANOVA table with the components required for calculation of significance is presented in Table 2.3. Note that the **sum of squares** adds up as follows:

$$SS_{treatment} + SS_{error} = SS_{total}. \tag{2.11}$$

Like sum of squares, the *d.f.* also add up:

$$(p-1) + (n-p) = n-1.$$

The MS (variances) are found by taking the appropriate SS and dividing by the *d.f.* Finally, the *F* statistic is calculated by dividing the two MS and compared to the F_{Table}. Table *F* values are given in Appendices III and IV.

A **one-way ANOVA** can be used for simple comparisons of treatments or factors. A **two-way ANOVA** can be used for combinations of treatments, where two factors (e.g., pH and temperature), for example, are applied in all possible combinations or levels. These are often termed factorial ANOVA. Picture a situation where two investigators are measuring the total lead concentration in wastewater samples. Table 2.4 provides the ANOVA report listing the probability that indicates whether or not there is a difference between the investigators' measurements. As shown, the Prob $> F$ (the p-value) is 0.0101, which confirms that there is a significant difference among their measurements. (See the Excel Spreadsheet Application I as follows for an example ANOVA calculation. This is the first application that expands on our discussion of Excel concepts introduced in Chapter 1.)

TABLE 2.4 ANOVA Report for the Analysis of Lead in Wastewater Samples

Source	*d.f.*[a]	Sum of Squares	Mean Square	*F*-Ratio	Prob>*F*
Investigators	1	0.53121	0.63112	11.997	0.0101
Error	6	0.40150	0.40991	–	–
Corrected Total	7	0.93271	–	–	–

[a]*d.f. = degrees of freedom for each source of variation.*

Excel Spreadsheet Application I

Four analysts performed five measurements each of the iron concentration (mg L^{-1}) in a given sample (Table 2.5). Perform a one-way ANOVA to determine whether there are significant differences among the analysts' measurements.

Step 1: Select the **Data Analysis** tab under **Tools**, then select **ANOVA: Single Factor** and click **OK**. Next, choose the **Input Range** in the window that appears and highlight cells C6 through F6. Note that an α value of 0.05 was chosen in the options window.

Step 2: Click **OK** and the ANOVA results will appear in a separate worksheet. Here, Excel automatically displays F_{Table} corresponding to the α level chosen earlier. Since F is greater than F_{Table} we can safely conclude that the measurements differ significantly between analysts. In addition, since the *p*-value (0.000299) < α (0.05), we can also reject the null hypothesis and conclude that the measurements differ significantly between analysts.

TABLE 2.5 Analyst Measurements of the Iron Concentration in a Given Environmental Sample

Measurements	Analyst 1	Analyst 2	Analyst 3	Analyst 4
1	12.2	11.6	12.9	11.4
2	11.8	10.9	13.1	10.9
3	12.6	11.8	12.0	11.2
4	12.4	11.5	11.9	10.9
5	11.9	12.0	12.1	10.9

(a)

	A	B	C	D	E	F
1						
2						
3						
4						
5		Measurement	Analyst 1	Analyst 2	Analyst 3	Analyst 4
6		1	12.2	11.6	12.9	11.4
7		2	11.8	10.9	13.1	10.9
8		3	12.6	11.8	12.0	11.2
9		4	12.4	11.5	11.9	10.9
10		5	11.9	12.0	12.1	10.9
11						
12						
13						
14						
15						
16						

(b)

	A	B	C	D	E	F	G	H
1								
2								
3	Summary							
4	Groups	Count	Sum	Average	Variance			
5	Column 1	5	60.9	12.18	0.112			
6	Column 2	5	57.8	11.56	0.173			
7	Column 3	5	62	12.4	0.31			
8	Column 4	5	55.3	11.06	0.053			
9								
10	ANOVA							
11	Variation Source	SS	d.f.	ms	F	p-value	Fcrit	
12	Between Groups	5.548	3	1.849	11.415	0.000299	3.2388	
13	Within Groups	2.592	16	0.162				
14								
15	Total	8.14	19					
16								

FIGURE 2.9 Goes with Excel Exercise I.

2.7. GRAPHICAL DATA ANALYSIS

The summary statistics discussed in section 2.3 provide unbiased estimates when the data are normally distributed. But as mentioned, environmental data sets are often skewed, and the use of summary statistics alone may hide the data structure and therefore be misleading. The use of graphical methods can reveal important features in such cases and are widely used in environmental data analysis efforts. Discussion in this section will concentrate on

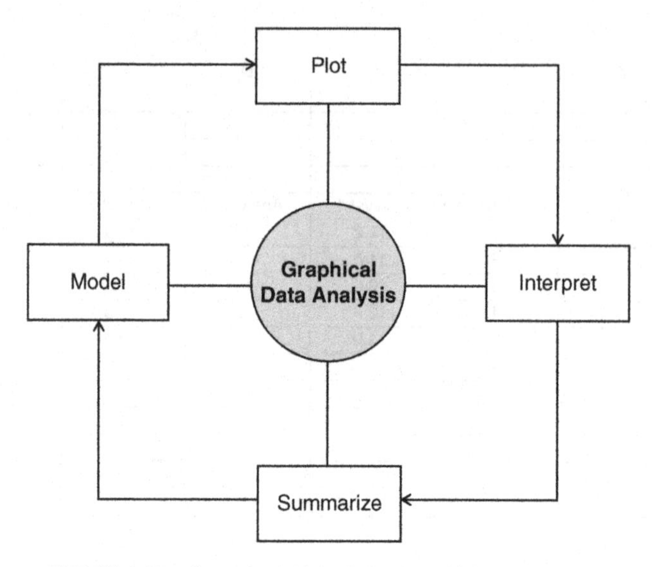

FIGURE 2.10 General principles behind graphical data analysis.

one key graphical display method: the **scatterplot**. The general principles of graphical data analysis that broadly apply to all methods are displayed in Figure 2.10.

2.7.1. The Use of the Scatterplot in Environmental Data Analysis

Scatterplots shows the relationship between two quantitative variables: the **response variable** (measures the outcome of a given study) and the **explanatory variable** (influences change in the response variable). Interpretation is described via direction, form, or strength of the relationships between the two variables. If the problem is roughly linear, one can summarize with correlation, means, and standard deviations. **Correlation** measures direction and strength of the linear relationship between two quantitative variables and will be defined shortly. Finally, a model (e.g., regression) can be generated if the relationship is roughly linear. **Regression** describes a group of methods that summarize the degree of association between one variable (or set of variables) and another variable (or set of variables).

Consider the scatterplot presented in Figure 2.11a, which shows the explanatory variable (horizontal x-axis, (river flow rate)) plotted against the response variable (vertical y-axis (total phosphorus (TP) concentration)) for an extended agriculturally-dominated watershed monitoring campaign. Assessing seasonal (e.g., monthly) trends in TP concentrations in rivers and streams is important given that large proportions of TP are delivered to waterbodies from such watersheds during high periods of rainfall (resulting in increased water-flow

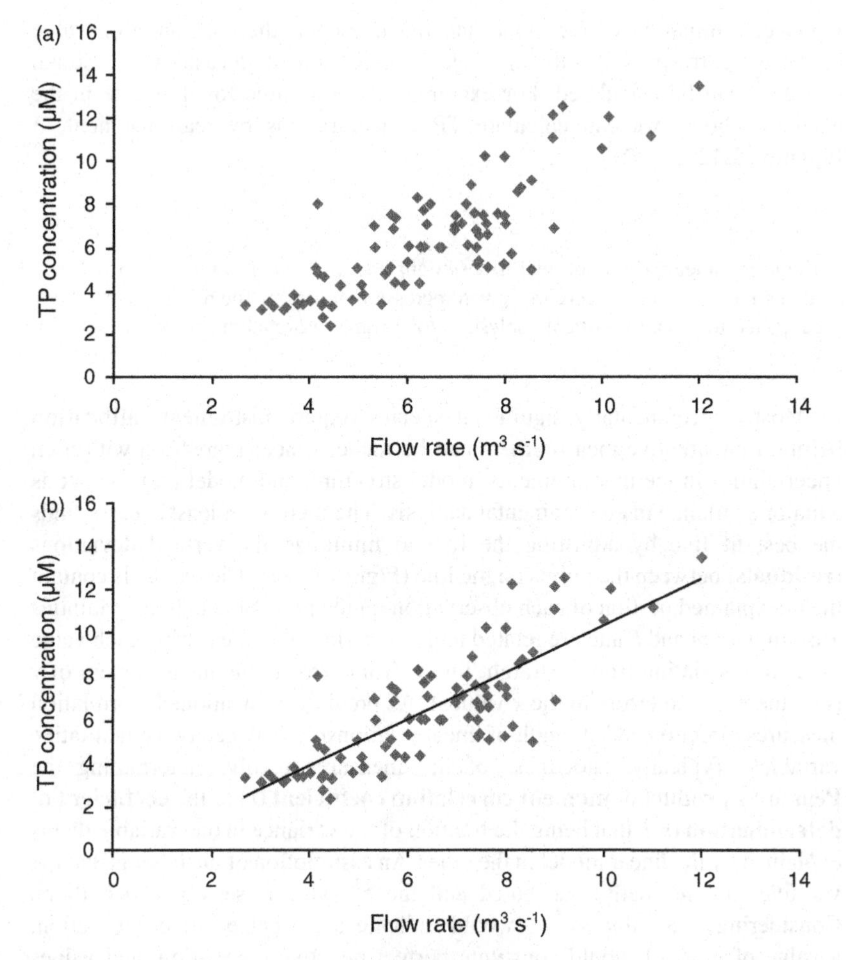

FIGURE 2.11 A scatterplot showing: a) river flow (explanatory variable) *versus* total phosphorus (TP) (response variable) and b) linear regression applied to the data set to determine the extent of correlation between the two variables.

rates). Applying a **linear regression** (also termed the least-squares best fit line) to the data set results in Figure 2.11b, which allows for the determination of the presence or absence of correlation between the two variables. The best fit line through these data is defined by:

$$y = mx + b, \tag{2.12}$$

where m = the slope of the line and b = the y-intercept (the point at which the line crosses the y-axis). The slope is defined by

$$m = \frac{y_2 - y_1}{x_2 - x_1}. \tag{2.13}$$

Upon determination of the slope and the intercept, the concentration value (x-value) corresponding to the relevant instrument measurement signal (y-value) can be calculated. For example, given a value for flow rate in the example above, we can calculate TP concentrations by rearrangement of Equation 2.12.

TEXTBOX 2.4

The term "**linear**" does not refer to a straight line, but rather to the way in which regression coefficients occur in a given regression equation. The results have to be subjected to proper statistical analysis before true interpretation can occur.

Most environmentally significant species require instrument **calibration** before a quantitative measurement can be made. Linear regression with high uncertainties in the measurements, model structure, and model permanence is a major challenge in environmental analysis. The method of least squares finds the best fit line by adjusting the line to minimize the vertical deviations (**residuals**) between the points on the line (Figure 2.12a). The residuals contain the unexplained portion of each observation on the plot. Note that uncertainties arise in both m and b and are related to the uncertainty in measuring each value of y. Any deviation from a straight line is from error in the measurement of y (i.e., there are no errors in the x values). As previously mentioned, correlation measures direction and strength of linear relationship between two quantitative variables. Typically, "goodness of fit" measures involve determining the **Pearson's product** or **moment correlation coefficient (*r*)** or the **coefficient of determination (r^2)**, that being the fraction of the variance in one variable that is explained by the linear model in the other. An assumption of the latter is that the variables are normally distributed and the r^2 value is sensitive to outliers. Considering r, a value $= +1$ would indicate a perfect positive correlation, a value of $r = -1$ would constitute perfect negative correlation, and values approaching $r = 0$ would be representative of no correlation between x and y. For Figure 2.12a, an $r = 0.68$ was observed, indicating positive linear association between increased flow rates and TP concentration. This trend confirms the original assumption that the highest concentrations of TP occur at the highest flow rate levels (for the watershed described). Ultimately, uncertainty can be quantified by displaying 95% confidence limits around the best fit line (gray shaded area in Figure 2.12b).

A full treatment of the standard deviation for results obtained from an instrument calibration curve, S_c, is given by

$$s_c = \frac{s_r}{m}\sqrt{\frac{1}{M} + \frac{1}{n} + \frac{(\bar{y}_c - \bar{y})^2}{m^2\left[\sum x_i^2 - \frac{(\sum x_i)^2}{n}\right]}}, \tag{2.14}$$

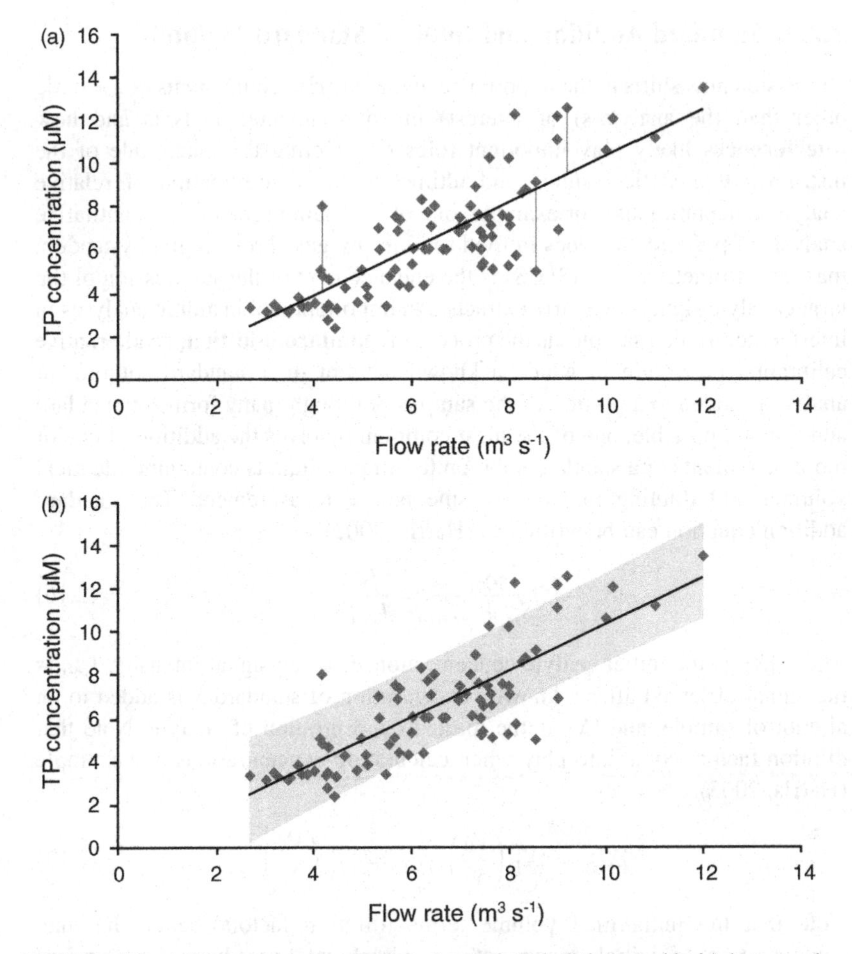

FIGURE 2.12 a) Linear regression showing the vertical deviations (residuals) between the points on the best fit line and b) 95% confidence limits applied (gray shaded area) to the best fit line.

where m is the value of the slope, s_r is the standard deviation about the regression, $\bar{y}_c$ is the average of M replicate analyses of the unknown value being determined from the calibration curve, and n is the number of calibration points. In the above equation, S_r is determined by

$$s_r = \sqrt{\frac{\sum_{i=1}^{n} (y_i - (b + mx_i))^2}{n - 2}}, \tag{2.15}$$

where two degrees of freedom are lost in the determination of m and b.

2.7.2. Standard Addition and Internal Standard Methods

Discussion now shifts to the importance of the **matrix** (components of a sample other than the analyte(s) of interest) in environmental analysis and how **interferences** likely play important roles in affecting the magnitude of the instrumental analytical signal and ultimately the determination of relative analyte concentrations. For example, one of the main problems in quantitative analysis of pesticide residues in food samples by gas chromatography-tandem mass spectrometry (GC-MS/MS) is the enhancement or the suppression of the target analyte signals in matrix extracts. An important tool in aiding analyses in interference-ridden samples is the process of **standard addition**, an alternative calibration procedure in which a known amount of a standard solution of analyte is added to a portion of the sample. Although many forms of standard addition are possible, one of the most common involves the addition of one or more increments of a standard solution to sample aliquots containing identical volumes and diluting to fixed volume before measurement. The standard addition equation can be written as (Harris, 2003)

$$\frac{[\mathrm{X}]_i}{[S]_f + [X]_f} = \frac{I_x}{I_{s+x}}, \tag{2.16}$$

where $[X]_i$ is the initial analyte concentration, I_x is the signal intensity, I_{s+x} is the signal observed after a known concentration of standard S is added to an aliquot of sample, and $[X]_f$ is the diluted concentration of analyte. Note that dilution factors come into play when calculating concentrations, for example (Harris, 2003),

$$[X]_f = [X]_i\left(\frac{V_0}{V}\right) \ [S]_f = [S]_i\left(\frac{V_S}{V}\right). \tag{2.17}$$

Note that the initial/final volume terms (dilution factors) relate the final concentration to initial concentrations. Graphical procedures for standard addition are commonly performed with a requirement of calibration graphs for each sample. This is in contrast to conventional calibration studies where one plot provides assessment of multiple samples.

Internal standards are in contrast to the standard addition method as they require the use of a different substance from the measured analyte of interest. In this method, a known amount of standard reference material is added to all samples, standards, and blanks with the intention of the reference material to aid in identifying and correcting for any run-to-run discrepancies. This substance can then be used for calibration by plotting the ratio of the analyte signal to the internal standard signal as a function of the analyte concentration of the standards, ultimately taking into consideration a given dilution factor (Harris, 2003):

$$\frac{A_x}{[X]} = F\left(\frac{A_s}{[S]}\right), \tag{2.18}$$

where F = a response factor and $[X]$ and $[S]$ are the concentrations of analyte and standard after mixed, respectively. Internal standards are especially beneficial for chromatographic analyses in which the instrument response varies widely between runs, or if loss of analyte occurs, for example, during sample processing. For example, liquid flow rates that differ by a few percent in a high performance liquid chromatography (HPLC) experiment could change the detector response (Kutnink, et al., 1987). A calibration curve is only accurate for the one set of conditions under which it is obtained. If, for example, the relative response of the detector to the analyte and standard increases by 7.5% due to a change in solvent flow rate, signal from the analyte typically also increases by 7.5%. Since the concentration of the internal standard is known, the accurate concentration of analyte can be determined.

2.8. MULTIVARIATE DATA ANALYSIS: A BRIEF INTRODUCTION

An important objective of environmental chemistry is to understand and characterize chemical and physical measurements of complex systems. The complexity of spatial and temporal dynamics in such systems has spurred interest in the use of **multivariate statistical techniques**. These methods are appropriate for meaningful data reduction and interpretation of multi-parameter investigations. In our discussion on simple regression, we learned that two variables can be graphically displayed. What if we have three or more variables? Does graphical representation still apply? Here, multivariate analysis can be fully explained by matrix algebra concepts. It is beyond the scope of this chapter to delve into such concepts, but a simple understanding of data matrices is assumed. This section highlights the use of common multivariate data analysis techniques and how they can be applied to real environmental situations.

2.8.1. Multivariate Correlation Analysis

As discussed, the aim of correlation analysis is to detect relationships among variables with the measure of linearity assessed by correlation. The importance of bivariate correlation was fully discussed, but considerations to multivariate data sets are needed. Here, the importance is the proportion of the response variable that can be attributed to the combined effects of all the independent explanatory variables acting together. A number of correlation techniques and visual tools can be used to assess relationships within multivariate data sets. Such techniques are discussed in detail as follows using multivariate lake data sets highlighting common measurable environmental parameters (Table 2.6).

A correlation matrix that summarizes the strength of the linear relationships from the sediment data housed in Table 2.6 is presented in Table 2.7. Typically, in such environmental data sets, coefficients from ±0.8 to 1.0 indicate a strong correlation, while values from ±0.5 to 0.8 indicate a moderate correlation.

TABLE 2.6 Descriptive Statistics for a Generic Lake Sediment Data Set

Parameter	Mean ($n = 15$)	Standard Deviation
Organic matter (OM)[a]	2.21	0.17
Total nitrogen (TN)[b]	1.56	0.41
Total phosphorus (TP)[b]	0.72	0.12
Cd[c]	12.1	1.00
Cu[c]	32.2	2.58
Fe[c]	7879	162
Mn[c]	724	47.1
Ni[c]	27.8	7.33
Pb[c]	75.1	15.6
Zn[c]	96.8	4.72

[a]*Units = %.*
[b]*Units = mg g^{-1} dry weight.*
[c]*Units = μg g^{-1} dry weight.*

TABLE 2.7 Correlation Matrix for the Generic Lake Sediment Data Set

	OM	TN	TP	Cd	Cu	Fe	Mn	Ni	Pb	Zn
OM	1.00	0.40	0.46	0.54	0.27	0.08	−0.05	0.71	0.44	0.39
TN	0.39	1.00	0.60	0.00	0.44	−0.08	0.09	0.57	0.50	0.27
TP	0.46	0.60	1.00	0.22	0.27	0.44	−0.15	0.63	0.32	0.30
Cd	0.54	0.00	0.22	1.00	0.07	0.04	−0.06	0.50	−0.02	0.30
Cu	0.26	0.44	0.27	0.07	1.00	0.17	0.01	0.20	0.65	0.50
Fe	0.08	−0.08	0.44	0.04	0.17	1.00	0.38	−0.09	−0.07	0.25
Mn	−0.05	0.09	−0.15	−0.06	0.01	0.38	1.00	−0.41	−0.19	0.15
Ni	0.71	0.57	0.63	0.50	0.20	−0.09	−0.41	1.00	0.29	0.39
Pb	0.43	0.50	0.32	−0.02	0.66	−0.07	−0.20	0.29	1.00	0.25
Zn	0.39	0.27	0.30	0.30	0.50	0.25	0.15	0.39	0.25	1.00

Finally, values less than ±0.5 indicate a weak correlation. Organic matter (OM) showed weak to moderate correlation with Pb (0.43), Cd (0.54), and Ni (0.71), confirming suspicion of higher OM levels retaining heavy metals. Intermetallic relationships as shown by Cd/Ni (0.50), Zn/Cu (0.50), and Cu/Pb (0.60) reveal similar behavior during transport in the aquatic system. In addition, total nitrogen (TN) showed a moderate correlation with total phosphorus (TP) (0.60), which could indicate a common source of nutrient input into the aquatic system.

2.8.2. Pattern Recognition Techniques

Pattern recognition techniques seek to identify similarities and regularities present in a given data set to attain natural classification (groupings). Two general areas are considered: **unsupervised** and **supervised learning methods**. In unsupervised methods, information on the individual groupings is likely unknown but not necessarily considered a prerequisite for data analysis. In supervised methods, the groupings of samples must be known to allow accurate predictions to be performed. More detailed explanations of these methods along with relevant applications are discussed as follows.

Cluster analysis (CA) is a common unsupervised learning technique for grouping objects of similar kinds into organized categories. More specifically, CA is an exploratory analysis tool that aims to sort different objects into groups in a way that the degree of association between two objects is maximal if they belong to the same group and minimal if not. The clusters formed with this tool should be highly internally homogenous (members are similar to one another) and highly externally heterogeneous (members are *not* like members of other clusters). Students should be aware that CA is not as much a typical statistical test as it is a "collection" of different algorithms that "put objects into clusters according to defined similarity rules."

Most clustering algorithms partition the data based on similarity; the more similar, the more likely that they belong to the same cluster. **Euclidean distance**, the square root of the sum of the square of the x difference plus the square of the y distance, is the most common distance measure. CA can accept a wide range of input data, and since one may be clustering items measured on different scales, some form of data pre-processing is likely. While these are generally called "similarity" measures, one may also see the terms "proximity," "resemblance," and "association." Combining of groups in CA is visually represented by a two-dimensional **dendrogram**—a hierarchical tree diagram that lists each observation and shows which cluster it is in and when it entered its cluster. As a result, we link more and more objects together and aggregate (amalgamate) larger and larger clusters of increasingly dissimilar elements. When the data contain a clear "structure" in terms of clusters of objects that are similar to each other, then this structure will often be reflected in the hierarchical tree as distinct branches. As the result of a successful analysis with the joining method, an investigator is able to detect clusters (branches) and interpret those branches.

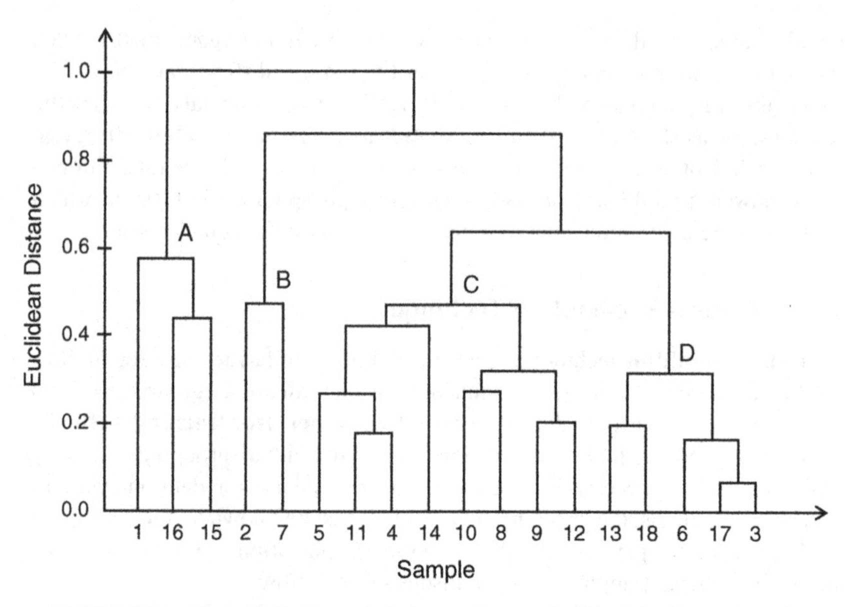

FIGURE 2.13 Hierarchical cluster analysis tree diagram showing four distinct clusters.

Let's now apply the theory above with a real-world application, considering, for example, the assessment of heavy metal distribution in an urban stream (surface sediments) affected by urban drainage. Urban streams are often heavily contaminated since they receive local runoff and the discharge of overflows from residential and industrial discharges. **Hierarchical cluster analysis** can be employed to quantify urban drainage impact and identify any specific areas of metal contamination. Refer to the representative hierarchical tree diagram for this study presented in Figure 2.13, which shows four distinct clusters (A, B, C, and D) with corresponding linkage distances. Cluster A contained the highest concentration levels (three times natural background levels) of Zn, Pb, and Cd in samples 1, 16, and 15. Cluster B comprises two profiles of samples collected (2 and 7), where cadmium concentrations were twice the averages of cluster A. High concentrations of cadmium could indicate, for example, a historical record of contamination by plating industries or precious metal processing plants. Cluster C (samples 5, 11, 4, 14, 10, 8, 9, and 12) represents intermediate contamination levels (two to three times natural background levels) of Zn, Pb, and Cd. Finally, Cluster D (samples 13, 18, 6, 17, and 3) represents low contamination levels (two times natural background levels) for Zn, Pb, and Cd. How were the clusters determined in the above example? A typical rule of thumb is to choose a place where the cluster structure remains stable for a long distance. Some other possibilities are to look for cluster groupings that agree with existing or expected structures, or to replicate the analysis on subsets of the data to see if the structures emerge consistently.

TEXTBOX 2.5

The majority of data clustering algorithms utilized in environmental data analyses are classified into **hierarchical** procedures. Hierarchical clustering creates a hierarchy of clusters that can be represented in a tree structure called a **dendrogram**.

Discriminant analysis (DA) is a supervised classification technique useful for investigating and classifying a set of observations into predefined classes. Unlike the unsupervised technique of CA discussed in the previous section, DA employs *a priori* knowledge in order to reach a solution. Investigators have the choice of modeling algorithms available to perform the necessary classification with ***k*-nearest neighbor** (*k*-NN) and **linear discriminant analysis** (LDA) being the most commonly used. In terms of applications, DA can be used to perform both analysis and classification. In the analysis process, variables that contribute most to the difference are determined. This allows for enhanced interpretation. The discriminant function can be used to classify unknown objects. Satisfactory separation aids in this process.

For demonstration purposes, consider a simplified example (Figure 2.14) with two classes (circles = contaminated sediments; squares = non-contaminated sediment). Each class can be characterized by multiple variables. Such a plot is a valuable tool in determining visual separation of the two classes of objects. It should be evident from the plot that the two groups form individual

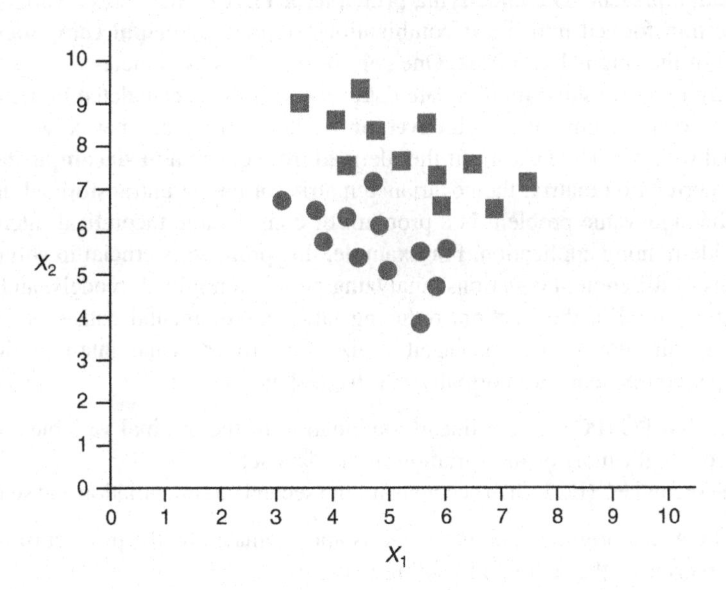

FIGURE 2.14 Discriminant analysis (DA) plot showing the separation of two classes of objects (circles = contaminated sediments; squares = non-contaminated sediments).

clusters. Unfortunately, complete differentiation of the classes is not possible by the sole use of this plot. Ultimately, the use of proper coefficients and the structure matrix accounts for full interpretation, with the larger the coefficient, the greater the contribution of the given variable to the discrimination among groups. Further examination of the structure matrix allows investigators to identify the explanatory variables that contribute to the discrimination among response variables.

2.8.3. Data Reduction

The ability to reduce the amount of data used to describe particular observations in environmental systems is appealing to investigators. The trick, however, is to do this effectively without the loss of vital information. As we learned, all real environmental data contain some form of noise. In addition, variables measured are rarely statistically independent and can be highly correlated. For example, suppose we are studying a string of sampling locations and measuring two environmental parameters at those locations. The second variable may be redundant due to its derivation from (and hence highly correlated with) variable one. One could rightly remove the second variable without losing vital information. A structured way to do this is through a technique called **principal component analysis** (PCA).

PCA is a variable reduction technique used to uncover the dimensions of a set of given variables. This technique is especially useful when there is correlation present. The underlying principle of PCA is that chosen variables can be transformed into linear combinations (termed **principal components** (PCs)) of the original variables. One can think of PCs as vehicles to map the structure of the original multivariate data set. Each PC is calculated by taking a linear combination of an **eigenvector** of the correlation matrix with an original variable. The PCs are further derived from **eigenvalue decomposition** of the correlation matrix, the covariance matrix, or on the untransformed data set. The eigenvalue problem is a problem of considerable theoretical interest and wide-ranging application. For example, this problem is crucial in solving systems of differential equations, analyzing population growth models, and as will be covered in this section, reducing large environmental data sets with varying units down to a manageable size for proper visual interpretation. Principal components are normally constructed as follows:

1. The first PC (PC1) is the linear combination of the original variables that accounts for most of the variation in the data set;
2. The second PC (PC2) then accounts for the second largest variation and so on.

In PCA, the original data matrix, **X**, is approximated by the product of two small matrices—the score and loading matrices:

$$\mathbf{X} = \mathbf{T}\mathbf{L}^{\mathrm{T}},$$

where **T** = the scores matrix and **L** = the loading matrix. The **component loadings** are the correlation coefficients between the variables (rows) and factors (columns) in a given data set. To obtain the percent of variance in all the variables accounted for by each factor, the sum of the squared factor loadings for the factor (column) is taken and divided by the number of variables present in the study. The **component scores** are the scores of each row on each column of data. To compute scores, one takes a standardized score of each variable, multiplies by the corresponding factor loading of the variable for the given factor, and finally sums these products. Note that the PCs are dependent on the units used to measure the original variables as well as the range they assume. Thus, standardization via data pre-processing should be done prior to performing PCA.

Examine the lake water physico-chemical data set presented in Table 2.8. Assume that with the exception of TP and total dissolved solids (TDS), water quality parameters exhibited a near normal distribution. Principal component analysis can be utilized in elucidating the physico-chemical make-up of this

TABLE 2.8 Physico-chemical Water Quality Parameters for PCA analysis

Parameter	Mean (n = 15)	Standard Deviation
pH	9.42	0.051
Dissolved oxygen (DO)[a]	8.71	0.171
Conductivity[b]	45.7	0.220
Temperature[c]	25.7	0.192
Total dissolved solids (TDS)[a]	690	9.441
Total nitrogen (TN)[a]	0.98	0.024
Total phosphorus (TP)[a]	0.09	0.010
Cd[d]	1.11	0.151
Cu[d]	4.93	0.199
Fe[d]	161	3.377
Mn[d]	10.1	0.126
Ni[d]	19.2	0.203
Pb[d]	2.51	0.140
Zn[d]	3.07	0.083

[a]*Units* = $mg\ L^{-1}$.
[b]*Units* = $\mu S\ cm^{-1}$.
[c]*Units* = °C.
[d]*Units* = $\mu g\ L^{-1}$.

aquatic environment and instructive in distinguishing between natural and anthropogenic sources of pollution in such systems. Therefore, PCA was performed on the correlation matrix to gain a more reliable display method and greater understanding about the relationships within the data set.

The resultant PCA report is highlighted in Table 2.9. As shown, the cumulative contribution of the first four PCs accounted for 71% of the cumulative variance in the lake water data set. The corresponding **screenplot** (Figure 2.15) shows the sorted eigenvalues as a function of the eigenvalue index. After the curve starts to flatten out, the corresponding components may be regarded as insignificant. The first four were retained for further analysis.

As shown in Table 2.9, PC 1 had higher loadings for conductivity, TDS, Cd, Cu, and Pb, and may represent metal accumulation with particulate matter. PC 2 is associated with Ni. PC 3 represents the carrier of micronutrients as higher loadings for TN, TP, Fe, and Mn were exhibited. PC 4 is associated with

TABLE 2.9 Principal components analysis report for the physico-chemical lake water parameters[a]

Vector	Eigenvalue	Total variance (%)	Cumulative (%)
1	4.2737	30.52	30.52
2	2.3285	16.63	47.15
3	1.7289	12.34	59.50
4	1.5604	11.14	70.65
5	1.1565	8.261	78.91
6	0.8222	5.873	84.78
7	0.7824	5.588	90.37
8	0.5177	3.698	94.07
9	0.4598	3.284	97.35
10	0.1576	1.126	98.48
11	0.1277	0.912	99.39
12	0.0668	0.477	99.87
13	0.0175	0.125	99.99
14	0.0004	0.003	100.0

[a]*Only the first four eigenvalues were retained for further analysis. See text for explanation.*

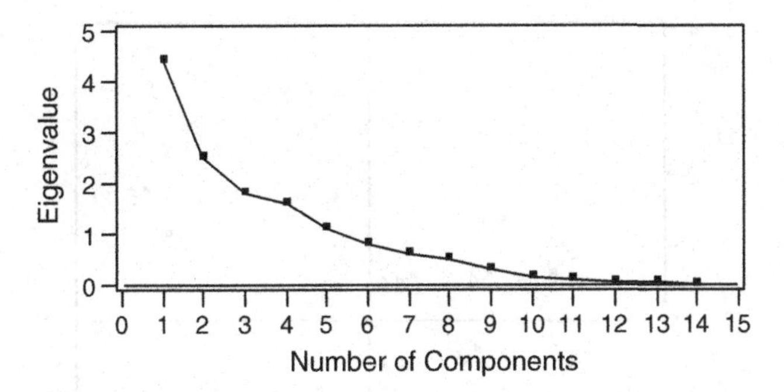

FIGURE 2.15 Screenplot for the 15 physico-chemical water quality parameters. Shown are the sorted eigenvalues as a function of the eigenvalue index.

dissolved oxygen (DO) and Zn. The corresponding **factor loading plot** for the first two PCs is shown in Figure 2.16. The loadings plot shows a matrix of two-dimensional representations of factor loadings. From this, information regarding the correlation between variables can be obtained. The representative **scoreplot** is shown in Figure 2.17. Scoreplots show a two-dimensional matrix representation of the scores for each pair of PCs. Such plots represent the linear projection of objects representing the main part of the total variance of the data. As shown, there are a number of clusters representing the 15 lake water samples.

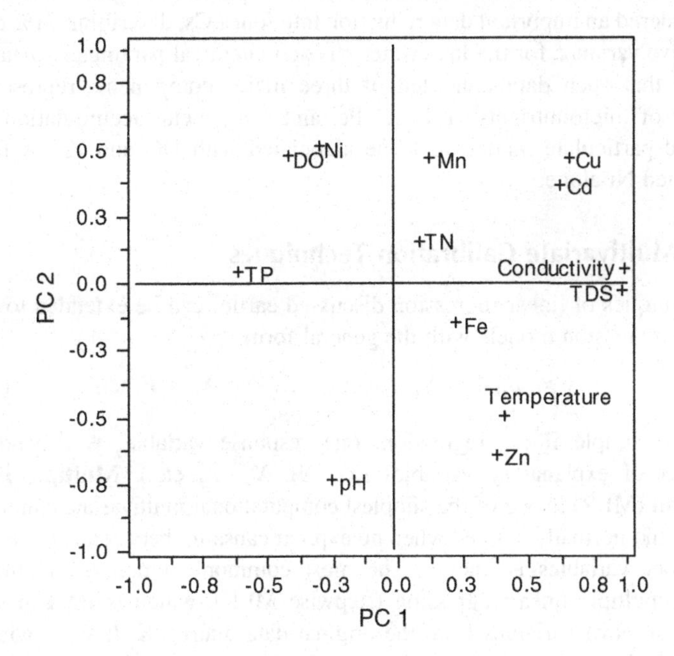

FIGURE 2.16 Factor loading plot for the first two PCs of the PCA water quality study.

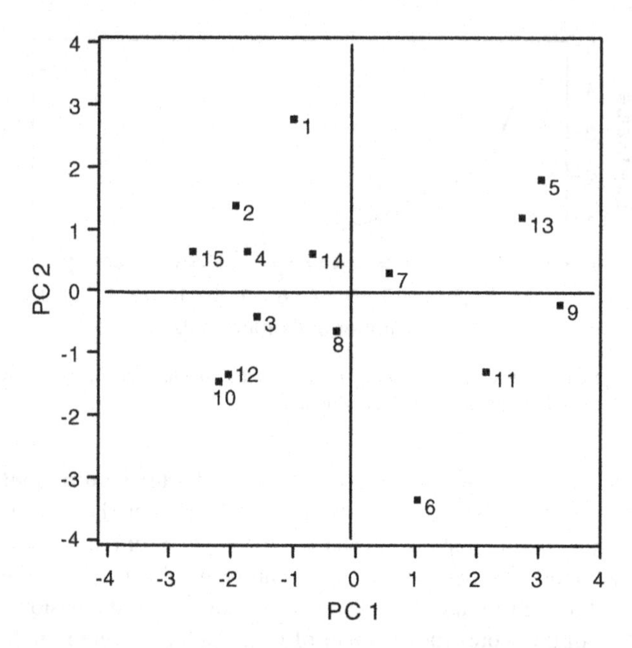

FIGURE 2.17 Scoreplot for the PCA water quality study.

What can we take away from this analysis? Of particular significance is how PCA rendered an important data reduction into four PCs, describing 71% of the cumulative variance for the lake water physico-chemical parameters. Analysis revealed that such data consisted of three major components representing a carrier of micronutrients (TN, TP, Fe, and Mn), metal accumulation with dissolved particulate matter, and one associated with DO and Zn. A fourth represented Ni alone.

2.8.4. Multivariate Calibration Techniques

The techniques of linear regression discussed earlier can be extended to form multiple regression models with the general form

$$y = \beta_0 + \beta_1 X_1 + \beta_2 X_2 + \ldots + \beta_n X_n + E. \tag{2.19}$$

Unlike simple linear regression, one response variable, y, depends on a number of explanatory variables (X_1, X_2, X_3 ..., etc.). **Multiple linear regression** (MLR) is one of the simplest computational multivariate calibration methods and normally applied when no explicit causality between response and explanatory variables is known. The most commonly employed method is stepwise multiple linear regression (stepwise MLR), which is used to select a small subset of variables from the original data matrix, **X**. It starts with the variable that has the strongest correlation with the response. If the chosen variable results in a significant regression, then the variable is retained and selection continues. Next, the variable that produces the largest significant increase of the regression sum of squares is added. This process is termed the **forward selection procedure**. After each forward selection step, the significance of the regression terms already in the model is tested, with the non-significant terms eliminated. This process is termed the **backward elimination step**.

As discussed above, scores on PCs can be used as descriptive variables and aid in finding structure within data sets. These characteristics can be used in MLR in a process termed **principal component regression** (PCR). Ultimately, PCR aids in removing the original variables containing **collinearity**, and thus helps guarantee more stable models. For PCR progression, the original data matrix, **X**, is approximated by a small set of PCs. Decisions based on the percentages of variance that each PC explains are then made to determine which ones to retain. This allows for the number of predictor variables to be reduced. A MLR model can then be built that relates the scores of the PCs to the response or property of interest.

Like PCR, **partial least squares (PLS) regression** is an extension of PCA and utilizes an algorithm that extracts linear combinations of the original data matrix, **X**. The PLS method uses variables u_i (matrix **U**) to model objects separately on the matrix of **Y** dependent data and x_i variables (matrix **T**) for modeling objects separately in the **X** matrix of independent data. The type of PLS model is distinguished by the response. When the response consists of a single variable, it is termed PLS1. When the response is multivariate in nature, it is termed PLS2. In terms of errors, PLS models calculate goodness of fit based on the error of the prediction. Variables that show higher correlation with the response variables are given extra weight due to their effectiveness during prediction.

Students wishing for more in-depth analysis of fundamental data matrix notation, associated calculations, and eigenvector analysis concepts are encouraged to examine Appendix V of this book. In addition, there are various software packages and Excel add-ins that provide complete multivariate statistical analysis. For example, XLSTAT is a complete analysis and statistics add-in for Excel. The use of Excel as an interface makes XLSTAT a student-friendly and highly efficient statistical and multivariate data analysis package that can be used in a wide variety of applications.

2.9. END OF CHAPTER PROBLEMS

2.1 Which of the following sampling techniques is most appropriate for sampling the diurnal fluctuations of nutrient concentrations in a stream? Provide justification for your answer.

a Random sampling

b Simple stratified sampling

c Systematic sampling

2.2 Provide logical reasoning as to why a single sample storage/preservation protocol is not appropriate for all investigations.
2.3 Describe the characteristics of normally distributed data.
2.4 Multiple determination results from monitoring benzene concentrations (mg L^{-1}) in a groundwater well are listed below.

Measurement	Benzene (mg L^{-1})
1	3.35
2	2.25
3	2.09
4	2.01
5	2.15

Calculate the sample mean and standard deviation for this study.
2.5 An analyst performed five replicate measurements of total lead concentration (mg L^{-1}) in a given wastewater sample. Use Excel to calculate the sample mean and standard deviation from the data below:

3.4 3.9 4.1 3.6 3.7 mg L^{-1}

2.6 An analyst has just performed a NO_3-N spectrophotometric calibration study to aid in determining the unknown NO_3-N concentration in a natural water sample. The relevant calibration data are given in tabular form below:

NO_3-N Standard (μM)	Mean Absorbance ($n = 3$)
0	0.0001
2.5	0.0137
5	0.0278
10	0.0500
15	0.0780

a Use the Excel **Analysis ToolPak Regression** tool to perform a linear least-squares analysis of your data. Be sure to include the table highlighting the **Summary Output**.
b Once plotted, use Excel functions to find the unknown NO_3-N concentration of a sample giving an absorbance of 0.0194.

2.7 Four analysts performed three determinations of the dissolved O_2 (DO) level using the same procedure. The results were as follows:

Analyst 1	Analyst 2	Analyst 3	Analyst 4
6.7	6.3	6.9	7.0
6.9	6.2	7.0	7.2
6.8	6.1	7.1	7.1

Use Excel and ANOVA analysis to determine whether there is a significant difference ($p = 0.05$) in the mean DO value measured by each analyst.

2.8 Scores on a recent Environmental Chemistry exam with an unknown population mean and population standard deviation of ± 3 points were reported. A total of 26 scores were considered, giving a sample mean of 68. Construct a 90% confidence interval for the true population mean of the Environmental Chemistry exam scores.

2.9 Suppose now that we wish to consider a two-tailed version of Example Problem 2.3 in the text. State the null hypothesis and work the problem appropriately.

2.10 Examine the results from the multi-determination of a given environmental variable:

70.15 69.62 69.70 69.65

Determine whether or not any of the above data should be rejected at 95% confidence.

2.11 Describe the importance of residual analysis in regression methods.

2.12 An environmental technician investigated the relationship between the pH of river water and its water hardness. In his/her notebook the following was written: "there was a very strong correlation of 1.22 between pH of the water and water's hardness." Was the correlation calculated appropriately?

2.13 A chemist employed the use of a single standard addition method to determine the concentration of analyte X in an unknown sample. Without standard addition, the response (absorbance) measured 0.230. To 10.0 mL of sample, 0.50 mL of a 100 mg/L standard was added and properly mixed. The new absorbance measured was 0.390. Determine the analyte concentration in the original (initial) sample.

2.14 In a given chromatography experiment, a solution containing 0.050 M of analyte (X) and 0.044 M of standard (S) gave peak areas of 500 and 450 for A_x and A, respectively. Calculate the response factor, F.

2.15 In order to determine the unknown concentration of analyte in X in the chromatographic experiment in Problem 2.13, 10.0 mL of 0.0190 M standard S was added to 10.0 mL of the unknown solution and diluted to 25.0 mL in a volumetric flask. Analysis of this mixture revealed chromatographic peak areas of 550 and 590 for A_x and A, respectively. Calculate $[S]$ and $[X]$.

2.16 Describe the conceptual difference between supervised and unsupervised classification methods.

2.17 Describe cluster analysis (CA) and compare how and when it is used in linear discriminant analysis.

2.18 List two distinct advantages of using principal component analysis (PCA) when faced with large, complex environmental data sets.

2.19 What do loading plots represent in PCA? Why are they useful in data analysis?

2.20 True or False. Multiple linear regression (MLR) is applied when a specific model between response and explanatory variables is known. Provide reasoning for your answer.

2.10 REFERENCES

Harris, D. C. (2003). *Quantitative chemical analysis*. New York: W. H. Freeman and Company.

Kutnink, M. A., Hawkes, W. C., Schaus, E. E., & Omaye, S. T. (1987). An internal standard method for the unattended high-performance liquid chromatographic analysis of ascorbic acid in blood components. *Analytical Biochemistry, 166*, 424–430.

Popek, E. P. (2003). *Sampling and analysis of environmental chemical pollutants*. New York: Academic Press.

World Health Organization. (2006). *Guidelines for drinking water*. 3rd ed., Geneva: WHO.

Chapter 3

Abstracts and Keywords

Abstract

This chapter begins by reviewing the chemical composition and physical properties of water, including the roles of hydrogen bonding, density, salinity, and temperature in governing aqueous systems. Next, the importance of a variety of weak acids, weak bases, and mineral assemblages in natural waters is covered in detail. This includes coverage of the carbonate system and species distribution. This is followed by material covering oxidation/reduction reactions, dissolution and precipitation, and adsorption concepts. Finally, a brief overview of Bjerrum plots is presented to complement material associated with the carbonate system.

Keywords

Adsorption, Hydrogen bonding, Density, Salinity, Oxidation/reduction reactions, Carbonate system, Bjerrum plots

Chapter 3

Aqueous Chemistry

Shallow lake and pond environments typically exhibit highly stochastic temporal dynamics in physico-chemical parameters due to hydrodynamic considerations. Such environments are further characterized by such qualities of low total nitrogen (TN) to total phosphorus (TP) ratios, low stratification, high primary productivity, and substantial internal transport of sediments by resuspension and sedimentation.

FIGURE 3.1 Lake Tuendae, a shallow, artificially-constructed aquatic habitat located in the Soda Basin of the Mojave Desert, California, USA. Two distinct aquifer systems influence this region: a carbonate rock-based system and the Mojave River Sink recharged from percolation.

Harsh desert conditions such as those observed in Lake Tuendae, Mojave Desert, USA, add to the complication of assessing and managing the health of aquatic species in such habitats. High evaporative desert processes are believed to play a major role in water quality dynamics. The Mojave Desert itself can undergo air temperature extremes (−6.7° to 40.6°C). Moreover, Lake Tuendae has limited protection from evaporation and is reported to sustain an annual evaporative loss of nearly 410,000 ft^3. Scientists are thus actively involved in studying this region for greater understanding of the physico-chemical makeup of shallow desert aquatic environments.

3.1. CHEMICAL COMPOSITION AND PROPERTIES OF WATER

Water has many unique properties and characteristics that contribute to its role as being nature's miracle worker. Given the complex nature of the water cycle and its role in many chemical, biological, and physical processes, and given the fact that such processes are crudely evaluated in lower division courses, reinforcement of the basic properties of water is warranted.

3.1.1. Atomic Structure and Hydrogen Bonding

Water's atomic structure is simple (Figure 3.2a), consisting of two hydrogen atoms bonded to one oxygen atom. The nature of the structure of water causes its molecules to possess unique molecular polarity with a partial negative charge (δ^-) near the oxygen atom due to the unshared pairs of electrons and partial positive charges (δ^+) near the hydrogen atoms. Electrostatic attractions between the partial positive charge near the hydrogen atoms and the partial negative charge near the oxygen results in the formation of **hydrogen bonds** as shown in Figure 3.2b. Water's polar nature is due to the unequal sharing of its electrons in a bent structure. This molecular polarity causes water to be a powerful **solvent** giving rise to various types of reactions in aqueous solutions—a main focus of this chapter. The **hydrophobic effect** (the exclusion of compounds containing carbon and hydrogen

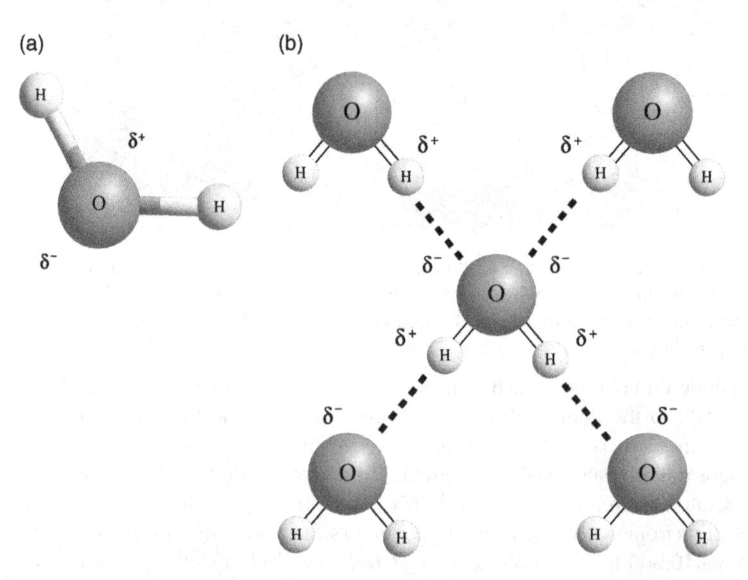

FIGURE 3.2 a) An electrically neutral water molecule. Note the partial negative charge, denoted as δ^- (the O atom), and the partial positive charge, denoted as δ^+ (the H atom). b) Hydrogen bonding between water molecules.

(non-polar compounds)) is another unique property of water caused by the hydrogen bonds. Other unique properties of water are described below.

Water is also the only substance on Earth that exists in all three physical states of matter: liquid, solid, and gas (Figure 3.3). In a liquid state, water molecules are constantly moving in relation to each other, and the hydrogen bonds are continually breaking and reforming at unprecedented timescales. In solid form (ice), motion is minimized and molecules become oriented to form a strong, open-lattice structure. When comparing the structure of ice and water in a liquid state, empty spaces within the ice structure result, allowing a more open structure. This structure takes up more volume than liquid water and hence ice is less dense. What unique role then does water's existence in different physical states have on the environment? Consider that massive amounts of heat exchange are incorporated in such changes of state. And as we learned in Chapter 1, oceans and other large water bodies have a noticeable influence on climate. They do this by acting as heat reservoirs and heat exchangers and the source of the majority of rain and snow that falls over land masses.

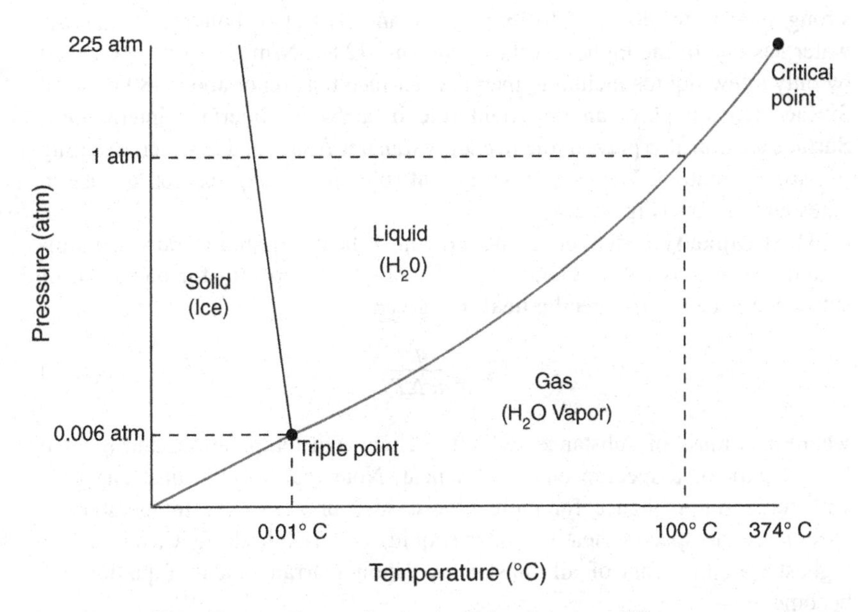

FIGURE 3.3 Phase diagram of water showing its preferred states at different temperatures and pressure. At typical temperatures and pressures on Earth, water is a liquid, but becomes solid if its temperature is lowered below 273 K and gaseous if its temperature is raised above 373 K, at the same pressure. A triple point is where three phase lines join. Here, three phases stably coexist but may abruptly change into each other, given a slight change in temperature or pressure. A critical point occurs at the end of a phase line where the properties of the two phases become indistinguishable from each other.

FIGURE 3.4 Wind-generated surface waves are controlled by the velocity of the wind, wind duration, and fetch. Surface tension is the direct result of the wind stress exerted on the sea surface. Most wind-generated waves eventually reach shorelines and dissipate.

3.1.2. Surface Tension and Heat Capacity

The hydrogen bonding exhibited by water molecules allows for the properties of strong surface tension and high heat capacity. **Surface tension** is the tendency of molecules of liquid to "stick" together at the surface. Water has strong surface tension due to its polarity and hydrogen bonding. In nature, water has one of the highest surface tensions, 72.8 mN/m (at 20°C), exceeded by only a few liquids including mercury (surface tension of about 480 mN/m). Surface tension plays an important role in air/water interface interactions. Surface tension also plays a role in energy transfer from wind to water, resulting in wave formation. Waves play important roles in helping oxygen diffuse in lakes and oceans (Figure 3.4).

Heat capacity is defined as the amount of heat required to raise the temperature of a mass of substance by 1°C. The heat capacity for one gram of substance is called its **specific heat, C_s**, given by

$$C_s = \frac{q}{m\,\Delta T}, \tag{3.1}$$

where m = mass of substance, ΔT = the change in temperature, and q = the loss or gain of a specific quantity of heat. Note that specific heat can vary with temperature; hence for more precise measurements the temperature is specified. The specific heat of water (liquid) is 4.184 Joules/g°C, one of the highest specific heats of all substances. Upon rearrangement, Equation 3.1 becomes

$$q = C_s m \Delta T. \tag{3.2}$$

Here, one can calculate the quantity of heat gained or lost by using its specific heat together with its mass and change in temperature. The ability of water to stabilize temperature depends on its relatively high specific heat, thus playing an important role in the Earth's climate through its cooling and transfer of heat effects in thermal and chemical processes. But in the world's oceans, heat capacities are affected by temperature, salinity, and pressure. For example, at a salinity of 35 parts per thousand (ppth), a temperature of 25°C, and under constant pressure conditions, the heat capacity of seawater is 3.995 Joules/g°C. Thus, 3,995 joules of energy are required to heat 1.0 kilogram of seawater by 1.0°C.

3.1.3. Water Density

The physical properties temperature, pressure, and salinity work together to determine water density (weight of water divided by the amount of space it occupies). The high density characteristic of liquid water is largely due to the cohesive nature of the hydrogen-bonded network discussed earlier. This reduces the free volume and ensures a relatively high density, partially compensating for the open nature of the hydrogen-bonding effect. Water's density changes with respect to temperature, but not on a linear scale as shown in Figure 3.5. Maximum density (at 3.984°C) is the result of the opposing effects of increasing temperature, causing both structural collapse that increases density and thermal expansion that lowers density. Increasing pressure shifts

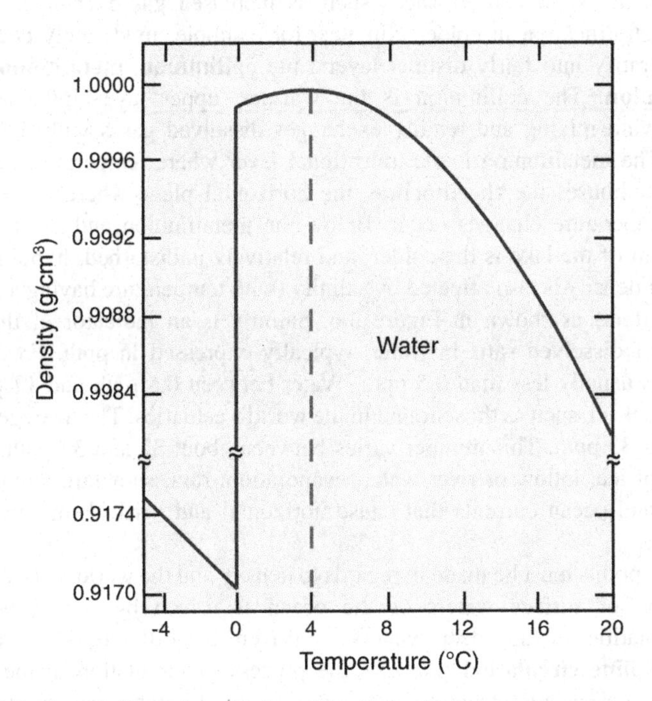

FIGURE 3.5 A graph illustrating a display of the relationship between temperature and density of water at normal atmospheric pressure.

the water equilibrium toward a more collapsed structure. Although pressure will increase the density of water at all temperatures (thus flattening the temperature density curve), there will be a disproportionate effect at lower temperatures.

Example Problem 3.1

As discussed, pressure will increase the density of water at all temperatures, with a disproportionate effect at lower temperatures. What would be the result of this occurrence? What is the effect at elevated pressures?

Answer: The result is a shift in the temperature of maximum density to lower temperatures. At high enough pressures the density maximum is shifted to below 0°C.

Density maximum ensures that the bottoms of freezing freshwater lakes generally remain at about 4°C and thus unfrozen. The change in density with temperature causes an inversion in cold water systems as the temperature is raised above about 4°C. Thus, in water below about 4°C, warmer water sinks whereas when above about 4°C, warmer water rises. As water warms up or cools down through 4°C, such **thermal stratification** causes considerable mixing with useful consequences such as increased gas exchange. During summer stratification in colder climates, for example, moderately deep lake waters stratify into fairly distinct layers: the **epilimnion**, **metalimnion,** and **hypolimnion**. The epilimnion is the warmer, upper layer influenced by surface wind mixing and readily exchanges dissolved gases with the atmosphere. The metalimnion is the transitional layer where temperatures rapidly change. It houses the **thermocline**, the horizontal plane where the greatest water temperature changes occur. Below the metalimnion and extending to the bottom of the lake is the colder, and relatively undisturbed, hypolimnion.

Water density is also affected by salinity (with temperature having a greater overall effect) as shown in Figure 3.6. Salinity is an indicator of the total amount of dissolved salts in water, typically expressed in ppth. Freshwater salinity is usually less than 0.5 ppth. Water between 0.5 ppth and 17 ppth is called brackish, such as those found in the world's estuaries. The average ocean salinity is 35 ppth. This number varies between about 32 and 37 ppth due to melting of ice, inflow of river water, evaporation, rain, snowfall, wind, wave motion, and ocean currents that cause horizontal and vertical mixing of the saltwater.

A few points must be made in regards to density and the world's oceans. The circulation of surface waters of the ocean is driven by winds, whereas the circulation of the deep waters is driven by differences in density. **Thermohaline circulation** refers to the process of circulation in the ocean depths. Water has a tendency to move horizontally throughout the deep ocean, moving along lines of equal density. One would thus expect that vertical

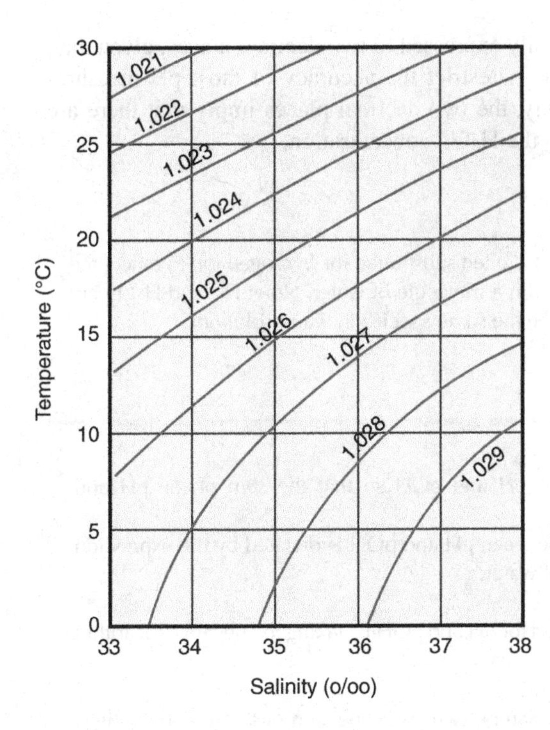

FIGURE 3.6 Variations in total salinity and in temperature cause variations in the density of seawater.

circulation is limited due to the ease of water to move along lines of constant density (**isopycnals**) than across them.

3.2. ACID-BASE PHENOMENA

In pure water, the majority of the water molecules remain intact. However, a small amount will collide with one another to cause the **self-ionization** reaction represented by

$$H_2O + H_2O \rightleftharpoons H_3O^+ + OH^- \; K_w = 1.0 \times 10^{-14} \text{ at } 25°C. \quad (3.3)$$

Here, $[H_3O^+]$ and $[OH^-]$ are 1.0×10^{-7} M. It is the 7 in the exponent of this number that gives neutral water a pH of 7.00 since $[H_3O^+]=[OH^-]$. For pH, we express the activity of the hydronium ion by means of its negative logarithm:

$$\text{pH} = -\log_{10}[H_3O^+]. \quad (3.4)$$

K_w is sensitive to both pressure and temperature; it increases when either increases. Similarly, to express the $[OH^-]$, the pOH scale can be used, where

$$\text{pOH} = -\log_{10}[OH^-]. \quad (3.5)$$

Measurements of pH are typically expressed to two decimal places, given that, on the whole, practical problems restrict the accuracy of most pH measurements to two places. Ostensibly, the two decimal places imply that there are only two significant figures in the H_3O^+ concentration.

TEXTBOX 3.1

The term *hydronium ion* (H_3O^+) is often substituted for *hydrogen ion* to reflect the association of a hydrogen ion with a molecule of water. Note: H^+ and H_3O^+ are used interchangeably to describe the same species in water solutions.

Example Problem 3.2

Derive a relationship between pH and pOH so that the sum of the pH and pOH = 14.00 (at 25°C).

Answer: The relationship between pH and pOH is dictated by the expression that describes the ionization of water:

$[H_3O^+][OH^-] = K_W.$

An expression can be derived for pH and pOH by taking the negative logarithm of both sides:

$-\log[H_3O^+][OH^-] = -\log(K_W).$

The logarithm of a product can be written as the sum of the two logarithms, therefore:

$-\log[H_3O^+] - \log[OH^-] = -\log(K_W).$

Realizing that the p-function = − log, we can rewrite

$pH + pOH = pK_W.$

Since the numerical value for $pK_W = 14$, the relationship can be written as

$pH + pOH = 14.00.$

For any acid-base system, a **mass balance** equation can be written that relates the concentrations of the various dissociation products of the substance to its **formal concentration** C_F. The mass balance equation expresses the conservation of mass in the form of atoms and groups of atoms that get distributed in a mixture or solution. Viewed another way, it states that the number of moles of an element placed in a solution will equal the total number of moles of that element in all the chemical forms it may adopt while in the solution. Let's consider a basic HCl solution with its mass balance expression shown as follows:

$$[HCl] + [Cl^-] = C_F.$$

But since HCl is a strong acid, we can neglect the first term and write the trivial mass balance equation:

$$[Cl^-] = C_F. \tag{3.6}$$

Note that the dissociation equilibrium of water must always be satisfied, where

$$[H_3O^+][OH^-] = K_w. \quad (3.7)$$

In any ionic solution, the sum of the positive and negative electric charges must be zero; in other words, all solutions are electrically neutral. This **charge balance** is expressed in terms of the concentration of positive and negative charges, taking into account not only the molar concentration of the cations and anions but of their individual charges as well.

$$[H_3O^+] = [OH^-] + [Cl^-] \quad (3.8)$$

In our approach, we must combine the above concepts into a single expression that relates $[H_3O^+]$ to C_F. By substituting Equation 3.6 into Equation 3.7, we can get rid of $[Cl^-]$:

$$[H_3O^+] = [OH^-] + C_F. \quad (3.9)$$

Using Equation 3.7 allows the $[OH^-]$ term to be eliminated:

$$[H_3O^+] = C_F + \frac{K_w}{[H_3O^+]} \quad (3.10)$$

It should be obvious from Equation 3.10 that the $[H_3O^+]$ will be the same as the formal concentration as long as the solution is not very dilute.

For proper discussion of natural waters we must consider the importance of a variety of weak acids, weak bases, and mineral assemblages characteristically present. Natural water composition is controlled by a combination of geochemical and biological processes and is largely affected by such parameters as pH and alkalinity. Alkalinity or Total Alkalinity (A_T) is a measure of the carbonate (CO_3^{2-}), bicarbonate (HCO_3^-), and hydroxide (OH^-) ions present in aqueous systems. Of additional consideration are borates, silicates, phosphates, ammonium, sulfides, and organic ligands. Ostensibly, alkalinity is a measure of the buffering capacity of water or the capacity of bases to neutralize acids. In the natural environment CO_3^{2-} alkalinity comprises the majority of the A_T due to the common occurrence and dissolution of carbonate rocks and the presence of CO_2 in the atmosphere. Such concepts will be discussed extensively in section 3.2.1.

Measuring alkalinity is important, for example, in determining a water system's ability to neutralize acidic pollution from rainfall or wastewater processes. Alkalinity of surface and groundwater is directly related to the underlying sediment and bedrock. Areas underlain by rocks and sediment rich in calcium from limestone, for example, are more resistant to changes in pH. Alkalinity also helps regulate the metal content of natural waters. For example, CO_3^{2-} and HCO_3^- in aqueous systems can remove toxic metals (e.g., lead, arsenic, barium, and cadmium) by precipitating the metals out of solution. For instance, barium forms an insoluble salt when reacting with CO_3^{2-}:

$$Ba^{2+} + CO_3^{2-} \rightleftharpoons BaCO_{3(s)}.$$

Barium is a common metal found in industrial wastewater streams and can be effectively removed to acceptable levels by precipitating the metal in the insoluble form above.

3.2.1. The Carbonate System in Aquatic Systems

Discussion in Chapter 1 touched upon CO_2 being the dominant end product of organic carbon degradation in most aquatic environments, with its variation being a measure of net ecosystem metabolism. Let's consider such a topic in greater detail. This section deals with the functioning carbonate system in aquatic environments in concurrence with atmospheric CO_2 and sedimental $CaCO_3$. We will learn how the carbonate system dominates pH control in most aquatic systems and thus regulates numerous global biogeochemical processes. Major discussion will be centered on the distribution of carbonate species in aqueous solutions with detailed reference to the carbonate system and its buffering source in the world's oceans.

Carbon dioxide is exchanged between the atmosphere and aquatic systems via equilibrium of CO_2 (g) and dissolved CO_2. Central to further discussion are four measureable parameters: pH, pCO_2, A_T, and total dissolved inorganic carbon (DIC). Dissolved CO_2 reacts with aqueous systems, ultimately becoming part of the carbonate system through various forms:

$$CO_2(aq) + H_2O \rightleftharpoons H_2CO_3 \rightleftharpoons HCO_3^- + H^+ \rightleftharpoons CO_3^{2-} + 2\,H^+. \tag{3.11}$$

Dissolved CO_2 in oceans occurs mainly in three inorganic forms: free aqueous carbon dioxide ($CO_{2(aq)}$), bicarbonate (HCO_3^-), and carbonate ion (CO_3^{2-}). A minor form is true carbonic acid (H_2CO_3) whose concentration is less than 0.3% of [$CO_{2(aq)}$]. Note that the sum of [$CO_{2(aq)}$] and [H_2CO_3] is denoted as [CO_2]. In thermodynamic equilibrium, gaseous carbon dioxide ($CO_{2(g)}$) and [CO_2] are related by Henry's law:

$$CO_2(g) \overset{K_0}{=} [CO_2],$$

where K_0 is the temperature and salinity-dependent solubility coefficient of CO_2 in seawater (Weiss, 1974).

The majority of dissolved inorganic carbon (DIC) in the world's oceans is in the form of HCO_3^- (>85%), with DIC described further in Textbox 3.2. Once dissolved, a small portion of the CO_2 reacts with water to form H_2CO_3:

$$CO_2 + H_2O \rightleftharpoons H_2CO_3 \tag{3.12}$$

Overall, the following equilibria are established in any carbonate-containing solution:

$$\frac{[H^+][HCO_3^-]}{[H_2CO_3]} = K_1 = 10^{-6.3} \quad (3.13)$$

$$\frac{[H^+][CO_3^{2-}]}{[HCO_3^-]} = K_2 = 10^{-10.3} \quad (3.14)$$

Considering the relationship $[H_3O^+][OH^-] = K_w$,

$$C_T = [H_2CO_3] + [HCO_3^-] + [CO_3^{2-}] \text{ (mass balance)} \quad (3.15)$$

$$[H+] - [HCO_3^-] - 2[CO_3^{2-}] - [OH^-] = 0 \text{ (charge balance).} \quad (3.16)$$

Note that C_T = the sum of all carbonate species concentrations.

TEXTBOX 3.2

The **partial pressure of carbon dioxide** (***p*****CO₂**) is the gas phase pressure of CO_2, which would be in equilibrium with the dissolved CO_2. pCO_2 measurements provide a relative measure of trophic status in the water column as a result of the balance between the capacity of an aqueous system to decompose organic matter and its capacity to take up CO_2 through photosynthesis. pCO_2 can be calculated from any two of the following parameters (together with temperature and salinity): pH, A_T, and DIC, are measured directly by equilibrating air (or another carrier gas) with water and then measuring the pCO_2 of the equilibrated air by either gas chromatography or infrared spectroscopy.

Dissolved inorganic carbon (**DIC**) is defined as

$$DIC = [CO_2] + [HCO_3^-] + [CO_3^{2-}].$$

DIC is an important parameter when performing measurements related to the pH of natural aqueous systems and CO_2 flux estimates. DIC is measured by the acidification of the sample, which drives the equilibria to CO_2. This gas is then **sparged** from solution and trapped, with the extracted quantity subsequently measured, typically by a coulometer or by an infrared CO_2 analyzer.

Like all acids, H_2CO_3 releases H^+ into the solution, leaving both HCO_3^- and CO_3^{2-} in the solution. The acidity of ocean waters is in turn determined by $[H^+]$, which is measured on the pH scale. What effect does this have on the world's oceans and biological processes? It has been estimated that over the last 200 years, oceans have taken up an excess of inorganic carbon from the atmosphere equivalent to approximately 25–30% of the total emissions of CO_2 from fossil-fuel burning, cement manufacturing, and land use change since the beginning of the industrial revolution (Sabine, et al., 2004). While this uptake potentially slows the growth rate in atmospheric CO_2, it is reported to lower surface water pH, with the potential for the resulting **ocean acidification** to disrupt marine ecosystems. For example, ocean acidification may limit corals and other coralline plants and animals to properly synthesize their carbonate skeletal materials (via the formation of biogenic calcium carbonate, $CaCO_3$) due to their ability to calcify being pH dependent. Even in today's saturated $CaCO_3$ oceans, calcification rates and $[CO_3^{2-}]$ continue to decline. As anthropogenic carbon emissions continue to accelerate, the CO_2 contained in these emissions will likely continue to reduce the alkalinity of the oceans.

3.2.2. Carbonate Species Distribution

In order to gain a true understanding of the distribution of carbonate species in aqueous systems we must consider pH in greater detail. To do this, determination of whether aqueous systems are in equilibrium with the atmosphere is paramount. Recall our Chapter 1 discussion of open and closed systems, which will aid in this determination. If aqueous systems are in equilibrium with the atmosphere, pCO_2 remains constant and the concentration of $[CO_2]$ is fixed. This is considered an open system as is the case for streams, shallow lake environments, and upper regions of the world's oceans where wind mixing is prevalent. However, even though the concentration of $[CO_2]$ is fixed, it does not mean that the total carbonate concentration is fixed since variable amounts of dissociated forms of carbonate, i.e., HCO_3^- and CO_3^{2-}, will be present depending on the pH. If we know the pH and the pCO_2 for an open system, we can calculate the concentrations of all relevant species in the system from the **mass-action expressions**, which include the Henry's law constraint (K_{CO_2}), the dissociation constants of carbonic acid (K_1, K_2), and K_w.

A **Bjerrum plot** can be constructed to show the concentrations of various inorganic carbon species as a function of pH in both an open and closed CO_2-H_2O system format. Such plots are typically used by marine chemists to track the response of an ocean to changes both in pH and of inputs in carbonate and CO_2. Consider Figure 3.7, which provides a plot of log concentration (molar) vs. pH for an open CO_2-H_2O system at 25°C with $pCO_2 = 10^{-3.5}$ atm. Upon inspection, this plot shows the pH ranges in which the three different carbonate species are predominant. At $pH < pK_1$, $[CO_2]$ is the predominant species. At this pH range it accounts for nearly all of the total carbonate, and in theory, a line for total carbonate could be drawn to coincide with the line showing $[CO_2]$. Notice that $[CO_2]$ is independent of pH (and so is the total carbonate concentration at this point). At $pH = pK_1$, $[CO_2]$ and HCO_3^- are at equal concentrations, which results in the total carbonate concentration being double the concentration of either of these species. This would result in the total carbonate curve rising 0.301 log units above the point at $pH = pK_1$. Here, carbonate ion contributes negligibly to total carbonate at this pH. At $pK_1 < pH < pK_2$, HCO_3^- is the predominant species, with its concentration approximately equal to total carbonate. Finally, at $pH > pK_2$, CO_3^{2-} will be

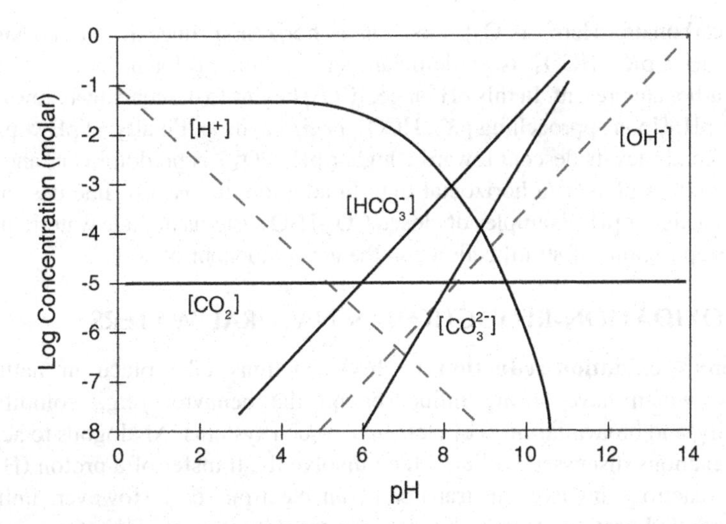

FIGURE 3.7 A Bjerrum plot of the log of concentrations (molar) of various inorganic carbon species as a function of pH in an open system at 25°C with $pCO_2 = 10^{-3.5}$ atm.

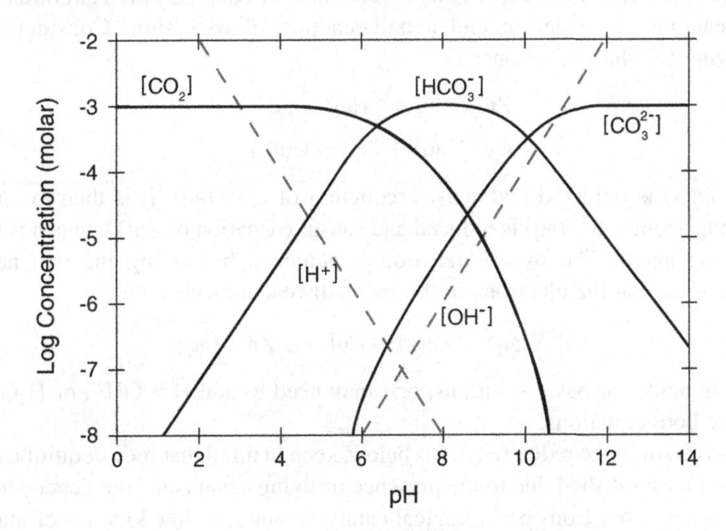

FIGURE 3.8 A Bjerrum plot of the log of concentrations (molar) of various inorganic carbon species as a function of pH in a closed system for a value of total inorganic carbon of 10^{-3} mol L^{-1}.

the predominant species. A theoretical total carbonate curve would thus follow CO_3^{2-}.

A Bjerrum plot of various inorganic carbon species as a function of pH in a closed system for a value of total inorganic carbon of 10^{-3} mol L^{-1} is shown in Figure 3.8. Let's examine this plot as we did for the open CO_2-H_2O system. Notice that at pH < pK_1, $[CO_2]$ is dominating and accounts for nearly all of the total carbonate. Here, $[CO_2]$ plots as a horizontal line. If we consider $pK_1 < pH < pK_2$, HCO_3^- is predominant and accounting for nearly all of the total carbonate present. In this pH range, $[CO_2]$ begins to descend away towards higher pH. Upon approaching pK_2, HCO_3^- begins to drop. Finally, at $pH > pK_2$, $[CO_2]$ continues its descent toward a higher pH, CO_3^{2-} is predominant, and its concentration plots as a horizontal line. In addition, the HCO_3^- line descends toward a higher pH. Examples of closed CO_2-H_2O systems include waterbodies with deep regions of stratification and the air component of soils.

3.3. OXIDATION-REDUCTION IN NATURAL WATERS

Numerous **oxidation-reduction (redox)** reactions take place in natural waters, which have strong influences on the behavior (e.g., solubility, mobility, and bioavailability) of elements in such systems. Analogous to acid-base reactions discussed earlier, which involve the transfer of a proton (H^+), redox reactions involve the transfer of an electron (e^-). However, unlike free hydrated protons, free hydrated electrons do not exist. Therefore, every reduction reaction must be accompanied by a corresponding oxidation reaction. A redox reaction can be considered a pair of coupled **half-reactions**—a half-reaction of oxidation and a half-reaction of reduction. Consider the following two half-reactions:

$$Zn(s) \rightarrow Zn^{2+}(aq) + 2e^-$$

$$Cu^{2+}(aq) + 2e^- \rightarrow Cu(s).$$

Here, Zn(s) is oxidized and causes reduction of $Cu^{2+}(aq)$. It is therefore the reducing agent. $Cu^{2+}(aq)$ is reduced and causes oxidation of Zn(s), thus it is the oxidizing agent. The overall reaction is achieved by adding the two half-reactions so that the electrons in the two half-reactions cancel:

$$Cu^{2+}(aq) + Zn(s) \rightarrow Cu(s) + Zn^{2+}(aq).$$

Note: In acidic or basic solutions, one may need to add H^+, OH^-, or H_2O to balance both equations.

As you study the redox reactions below, keep in mind that redox equilibrium is not well established due to the presence of living organisms, the dependence of most redox reactions on biological catalysis, and the slow kinetics of many oxidation and reduction reactions. One should be cautious when interpreting redox information, since reactions are very slow in the natural environment. In addition, some redox species may not be electroactive in environmental systems. Potential limitations to redox potential measurements in natural waters will also be discussed.

As an example half-reaction of environmental significance, organic pollutants (e.g., non-biodegradable organic matter in pigment wastewater) and other wastes (e.g., pesticides) can be mineralized upon oxidation by powerful oxidizing agents like hydrogen peroxide (H_2O_2), which through the following reaction

$$Fe^{2+} + H_2O_2 \rightarrow Fe^{3+} + OH^- + \bullet OH$$

produces an oxidizer intermediate: the hydroxyl radical ($\bullet OH$). The efficiency of this process depends on H_2O_2 and Fe^{2+} concentrations and pH (optimal range: pH 3.0–5.0) of the reaction. The initial concentrations of the pollutant and temperature also have a substantial influence on the final efficiency.

U.S. Air Force scientists have recently acknowledged the role of hydrogen as a direct electron donor in the anaerobic dechlorination of chlorinated aliphatic hydrocarbons (CAHs). Consider the reduction of perchloroethene (PCE) in groundwater. Initially, molecular hydrogen disassociates in the following half-cell reaction:

$$H_2 \rightarrow 2H^+ + 2e^-. \tag{3.17}$$

Next, PCE is reduced by the substitution of a chloride ion with a hydrogen ion and the transfer of one electron:

$$C_2Cl_4 + 2H^+ + 2e^- \rightarrow C_2HCl_3 + H^+ + Cl^- \tag{3.18}$$

The hydrogen ion and chloride ion produced likely form HCl as follows:

$$H^+ + Cl^- \rightarrow HCl. \tag{3.19}$$

Combining and balancing Equations 3.17, 3.18, and 3.19, the dechlorination of PCE using hydrogen as the electron donor can be written as follows:

$$H_2 + C_2Cl_4 \rightarrow C_2HCl_3 + HCl. \tag{3.20}$$

Anaerobic reductive dechlorination is performed by limited metabolic classifications of bacteria. These groups may act differently from one another, and include methanogens, sulfate-reducing bacteria, and dechlorinating bacteria.

3.3.1. Redox Variables

Many variables are commonly employed to express redox conditions: Eh, pε, f_{O_2} (fugacity of oxygen), and f_{H_2} (fugacity of hydrogen). They are all more or less useful depending on a specific problem, or an approach to measurements or calculations, and can be readily converted from one to the other given the appropriate conversion factors. Fugacity of hydrogen, f_{H_2}, is a convenient parameter because any redox reaction can be written to include hydrogen as a product or reactant. High values of log f_{H_2} indicate reducing conditions whereas low log f_{H_2} values indicate oxidizing conditions. Some of the most important redox reactions in natural waters are the oxidation of organic matter and the corresponding reduction reactions: reduction of oxygen to H_2O, nitrate to elementary nitrogen N_2, manganese(III/IV) to Mn(II), Fe(III) to Fe(II), sulfate to sulfide, and CO_2 to methane. A few specific examples are provided below. Diagrams that show the dominant ionic or solid phase forms of redox reaction participants as a function of pH and Eh (or of pH and pε or f_{H_2}) are useful ways of summarizing equilibrium speciation for a variety of conditions.

TEXTBOX 3.3

Eh is the redox potential of a solution and a measure of the relative intensity of oxidizing or reducing conditions in a given system. It is expressed in volts and at equilibrium can be expressed by standard equations of chemical thermodynamics. **Electron activity, pε,** indicates the tendency of a solution to donate or accept a proton. If pε is low, the solution is reducing. Alternatively, if pε is high, the solution is oxidizing.

Fugacity is a calculated property that is intrinsically related to chemical potential. Fugacity is a measure of how much the chemical potential of the component in the gas deviates from the chemical potential of some reference, namely, the standard state, due to changes in pressure and/or the mole fraction of the component.

3.3.2. Aqueous Nitrogen Species

Consider the representative N-O-H system at 25°C and 1 atm depicted in Figure 3.9. Shown is a f_{H_2} – pH plot that indicates both oxidation-reduction and acid-base conditions. The horizontal blue lines in the plot correspond to reducing (upper) and oxidizing (lower) boundaries in terms of log f_{H_2}, where H_2O is stable. The dissociation reaction of water is

$$H_2O \rightleftharpoons \frac{1}{2}O_{2(g)} + H_{2(g)},$$

where, at equilibrium,

$$\log K = \log f_{H_2} + \frac{1}{2}\log f_{O_2} \tag{3.21}$$

and

$$\log f_{H_2} = \log K - \frac{1}{2}\log f_{O_2}. \tag{3.22}$$

The usual atmospheric O_2 defines the oxidizing boundary of water stability, and we may use 0.21 atm (or $\log f_{O_2} = -0.68$) since oxygen makes up about 21% of the atmosphere by volume. Combining with the equilibrium constant of reaction 14 in Table 3.1, this corresponds to $\log f_{H_2} = -41.21$. The reducing boundary occurs at 1 atm $H_{2(g)}$ since the pressure of hydrogen gas in surface environments cannot exceed the atmospheric pressure. This sets the upper boundary at $\log f_{H_2} = 0$.

For the vertical line separating NH_4^+ and $NH_{3(aq)}$,

$$NH_4^+ \rightleftharpoons NH_{3(aq)} + H^+$$

$$\log K = -pH + \log a_{NH_3} - \log a_{NH_4^+}. \tag{3.23}$$

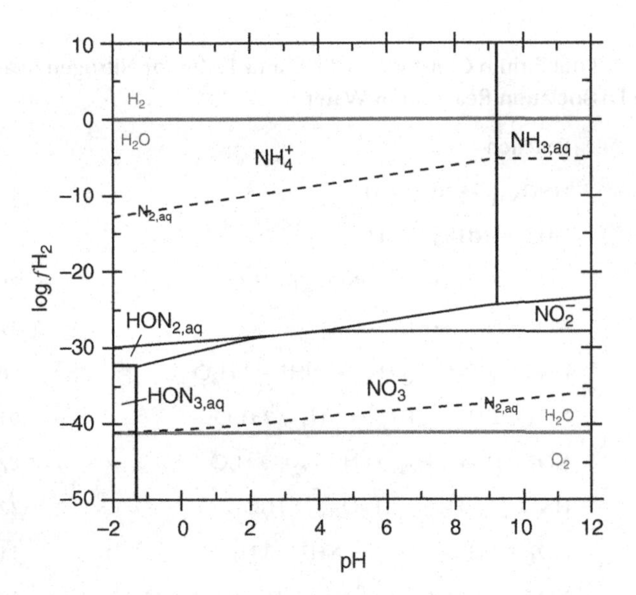

FIGURE 3.9 A representative N-O-H system at 25°C and 1 atm. Shown is a f_{H_2}-pH plot that indicates both oxidation-reduction and acid-base conditions.

If we consider $a_{NH_3} = a_{NH_4^+}$, the equation reduces to $pH = -\log K$, or $pH = 9.24$. So this line corresponds to the pH where the activity of NH_4^+ equals the activity of $NH_{3(aq)}$. At $pH > 9.24$, the ratio $a_{NH_3}/a_{NH_4^+}$ is greater than 1; at any pH values left of this line, less than 1.
Next, consider the redox couple between nitrite and ammonium ion:

$$NO_2^- + 2H^+ + 3H_{2(g)} \rightleftharpoons NH_4^+ + 2H_2O.$$

The equilibrium constant expression gives

$$\log K = \log a_{NH_4^+} - \log a_{NO_2^-} + 2pH - 3\log f_{H_2}. \quad (3.24)$$

Setting the activity of NH_4^+ equal to the activity of NO_2^-, rearrangement gives $\log f_{H_2}$ as a function of pH:

$$\log f_{H_2} = \frac{2}{3}pH - \frac{1}{3}\log K. \quad (3.25)$$

This equation has a slope of 2/3 and its intercept can be evaluated from the log K of reaction 6 in Table 3.1, and it is plotted in Figure 3.9 as a line that separates the predominance areas for NH_4^+ and NO_2^-. Using these methods, similar lines can be constructed for the other aqueous nitrogen species. The appropriate reactions and their equilibrium constants are listed in Table 3.1. Also overlaid on the figure are dashed lines, derived from reactions 10–13, that illustrate the area of nitrogen, $N_{2(aq)}$, dissolved in aqueous solutions in contact with the atmosphere having 78% N_2.

TABLE 3.1 Equilibrium Constants at 25°C and 1 Atm for Nitrogen Reactions and the Dissociation Reaction of Water

	Reaction	log *K* [a]
1	$HNO_{3(aq)} \leftrightarrow NO_3^- + H^+$	1.303
2	$NH_4^+ \leftrightarrow NH_{3(aq)} + H^+$	−9.241
3	$NO_3^- + H^+ + H_{2(g)} \leftrightarrow HNO_{2(aq)} + H_2O$	30.992
4	$NO_3^- + H_{2(g)} \leftrightarrow NO_2^- + H_2O$	27.767
5	$HNO_{2(aq)} + H^+ + 3\ H_{2(g)} \leftrightarrow NH_4^+ + 2\ H_2O$	88.155
6	$NO_2^- + 2\ H^+ + 3\ H_{2(g)} \leftrightarrow NH_4^+ + 2\ H_2O$	91.381
7	$NO_2^- + H^+ + 3\ H_{2(g)} \leftrightarrow NH_{3(aq)} + 2\ H_2O$	82.140
8	$HNO_{3(aq)} + H_{2(g)} \leftrightarrow HNO_{2(aq)} + H_2O$	32.295
9	$NO_3^- + 2\ H^+ + 4\ H_{2(g)} \leftrightarrow NH_4^+ + 3\ H_2O$	119.147
10	$NO_3^- + H^+ + \frac{5}{2}H_{2(g)} + \frac{1}{2}N_{2(g)} \leftrightarrow N_{2(aq)} + 3\ H_2O$	102.041
11	$N_{2(aq)} + H^+ + \frac{3}{2}H_{2(g)} \leftrightarrow NH_4^+ + \frac{1}{2}N_{2(g)}$	17.106
12	$HNO_{3(aq)} + \frac{5}{2}H_{2(g)} + \frac{1}{2}N_{2(g)} \leftrightarrow N_{2(aq)} + 3\ H_2O$	103.344
13	$N_{2(aq)} + \frac{3}{2}H_{2(g)} \leftrightarrow NH_{3(aq)} + \frac{1}{2}N_{2(g)}$	7.865
14	$H_2O \leftrightarrow \frac{1}{2}O_{2(g)} + H_{2(g)}$	−41.552

[a]*All values of log K were calculated with* ΔG_f^o.

3.3.3. Reduced Phosphorus Species Reconsidered

While other biologically relevant elements have well-defined redox mechanisms (e.g., denitrification for the conversion of nitrate to nitrogen gas), thus far, it has been assumed that phosphorus species do not undergo redox reactions in the environment and exist solely in the phosphate, organic phosphorus esters (P-O-C bonds), and colloidal complex forms in aqueous systems. But recall our discussion in section 1.5.4 describing recent developments in the field of microbiology and research on the origin of life suggesting a possibly significant role for reduced, inorganic forms of phosphorus in bacterial metabolism and as evolutionary precursors of biological phosphate compounds.

Figure 3.10 depicts the calculated equilibrium distributions of the various species of phosphate, hypophosphite, and phosphite as a function of pH. At the circumneutral pH of most natural waters and soils, the dominant P species according to equilibrium calculations are $H_2PO_4^-$ and HPO_4^{2-} for phosphate, $H_2PO_3^-$ and HPO_3^{2-} for phosphite, and $H_2PO_2^-$ for hypophosphite. This

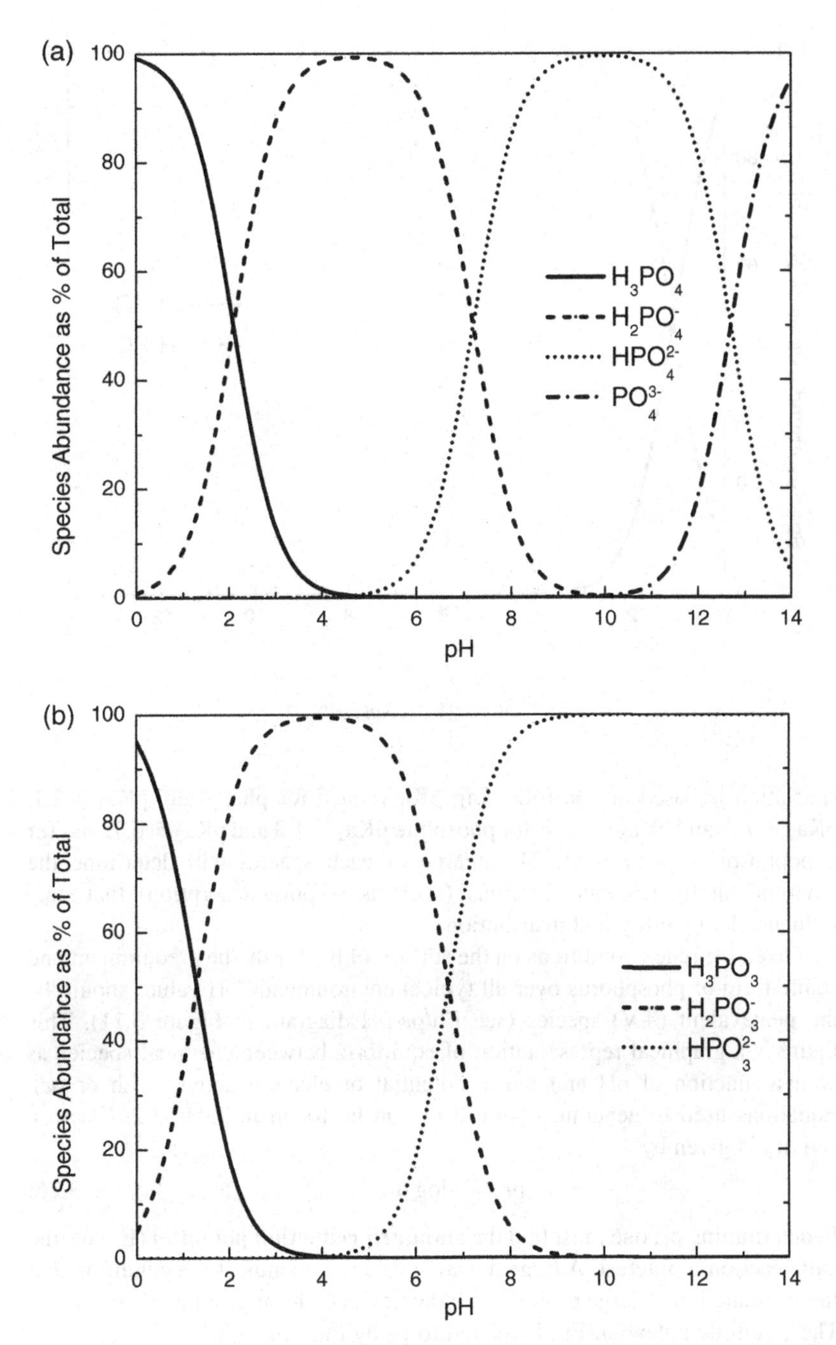

FIGURE 3.10 Species abundance, as a percentage of the total concentration, is shown for a) phosphate, b) phosphite, and c) hypophosphite species. At circumneutral pH (typical of surface waters), the dominant species are $H_2PO_4^-$ and HPO_4^{2-} for phosphate, $H_2PO_3^-$ and HPO_3^{2-} for phosphite, and $H_2PO_2^-$ for hypophosphite.

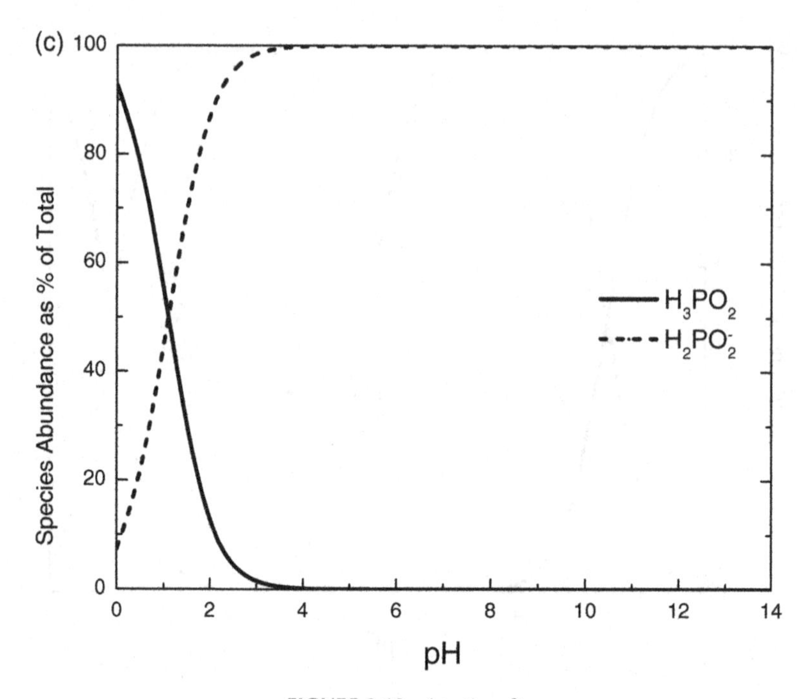

FIGURE 3.10 *(continued).*

speciation is based on the following pKa values: for phosphate, $pKa_1 = 2.1$, $pKa_2 = 7.2$, and $pKa_3 = 12.7$; for phosphite $pKa_1 = 1.3$ and $pKa_2 = 6.7$; and for hypophosphite $pKa_1 = 1.1$. The charge of each species will determine the environmentally relevant reactions (such as sorption/desorption) that may influence its mobility and distribution.

Given the redox conditions on the surface of the Earth, the predominant and stable form of phosphorus over all typical environmental pH values should be the pentavalent (+V) species (see Eh/pε-pH diagram in Figure 3.11). This figure is a graphical representation of equilibria between chemical species as both a function of pH and redox potential or electron activity (Eh or pε). Equations used to generate Figure 3.11 can be found in Table 3.2. Electron activity is given by

$$p\varepsilon = -\log a(e^-). \tag{3.26}$$

In determining pε, one must find the **standard reduction potential (E°)** for the half-reaction of interest. A large negative value of pε indicates a condition that favors reduction. A large positive pε indicates a condition that favors oxidation. The electrode potential, Eh, is related to pε by the equation

$$p\varepsilon = \frac{F}{2.30\,RT}\text{Eh}, \tag{3.27}$$

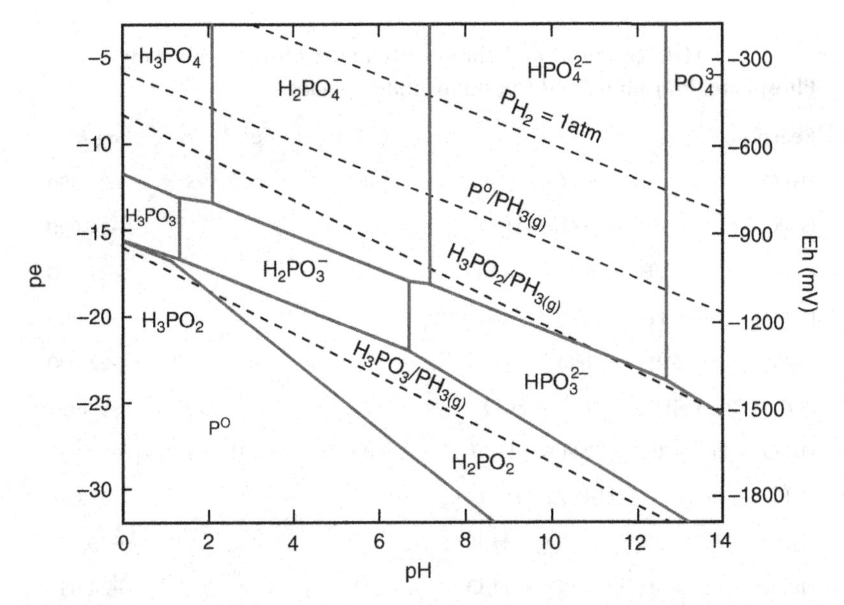

FIGURE 3.11 Eh/pε-pH diagram for phosphorus species in water. For calculations involving equilibrium between reduced phosphorus species and phosphine (i.e., dashed lines including PH_3), an equilibrium concentration of 10^{-6} M for the reduced phosphorus compound was assumed. The dashed line with P_{H_2}=1 atm is shown as a reference for a reducing environment on Earth.

where F = Faraday's constant (96.42 kJ volt^{-1} eq^{-1}), R = gas law constant (8.314 J/(k mol)), and T = temperature (K). At 25°C, this becomes

$$p\varepsilon = 16.9\ \text{Eh, or Eh} = 0.059\ p\varepsilon.$$

Note that the Eh and pH of an aqueous solution are related. For a given half-cell equation

$$aA + bB + n\,e - +h\,H^+ = cC + dD.$$

The half-cell standard potential E° is given by

$$E^\circ(V) = \frac{-\Delta G}{nF}, \tag{3.28}$$

where ΔG is the Gibbs free energy change, n is the number of electrons involved, and F is Faraday's constant. The Nernst equation then relates pH and Eh:

$$Eh = E^\circ + \frac{0.0592}{n} \times \log\left\{\frac{[A]^a[B]^b}{[C]^c[D]^d}\right\} - \left(\frac{0.0592h}{n}\right)pH. \tag{3.29}$$

TABLE 3.2 Half-Reactions and Thermodynamic Constants Involving Phosphate, Phosphite, and Hypophosphite Species

Reaction	E° (V)*	pε°*	log *K*
$H_3PO_4 + 2e^- + 2H^+ \rightarrow H_3PO_3 + H_2O$	−0.69	−11.695	−23.390
$H_3PO_4 + 2e^- + H^+ \rightarrow H_2PO_3^- + H_2O$			−24.690
$H_2PO_4^- + 2e^- + 2H^+ \rightarrow H_2PO_3^- + H_2O$			−22.590
$H_2PO_4^- + 2e^- + H^+ \rightarrow HPO_3^{2-} + H_2O$			−29.290
$HPO_4^{2-} + 2e^- + 2H^+ \rightarrow HPO_3^{2-} + H_2O$			−22.090
$PO_4^{3-} + 2e^- + 3H^+ \rightarrow HPO_3^{2-} + H_2O$			−9.390
$H_3PO_3 + 2e^- + 2H^+ \rightarrow H_3PO_2 + H_2O$	−0.913	−15.475	−30.95
$H_3PO_3 + 2e^- + H^+ \rightarrow H_2PO_2^- + H_2O$			−32.05
$H_2PO_3^- + 2e^- + 2H^+ \rightarrow H_2PO_2^- + H_2O$			−30.75
$HPO_3^{2-} + 2e^- + 3H^+ \rightarrow H_2PO_2^- + H_2O$			−24.05
$H_3PO_2 + e^- + H^+ \rightarrow P^\circ + 2H_2O$	−0.922	−15.627	−15.627
$H_2PO_2^- + e^- + 2H^+ \rightarrow P^\circ + 2H_2O$			−14.527
$P^\circ + 3e^- + 3H^+ \rightarrow PH_{3(g)}$	−0.525	−8.898	−26.7
$H_3PO_2 + 4e^- + 4H^+ \rightarrow PH_{3(g)} + 2H_2O$			−42.322
$H_3PO_3 + 6e^- + 6H^+ \rightarrow PH_{3(g)} + 3H_2O$			−73.271

* *pε° = pε in standard state.*

In spite of the importance of the Eh its experimental measurement in a natural environment is still not entirely resolved. However, if correct thermodynamic data are available, it is possible to propose viable models to determine conditions of natural redox systems. Nonetheless, natural waters range between pH values of 4 and 10 and between Eh values of +1000 and −600 mV, suggesting that based on equilibrium considerations, reduced phosphorus should not be present in natural waters. However, despite the low Eh environments in which reduced phosphorus species are expected to exist, phosphorus in the −III oxidation state has been detected in the environment and has been shown to be relatively stable in the presence of oxygen and absence of other catalysts. If we, however, extend our definition of "natural waters" to include unusual and acidic aquatic environments such as those associated with acid mine drainage, the lower limit of Eh values in acidic environments (pH = 1 or lower) can be below −800 mV. In such environments, the fully protonated form of phosphite, H_3PO_3, is thermodynamically stable.

These observations suggest that the kinetics of chemical oxidation of these species may be relatively slow with respect to the equilibrium conditions depicted in Figure 3.11. Thus, given potentially slow redox kinetics, it would be expected that reduced forms of phosphorus may exist in highly reducing microcosms in nature but that, thus far, have circumvented detection due to limitations in detection capabilities. Moreover, microorganisms (rather than chemical oxidants) are likely to be the catalysts or mediators of redox-reactions that rapidly cycle reduced phosphorus to phosphate in the environment. The strongest evidence for the existence of a detectable redox cycle for phosphorus is in the microbiology literature. The idea of the biological formation of reduced gaseous phosphorus compounds dates back more than 100 years.

3.3.4. Trace Metals in Natural Waters

The study of metals in aquatic ecosystems is an extremely important and complex environmental issue. The complexity stems partially from the fact that trace metals in natural waters are present in a variety of forms having different reactivities and bioavailabilities. The determination of oxidation states is an important consideration for metal toxicity, especially for elements such as arsenic (As), selenium (Se), and chromium (Cr). Let's consider the former in greater detail. In natural waters, As exists in a variety of forms including the inorganic species arsenite [As(III)] and arsenate [As(V)] as well as those bound to particulates and organic matter. As(III) and As(V) species are considered the most abundant species, with As(V) being thermodynamically stable under **oxic** conditions. In reducing conditions, such as those in naturally occurring geothermal systems, As(III) dominates (Wilkie and Hering, 1998). As(III) is more toxic than As(V), with both affecting the health of millions of people worldwide. Ultimately, the concentrations and speciation of As(III) and As(V) are determined by various abiotic and biologically mediated biogeochemical processes. (See Research Application II for a greater discussion on redox-sensitive As species.)

TEXTBOX 3.4

Oxic waters are defined as having chemically active dissolved oxygen levels. In contrast, **anoxic** waters are areas that are depleted of dissolved oxygen. Typically, such conditions occur if the rate of oxidation of organic matter by bacteria is greater than the supply of dissolved oxygen.

Organometallic species (covalently bonded molecules) are also of great importance when studying aquatic ecosystems. For example, methylmercury (CH_3Hg^+) is of immense environmental importance because its toxicity is even greater than Hg^{2+}. Such a high level of toxicity stems from the fact that it is highly membrane permeable and has a strong tendency to bind to tissues. In acidified aqueous solutions, the mechanism of the electrochemical reduction of CH_3Hg^+ proceeds as follows:

$$CH_3Hg^+ + e^- = CH_3Hg^\bullet$$
$$2CH_3Hg^\bullet \rightarrow (CH_3Hg)_2$$
$$(CH_3Hg)_2 \rightarrow (CH_3)_2Hg^+ + Hg$$

The above mechanism suggests that the reduced methylmercury radicals can quickly undergo a fast **dimerization reaction**, with the reactions eventually leading to the formation of dimethyldimercury. As shown from the example above, organometallic species can be produced naturally from inorganic precursors, i.e., by chemical or biological methylation. Note that they are also formed via industrial processes and introduced into the environment as biocides in agriculture, fuel additives, or antifouling paints. Ultimately, the bioavailability and toxicity of metals are influenced by the physico-chemical characteristics of the forms in which they are present (termed speciation) and the physiological characteristics of a given organism. Note that biological availability of a trace metal is not a function of total metal concentration, but rather of particular species of the metal that either can interact directly with an organism or can readily convert to species that can interact (the kinetically labile metal concentration). The interactions between chemical species in the environment and aquatic organisms are complex, and their elucidation requires interdisciplinary approaches and detailed knowledge of relevant chemical, physical, and biological processes. Such processes will be given full consideration throughout this book.

Research Application II: Redox Conditions and Arsenic Speciation and Transport

To extend our discussion of redox-sensitive environmental species consider the following research application highlighting the study of arsenic mobility in groundwater in China (Xie, et al., 2009). In addition to selected regions of China, high arsenic groundwater concentrations have been documented in Argentina, Bangladesh, Chile, India, Mexico, and the United States. Such occurrences are significant and timely given the current state of global drinking water supplies and lack of treatment facilities, especially in regards to developing nations. The selected application focused on the importance and control of arsenic mobilization along three flow pathways in the northern Datong Basin—an area of China home to numerous rural villages and high population densities. Concentrations ranged from 68 to 670 $\mu g\ L^{-1}$ in the central basin and 3.1 to 44 $\mu g\ L^{-1}$ in the western and eastern regions. In the latter, highly oxidized waters were evident with low contents of As(III) and high proportions of As(V). In contrast, the central basin contained highly reducing waters with high proportions of As(III) among As species.

In order to mitigate the health hazards that arsenic may pose in these and other communities, we must better understand the processes that lead to the mobilization of arsenic within groundwater systems. The authors discuss a number of mechanisms that likely control arsenic mobilization, including arsenic being sorbed and sequestered on Fe/Mn oxides, reductive dissolution of Fe(III) hydroxides, and sulfide oxidation. While the latter may be a source of arsenic, microbial sulfate reduction could result in the reduction of arsenic concentrations in groundwater through the formation of pyrite, which could precipitate/sorb arsenic. As a result, sulfur isotope values of dissolved sulfate may indicate sulfate reduction and that in turn may provide insights into reaction pathways and conditions that mobilize or retard arsenic. Ultimately, the authors used an integrated approach to studying arsenic distribution and mobilization involving the following processes and factors: Eh measurements, Fe-hydroxide sorption studies, sulfur cycle link considerations, and study of the behavior of other redox-sensitive species (Mo, U, and nitrate). Their study area included 15 groundwater wells and two springs located in the central and outer regions of the Datong Basin.

In terms of results, the authors noted that the mobility of sulfate, U, and Mo was controlled by the change in redox conditions as the groundwater flowed toward the central basin. Reducing conditions along the pathway could not account for the occurrence of high arsenic concentrations but does provide insight into arsenic speciation. As expected, all groundwater wells with As(III) as the major As species showed low Eh values. In contrast, those with As(V) demonstrated high Eh values. Reductive dissolution of Fe-oxyhydroxides or reduction of As(V) were consistent with these observations. In terms of sulfur isotope data, no significant correlations were observed between sulfur isotope values and total arsenic concentrations. The authors stressed that limited evidence for sulfate reduction was observed, thus indicating that sulfate reduction did not sequester arsenic. Overall, the data were consistent with traditional models of arsenic mobilization (see Figure 3.13): those that considered reductive dissolution of Fe-oxyhydroxides, reduction of As(V) to more mobile As(III), and bacteria-mediated reactions (due to the presence of organo-arsenic compounds) as active processes. This application provided further information in regards to the study of arsenic mobilization in groundwater processes. Students and faculty are encouraged to explore this paper and other relevant papers for a greater appreciation of this topic and its importance in global drinking water considerations.

3.4. DISSOLUTION AND PRECIPITATION IN AQUATIC ENVIRONMENTS

A good understanding of the processes of **dissolution** and **precipitation** is essential for a quantitative assessment of contaminant fate and transport in aquatic environments. Unlike our previous discussions of dissolved species, dissolution and precipitation reactions are typically slower, although rates are difficult to generalize. For a greater understanding we will first consider equilibrium relations. The equilibrium governing the dissociation of solids in water is given by the general form $A(s) \rightleftharpoons A(aq)$, and its equilibrium constant, K_{sp}, is given the name "**solubility product**." In the case of ionic compounds, the solubility product equilibrium is usually written in the form

$$M_mA_n(s) \rightleftharpoons m\ M^{n+}(aq) + n\ A^{m-}(aq),$$

where M^{n+} is the metal cation of charge n^+ and A^{m-} is the anion of charge m^-. It is important to note the position of the undissociated compound on the left side (i.e., as the reactant) in the K_{sp} equilibrium equation, and the dissociated ions on the right (i.e., as the products). The corresponding general expression for K_{sp} is

$$K_{sp} = [M^{n+}]^m [A^{m-}]^n. \tag{3.30}$$

Calculations involving the K_{sp} are usually for the cases of sparingly-soluble or "insoluble" ionic compounds. Values of K_{sp} are typically very small numbers several orders of magnitude smaller than 1. In such cases, one can view the dissociation as being very unfavorable and the concentrations of the dissociated species as being very low in concentration. The process of precipitation (the reverse of dissociation) occurs when ions leave solution to regenerate an ionic solid. The K_{sp} dictates the prediction of formation of precipitates from aqueous solution.

Example Problem 3.3

Consider the reaction below:

$$Mg(OH)_2(s) \rightleftharpoons Mg^{2+}(aq) + 2OH^-(aq) \qquad K_{sp} = 1.8 \times 10^{-11}.$$

Describe the solubility behavior of $Mg(OH)_2(s)$ as the acidity of the solution increases.

Answer: As mentioned, the pH of a solution will affect the solubility of any substance whose anion is basic. The solubility of $Mg(OH)_2(s)$ greatly increases as the acidity of the solution increases:

$$Mg(OH)_2(s) + 2H^+(aq) \rightleftharpoons Mg^{2+}(aq) + 2H_2O(l).$$

What factors affect solubility and how does this relate to metal ion behavior in aquatic systems? The **common ion effect** states the solubility of insoluble substances can be decreased by the presence of a common ion. In addition, we can consider **complex ion formation** due to the presence of complexing agents. Here, the complexing agent that reacts with the anion or cation of the precipitate increases the solubility of the precipitate. For example, silver chloride, AgCl, is reasonably insoluble in water. However, upon the addition of aqueous ammonia, it dissolves appreciably. Consider the equlibria:

$$AgCl(s) \rightleftharpoons Ag^+ + Cl^-$$

$$Ag^+ + NH_3 \rightleftharpoons AgNH_3^+$$

$$AgNH_3^+ + NH_3 \rightleftharpoons [Ag(NH_3)_2]^+.$$

As shown, the ammonia combines with silver ions to form a complex diamminesilver(I) ion. Although this is a reversible reaction, the position of the equilibrium lies well to the right, thus forming a fairly stable complex. In addition, the solubility of a given substance will be affected by the pH of solution if the cation of the substance is a weak acid or the anion is a weak base. For example, metal ions can be considered multi-protic weak acids, where the relative concentrations of the different **hydrolysis** species are dependent on pH. Upon hydrolysis, the free ions will react with other solutes to form more complex ions. The solubility of hydroxides (and oxides) can be expressed (in general terms) in equilibrium with free ions. Considering divalent metal ions:

$$Me(OH)_2(s) \rightleftharpoons Me^{2+} + 2OH^-$$

$$MeO(s) + H_2O \rightleftharpoons Me^{2+} + 2OH^-$$

The conventional solubility product is given by (Stumm and Morgan, 1981)

$$^{c}K_{s0} = [Me^{2+}][OH^-]^2 \ mol^3 \ L^{-1}.$$

TEXTBOX 3.5

Hydrolysis is a chemical reaction in which the action of water (or its ions) breaks down a substance into smaller molecules. The most common hydrolysis occurs when a salt of a weak acid or weak base (or both) is dissolved in water.

This constant is corrected for ionic strength and temperature. Note that it is often appropriate to express the solubility in terms of reactions with protons, since the equilibrium concentrations of OH^- ions may be exceedingly small. The solubility equation can then be defined as (Stumm and Morgan, 1981)

$$^{c^*}\mathbf{K}_{s0} = \frac{[me^{2+}]}{[H^+]^2} mol^{-1} \ L \quad (3.31)$$

And when considering the ion product of water, $\mathbf{K}_w$, we can write

$$^{c^*}\mathbf{K}_{s0} = \frac{^{c}\mathbf{K}_{s0}}{\mathbf{K}_w}. \quad (3.32)$$

The logarithm of the resulting **mass action equilibrium** for ion concentrations gives a linear equation relating ion concentration and pH. Note that the subscript zero in Equations 3.31–3.32 indicates that the equilibrium of the solid with the uncomplexed species Me^{2+} and OH^- is considered. Finally, the solubility of metal ions in aqueous solutions is affected by initial metal concentration (impacts the formation of polynuclear complexes), time (affects crystal and polynuclear complex formation), and particle size.

TEXTBOX 3.6

Concerning the behavior of metals in aqueous solutions, if no complexing ligands other than OH^- are present, the metal cation (M^{z+}) will tend to hydrolyze to form complexes with OH^- as a function of pH. Such complexes can be anions, cations, and neutral molecules.

Example Problem 3.4

Hydrogen sulfide H_2S often occurs naturally in well water, or can be caused by the presence of sulfate-reducing bacteria in a well or water system. It dissolves in water to make a solution that is weakly acidic. Consider a 0.10 M saturated solution of H_2S and calculate the $[H^+]$.

Answer: Write the associated equlibria and equilibrium constant expressions:

$$H_2S \rightleftharpoons H^+ + HS^- \qquad K_1 = \frac{[H^+][HS^-]}{[H_2S]} = 1.0 \times 10^{-7}$$

$$HS^- \rightleftharpoons H^+ + S^{2-} \qquad K_2 = \frac{[H^+][S^{2-}]}{[HS^-]} = 1.0 \times 10^{-14}$$

$$H_2O \rightleftharpoons H^+ + OH^- \qquad K_w = [H^+][OH^-] = 1.0 \times 10^{-14}.$$

Next, we write the charge balance equation: $[H^+] = [HS^-] + [OH^-] + 2[S^{2-}]$.
And the mass balance equation: $C = [H_2S] + [HS^-] + [S^{2-}]$.

A system of five equations with a total of five unknowns ($[H^+]$, $[HS^-]$, $[OH^-]$, $[S^{2-}]$, and $[H_2S]$). From this system we can make the assumption that $[OH^-]$ is negligible, and since $K_2 < K_1$, we can neglect the second dissociation step above. Here, $[S^{2-}]$ is considered negligible and $[H^+]$ is roughly equal to $[HS^-]$. Next, we can simplify our mass balance and charge balance equations from above:

$$C = [H_2S] + [HS^-] \qquad [H^+] = [HS^-].$$

If we combine our first dissociation step with the simplified charge and mass balance equations above, the following relationship results:

$$K_1 = 1.0 \times 10^{-7} = \frac{[H^+]}{C - [H^+]}.$$

We can then solve for $[H^+]$:

$$[H^+] = \sqrt{1.0 \times 10^{-7} \times 0.10} = 1.0 \times 10^{-4}.$$

3.5. ADSORPTION IN AQUATIC ENVIRONMENTS

As discussed, the behavior of many elements present in macroconcentrations in aquatic systems is controlled by rules of acid-base equilibria, solubility, and complexation with the extent of elements participating in such reactions determined by the kinetics (rates) and thermodynamics of involved reactions. The behavior of trace elements, however, is largely influenced by the **adsorption** process. Adsorption refers to the process leading to the concentration of a trace element from an aqueous phase onto the surface of a solid phase. Note that this does not apply to coprecipitation and biological uptake activities. In contrast, **desorption** refers to the release into the aqueous phase of previously adsorbed trace elements. **Ion exchange** is considered to be one of the most important adsorption mechanisms for inorganic ions. We will consider this process in greater detail below. The extent of adsorption of inorganic solutes through empirical adsorption equations will be covered in Chapter 8.

Consider our earlier discussion of phosphorus distribution and transformation in section 1.5.4. There has been considerable attention paid to phosphorus adsorption/desorption processes on natural sediments and mineral surfaces. As shown in Figure 1.9, phosphorus will be exchanged on the sediment-water interface until a dynamic equilibrium is reached. Sediments are those mineral and organic materials situated beneath an aqueous layer. Sediment composition varies dramatically and includes such components as metal oxides, iron sulfides, clay minerals, ferric hydroxide, calcite, interstitial microorganisms, and organic matter. Sedimentary organic matter, for example, acts as an important source of nutrients to the overlying water. As discussed in Chapter 1, release of nutrients to the system in sufficient quantities can promote algal blooms, and thus primary productivity is greatly enhanced.

The exchange of phosphorus between water and sediment is driven by both **sorption** reactions and biologically-controlled processes. For example, studies in estuarine environments have shown a two-step ion exchange between clay minerals and water, plus exchange between interstitial microorganisms and water. In lake sediments, the phosphate ion has been shown to be bound by ferric hydroxide ($Fe(OH)_3$). Here, the processes involving adsorption and subsequent transformation of the phosphate ion become complex. Inorganic phosphate is sorbed by $Fe(OH)_3$, which is present in the upper oxidized sediment layer. As sediments become saturated with organic matter (devoid of oxygen excess), Fe(III) is reduced. Since Fe(II) salts possess high solubility, Fe^{2+} and inorganic phosphate begin to diffuse. The Fe^{2+} migrates to the boundary of the oxidized zone where $Fe(OH)_3$ is again precipitated. Inorganic phosphate migrates with Fe^{2+} and eventually becomes bound by the $Fe(OH)_3$ recently precipitated.

The above processes are complicated by chemical transformations driven by microorganisms. We must also consider ancillary measurements of pH, redox potential, dissolved Ca, Fe, Mn, and alkalinity during our study of the understanding of reactions controlling, for example, phosphorus in solutions in contact with sediments. What role does **seasonal turnover** play in phosphorus exchange? In spring, for example, increasing temperatures, enhanced sedimentation, and turnover of organic matter occur in shallow lake systems. As a result, the thickness of the oxidized sediment layer decreases and hence the phosphorus binding capacity. What results from this occurrence? The result is release of phosphorus from the surface layer where it was retained during the winter months.

TEXTBOX 3.7

Sorption is a common term used for both **absorption** and **adsorption.** Don't confuse the latter two terms. Absorption is the incorporation of a substance in one state into another of a different state (e.g., liquids being absorbed by a solid or gases being absorbed by water). Recall that adsorption refers to the process leading to the concentration of a trace element from an aqueous phase onto the surface of a solid phase.

Although phosphorus was used extensively in our discussion, the adsorption of metals from aqueous solutions is of strong interest and environmental significance. Metal ions form soluble complexes with a variety of dissolved ions, including hydroxide, organic acids, carbonate, fluoride, chloride, and sulfate. As a demonstration, consider hydroxide complexes involving zinc (Zn). Refer to the species distribution diagram in Figure 3.12 showing how the solubility of

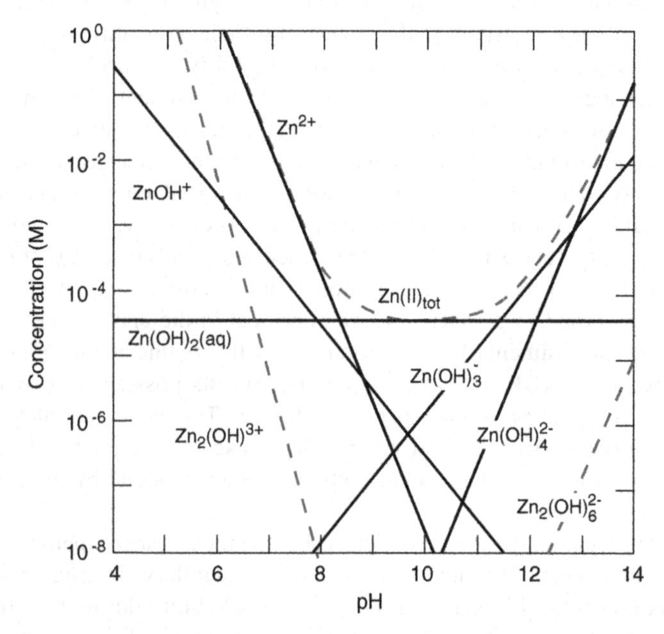

FIGURE 3.12 Species distribution diagram for $Zn(OH)_2$. This diagram depicts how the solubility of $Zn(OH)_2$ (s) is determined by the sum of all concentrations of all zinc species across the full pH scale.

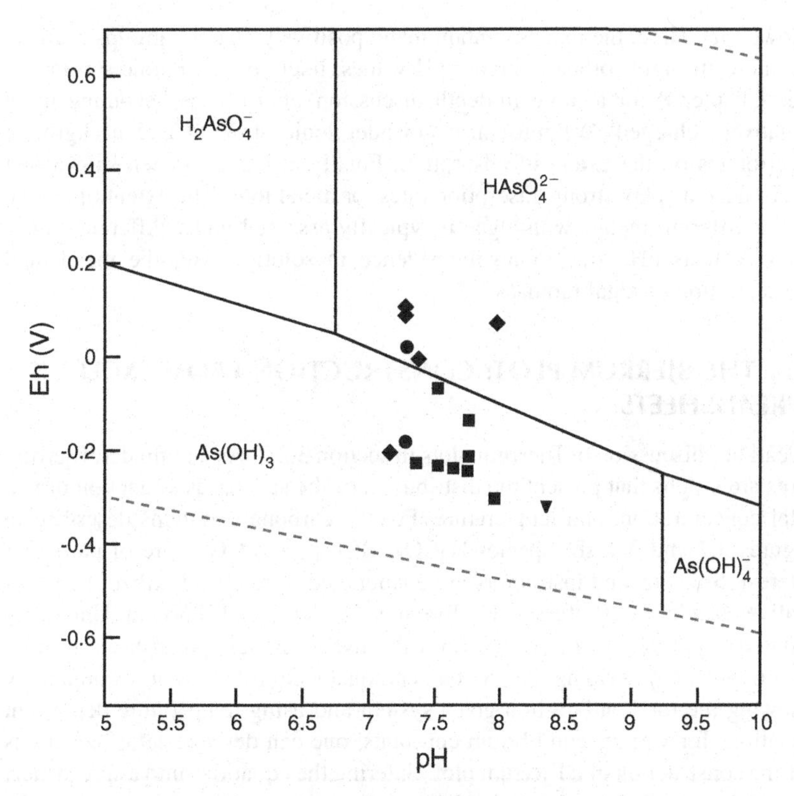

FIGURE 3.13 Plot of observed pH and Eh illustrating the speciation of arsenic. Shown are groundwater samples from the Datong Basin, northern China, designated as symbols. A temperature of 20°C and pressure = 1 atm was assumed. The dashed lines represent the limits of the water stability field. The plot illustrates that arsenic speciation in this study followed similar patterns as those observed in traditional aquatic modeling studies. *Modified from (Xie et al., 2009) with permission from Elsevier.*

$Zn(OH)_2$ (s) is determined by the sum of all concentrations of all zinc species across the full pH scale. Pay particular attention to the dashed lines, which represent the total solubility of Zn(II) species. If we again consider the thermodynamic law of mass action, the solubility for $Zn(OH)_2$ (s) shows that $[Zn^{2+}]$ is fixed by pH and the conventional solubility constant (defined above). Figure 3.12 assumes that no complexing ligands other than OH^- are present, and neglects effects that may arise if multiple metals were present in solution. Overall, dissolution and precipitation of metal hydroxides and the carbonate mineral family partially regulate pH and alkalinity of natural waters, ultimately affecting the fate and transport of organic and inorganic contaminants.

The influence of pH on metal adsorption is significant. A primary reason for this is the variability in particle charge, with electrostatic attraction varying depending on the pH value. For example, anions are adsorbed more strongly at a lower pH. Here, the oxides contain many positively charged groups. Cations are more strongly sorbed at higher pH values, likely due to humic substances (see Chapter 8 for a more in-depth discussion) and oxides becoming more negatively charged. We must also consider ionic strength and background electrolytes on the extent of adsorption. Finally, it has also been shown that microalgae display strong adsorption rates for metal ions. The extent of uptake of the different metals with algae is typically assessed under different conditions such as pH, time of algal residence in solution with the metal, and concentration of algal biomass.

3.6. THE BJERRUM PLOT: CONSTRUCTION FROM EXCEL SPREADSHEETS

Recall our discussion of Bjerrum plots in section 3.2.2. As a reminder, Bjerrum plots are graphs that present the distribution of the species as a function of pH, total concentration, and temperature. For the carbonate systems described in Figures 3.7 and 3.8, the species H_2CO_3, HCO_3^-, and CO_2^{2-} are of particular interest. Students and instructors are encouraged to read and utilize the informative article by Professor C. Brannon Andersen of Furman University (Anderson, 2002). In it, he describes the use of Excel spreadsheets in constructing and describing theoretical carbonate equilibria. For example, by knowing the total carbon in a given system and using temperature-dependent equations for various equilibrium constants, one can derive a set of equations for the construction of a Bjerrum plot. Entering the equations into a spreadsheet allows the calculation of the log of the species concentration for any given pH. Finally, the plot is constructed by plotting pH versus the log of each of the species concentrations. This article contains valuable content for students learning equilibria concepts (e.g., the concept of carbonate equilibria) and provides an exercise set with supplemental information (e.g., example Excel spreadsheet data) for students and instructors to follow in the course of studying this chapter. Full derivation of the equations presented can be found in cited references within the paper.

3.7. END OF CHAPTER PROBLEMS

3.1 What can water's polar nature be attributed to? Describe why water is a powerful solvent giving rise to various types of reactions in aqueous solutions.

3.2 Define the term *heat capacity* and discuss its importance in natural water systems.

3.3 How much energy is required to heat 1,000 g of seawater from 18.0°C to 22.5°C?

3.4 How would negative pressure affect the temperature of maximum density of water?

3.5 As discussed, water at 0°C becomes ice and is less dense than its liquid counterpart. As a result, newly formed ice is able to float on the surface of waterbodies, even though it is in a solid state. Describe the significance of this effect in terms of aquatic life.

3.6 Define the term alkalinity. Discuss the main sources of alkalinity in natural water systems.

3.7 Acid mine drainage is acidic water laden with iron, sulfate, and other metals that forms under natural conditions when strata containing pyrite are exposed to the atmosphere or oxidizing environments. Calculate both the $[H_3O^+]$ and $[OH^-]$ of a drainage sample that has a pH of 3.50.

3.8 Using the equilibrium constant expression for the ionization of water, provide an alternative approach to solving the $[OH^-]$ in problem 3.7.

3.9 Write the charge and mass balance equations for a solution of sodium hydrogen carbonate, $NaHCO_3$, noting that it dissociates into Na^+ and HCO_3^-. The anion is also known to ionize to CO_3^- as well as form carbonic acid, H_2CO_3.

3.10 Like CO_3^{2-}, phosphate (PO_4^{3-}) is known to form insoluble precipitates with metal ions in solution. Provide a complete and balanced equation for the reaction of Ba^{2+} with PO_4^{3-}.

3.11 Carbon dioxide dissolved in water is in equilibrium with carbonic acid. a) Write the equilibrium expression for this reaction. b) Write two equilibrium expressions for this product in water and discuss the potential consequences of this product formation.

3.12 Would you expect seasonal variations in pCO_2 to exist in seawater? What would be some of the key controlling factors in determining the occurrence and extent of possible variations?

3.13 Describe an open system in terms of pCO_2 and $[CO_2]$ levels.

3.14 Explain ocean chemistry changes that could likely result in significant changes in the global carbon cycle (e.g., increased atmospheric CO_2 levels).

3.15 Discuss why determining dissolved inorganic carbon (DIC) concentrations is vitally important in aquatic systems.

3.16 Discuss potential limitations in determining redox conditions in natural waters.

3.17 During methanogenesis (the formation of methane by microbes), CO_2 is used as an electron acceptor in a reduction half-reaction. Complete and balance this half-reaction:

$$4H_2 + CO_2 \rightarrow - + -.$$

3.18 Consider the following reaction in a basic solution:

$$ClO_2 + OH^- \rightarrow + ClO_2^- + ClO_3^-.$$

a Show the two half-reactions and label as reduction and oxidation as appropriate.

b Provide the overall balanced reaction.

3.19 What is meant by the term metal "speciation" and why is it important in aqueous systems?
3.20 Name three factors that affect the solubility of a substance in aqueous solutions.
3.21 What role does pH play in metal adsorption?
3.22 Based on your study of Research Application II, and your newfound knowledge of adsorption processes, would you expect greater arsenic adsorption to occur in highly reducing or oxidizing environments? Justify your answer.

3.8 REFERENCES

Anderson, C. B. (2002). Understanding carbonate equlibria by measuring alkalinity in experimental and natural systems. *Journal of Geoscience Education, 50*, 389–403.

Sabine, C. L., Feely, R. A., Gruber, N., Key, R. M., Lee, K., Bullister, J. L., Wanninkhof, R. R., Wong, C. S., Wallace, D. W. R., Tilbrook, B., Millero, F. J., Peng, T.-H., Kozyr, A., Ono, T., & Rios, A. F. (2004). The oceanic sink for anthropogenic CO_2. *Science, 305*, 367–371.

Stumm, W., & Morgan, J. L. (1981). *Aquatic chemistry: An introduction emphasizing chemical equilibria in natural waters* (2nd ed.). Hoboken, NJ: John Wiley & Sons.

Weiss, R. F. (1974). Carbon dioxide in water and seawater: The solubility of a non-ideal gas. *Marine Chemistry, 2*, 203–215.

Wilkie, J. A., & Hering, J. G. (1998). Rapid oxidation of geothermal arsenic (III) in streamwaters of the eastern Sierra Nevada. *Environmental Science & Technology, 32*, 657–660.

Xie, X., Ellis, A., Wang, Y., Xie, Z., Duan, M., & Su, C. (2009). Geochemistry of redox-sensitive elements and sulfur isotopes in the high arsenic groundwater system of Datong Basin, China. *Science of the Total Environment, 407*, 3823–3835.

Chapter 4

Abstracts and Keywords

Abstract

In this chapter, detailed understanding of the hydrologic cycle presented in Chapter 1 will be beneficial in understanding surface and groundwater concepts and the interactions that occur within such systems. First, hydraulic principles and groundwater flow concepts will be exposed. This material will be instrumental in covering subsequent material on chemical contaminants and pollutant transport. Next, a review of typical concentration units encountered in aqueous chemistry is presented. This is followed by information on drinking water standards, physico-chemical indicators, and selected environmental regulations relative to chemical, biological, and physical contaminants. Finally, a brief introduction to environmental modeling in relation to aquatic systems is presented.

Keywords

Groundwater, Contaminant transport, Chemical concentration units, Drinking water standards, Physico-chemical indicators, Developmental toxicants

Chapter 4

Surface/Groundwater Quality and Monitoring

Since the late 1970s, methyl *tert*-butyl ether (MTBE) and other oxygenates have replaced lead and other toxic chemicals as octane boosters to gasoline in the United States. Oxygenates have been shown to optimize the oxidation during combustion, reduce tailpipe emissions, and help boost the anti-knocking capacity of motor vehicles. MTBE is produced by reacting methanol with isobutylene (2-mehtyl-1-propene) using an acidic catalyst at 100°C (Figure 4.1):

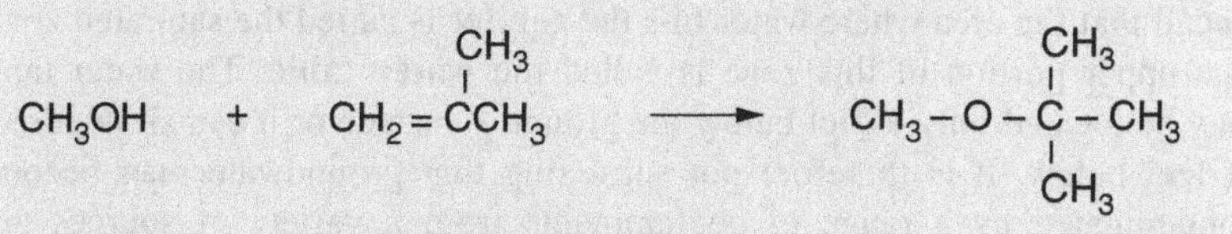

$$CH_3OH + CH_2{=}C(CH_3)CH_3 \longrightarrow CH_3{-}O{-}C(CH_3)_2{-}CH_3$$

FIGURE 4.1

But as we will learn, MTBE and other fuel oxygenates have been associated with groundwater contamination through leaking underground storage tanks (USTs), surface spills, broken or cracked pipelines, refueling facilities, and related sources. In this chapter, we will spend a great deal of time reviewing common pollutants in groundwater, including oxygenates and related aqueous degradation products. Implications for drinking water supplies and aquatic life will be assessed. Similarly, we will cover available remediation technologies and the formulation of appropriate laws to regulate common pollutants.

4.1. SURFACE AND GROUNDWATER RELATIONSHIPS

Detailed understanding of the hydrologic cycle (Chapter 1) is needed for comprehending hydrological processes, groundwater/surface water interactions, and how to manage water resources. Recall that precipitation, which is the source of effectively all freshwater in the hydrologic cycle, has an unpredictable pattern of distribution. And except for during (and shortly after) rainfall events, the majority of surface water in streams and rivers is furnished by the gradual release of groundwater. This is termed the **base flow**. Similarly, evaporation and transpiration return water to the atmosphere nearly universally, but such rates vary significantly according to climatic conditions. As a result, much of the precipitation never reaches the oceans as runoff before the water is restored to the atmosphere. The comparative enormity of the individual components of the hydrologic cycle, such as evapotranspiration, may differ significantly even at smaller scales. As we will learn shortly, hydraulic principles and the force of gravity dictate which direction the water travels in all surface water/groundwater interactions.

4.1.1. Flow in Groundwater Systems

As a reminder, groundwater is stored in—and moves rather slowly through—layers of soil, sand, and rocks called aquifers. Aquifers typically consist of gravel, sand, sandstone, or fractured rock, like limestone. These materials are permeable because they have large connected spaces that allow water to flow through relatively freely. The speed of groundwater flow is important when dealing with potential contaminants and depends on the size of the spaces in the soil or rock (and how well the spaces are connected). Recall that the area where water fills the aquifer is called the saturated zone. The upper portion of this zone is called the **water table**. The water table may be located only a foot below the ground's surface or it can sit hundreds of feet below. It is therefore not surprising that groundwater can become contaminated by a range of contaminants from a variety of sources and pathways.

TEXTBOX 4.1

Groundwater is the world's largest and most reliable source of freshwater, with nearly 80% of global inhabitants relying on it as a source of drinking water.

To understand the movement of contaminants in groundwater, we must first understand the flow of water to, and through, a groundwater system. This requires a basic understanding of all physical factors controlling sources of water and water movement, and therefore advective transport of potential contaminants (more on this concept a little later). As an extension to Figure 1.5, we can detail the groundwater flow system. Groundwater can be modeled to show movements along flow paths from areas of recharge to areas of discharge (Figure 4.2). For example, recharge can occur as precipitation falls on the land surface or from water in surface waterbodies moving into the ground (infiltration). Some quantity of water that infiltrates will remain in the shallower soil layer, where it will progressively move vertically and horizontally through the soil and subsurface material. Water will likely continue to move deeper (percolate), recharging the local groundwater aquifer. The capillary fringe is

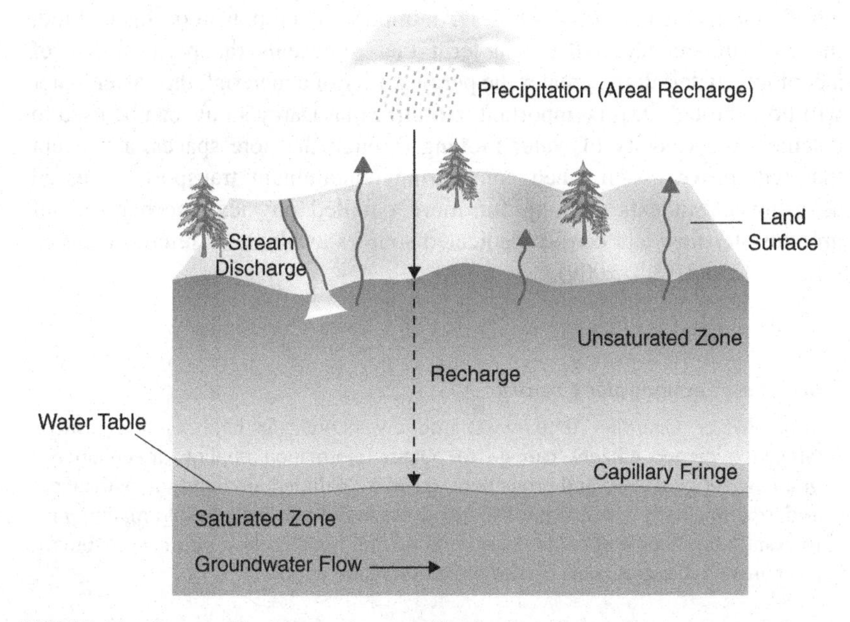

FIGURE 4.2 Conceptualized figure of water infiltration through the soil-water unsaturated zone and into the water table. Note that the capillary fringe refers to that zone of soil immediately above the water table. Most of the soil pores are completely filled with water at the base of the capillary fringe. In contrast, only the smallest soil pores are filled with water at the top of the capillary fringe.

a region that acts as a critical connection between the unsaturated zone and aquifers. Flow within the capillary zone has been shown to demonstrate potential impact on the hydraulics in the subsurface, particularly in regards to water table dynamics and infiltration events (Nielsen and Perrochet, 2000). Understand that the rate of infiltration is limited by the capacity of the soil and the rate at which water is applied to the surface itself. This is a volume flux of water flowing into the profile per unit of soil surface area, commonly expressed as velocity. Discharge from the saturated subsurface material occurs at springs, streams, lakes, wetlands, the ocean and coastal areas, and as transpiration by plants.

How do we now go about improving our understanding of the susceptibility of groundwater systems to contamination? To answer this, we need to consider two important sets of factors controlling groundwater movement: those that define the movement of water within porous media (e.g., sand and gravel), and those that define the movement of water along the boundary of the porous media (areas where water enters or leaves the groundwater system). Porous medium is composed of a mixture of inorganic and organic natural materials, with the surface properties of the solid phase being major factors that control the behavior of chemicals (Berkowitz, et al., 2008). Henri Darcy, a French engineer, developed a method to describe the flow of groundwater within porous media (Darcy's Law). An important component of this method, the hydraulic conductivity, is a term that represents the permeability of a geologic material; the greater the permeability of a material, the easier water will flow through it. It is important to realize that Darcy's Law can be used to calculate the velocity of water moving through the pore spaces, a concept that will prove useful when considering contaminant transport discussed as follows. Students wishing for more detailed physical mechanisms of groundwater flow can consult dedicated sources available as references (e.g., Fitts, 2002 and Tóth, 2009).

TEXTBOX 4.2

Become a Groundwater Guardian

Groundwater Guardian (http://www.groundwater.org/gg/gg.html) is a website where concerned citizens can go to obtain information on how to enhance groundwater education and protection activities. Included are hands-on activities and experiments in which to participate, groundwater terminology, information on relevant workshops and conferences, and helpful hints on how you can become more involved in grassroots and larger conservation efforts.

4.1.2. Contaminant Retention and Release

The contaminants discussed above typically enter a groundwater system from the land surface, infiltrating and percolating down through the aerated soil and unsaturated zone. This movement is affected by several variables including properties of the contaminant itself (water solubility and persistency), soil and unsaturated zone conditions (infiltration characteristics, pore size distribution, microbial population density and diversity, organic matter content, total porosity, ion exchange capacity, hydraulic properties, pH, and oxygen status), climatic factors (temperature, wind speed, solar radiation, and intensity and duration of rainfall), and vegetation cover. These combinations of factors make the likelihood of groundwater contamination a very site-specific endeavor of scientific study. Nevertheless, a thorough understanding of these processes and variables is critical to effectively managing potential groundwater contamination.

Specific geochemical subsurface processes involved in contaminant retention and release are highlighted in Figure 4.3. Many of these processes that occur in natural waters, including acid-base reactions, complex formation, dissolution/precipitation, redox reactions, and adsorption and ion interactions, were covered in detail in Chapter 3. Consider redox processes and their affects in the subsurface further. For example, the complete oxidation of a given organic compound to CO_2 and water is considered thermodynamically favorable. But is this process favored kinetically? Recall that redox reactions in the subsurface are fairly slow compared to other

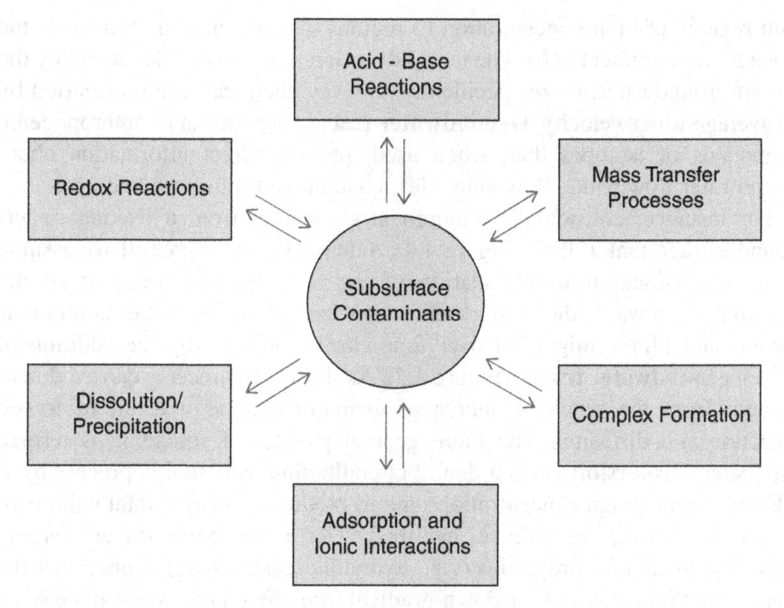

FIGURE 4.3 Geochemical subsurface processes involved in contaminant retention and release.

reactions due to the sluggish species diffusion from one environmental compartment to the next. Many of these are dependent upon catalytic processes; thus, we can say that they are not considered kinetically favorable. And what role does adsorption have on contaminant retention and release? As defined, adsorption is the net deposition of matter on the solid phase at the boundary with an aqueous solution or gaseous phase. If the adsorption process removes a given compound from the bulk phase, it will have an effect on its behavior in the subsurface. Moreover, the chemical form of a contaminant will largely affect its sorption. For example, most inorganic contaminants in aqueous solution occur as ionic or charged species. However, in many contaminated sites, metals and other inorganic constituents can exist as polar or non-polar neutral species. This will change the sorption characteristics, and therefore mobility.

4.1.3. Contaminant Transport Processes and Plume Behavior

Chemical contaminants can migrate directly into the groundwater from below-ground sources or from the surface by vertical leakage. Mechanistically, chemical contaminants can enter, and be transported in, the subsurface via numerous processes. If we consider the transport of passive contaminants, those not involving an outlay of chemical energy, two main processes are implicated: diffusion and advection. **Diffusion** (e.g., molecular diffusion) is a process in which chemical species move under the influence of thermal-kinetic energy from regions of high concentration to regions of low concentration (until the concentration gradient = 0). The term **advection** refers to the movement by the flow of groundwater; more specifically, the way chemicals are transported by the average water velocity. **Groundwater tracers** are natural or anthropogenic compounds or isotopes that, when used, provide direct information about groundwater flow paths, flow velocities, contaminant transport, and age.

For instance, consider a contaminant emanating from a leaking underground storage tank (UST) (Figure 4.4), a defined point source. If we assume a one-dimensional flow, elucidation of the behavior and transport of the contaminant toward the water table can be examined. The continuous contaminant plume migration over time can be pictured by the addition of a given **groundwater tracer** (Figure 4.5). As shown, the tracer is carried down-gradient from the potential source of contamination as a result of forced advection and diffusion. The more general process of spreading is termed dispersion. **Dispersion** plays a dual, but conflicting, role in this process by i) reducing contaminant concentrations and ii) producing a larger total volume of contaminated water to remove and treat. The above processes are largely controlled by aquifer properties (e.g., hydraulic conductivity). Notice that the contaminant plume is moving down-gradient from the source, while at the same time being diluted (as a result of uncontaminated groundwater and travel through soil and rock within the aquifer) and dispersing over relative time periods (e.g., Time 1 (T_1), Time 2 (T_2), etc.).

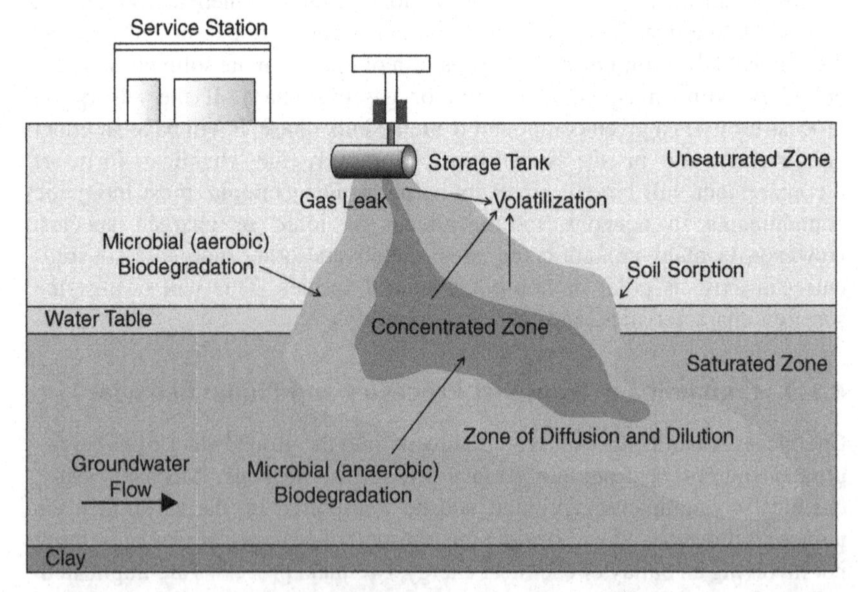

FIGURE 4.4 Groundwater contamination from the surface by vertical leakage through an underground gasoline storage tank.

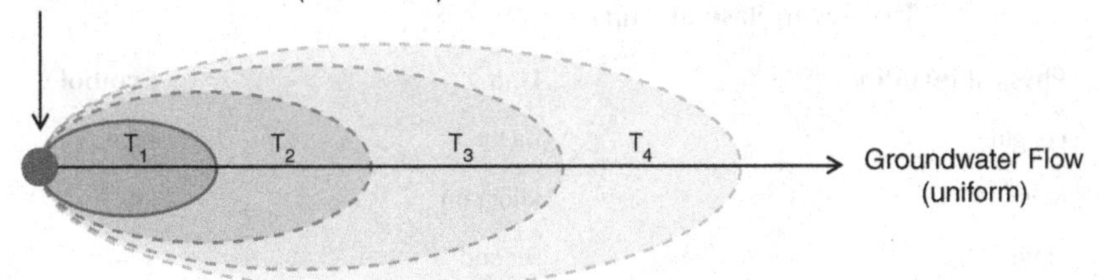

FIGURE 4.5 A non-reactive tracer plume moving down-gradient from a point source of potential contamination. This is assuming that groundwater flow is uniform. The tracer can provide direct information on potential transport patterns and mechanisms in groundwater. Figure based partially on an original diagram by (Freeze and Cherry, 1979).

However, at the process level, evaluation of plume development and movement requires more detailed knowledge. For example, the same redox conditions mentioned above dictate the identification of electron acceptors (recall our discussion on the degradation of organic pollutants), organic carbon solutes, their concentration patterns, and relative rates of supply (Rees, 2007). If we consider a plume of dissolved organic contaminants, strong reducing environments will develop shortly after their emanation from a point source, with electron acceptors being depleted near the source and within the plume core downstream of the source area. Rates of biodegradation through microbial respiration will thus be limited by the rate of electron acceptor supply at the plume fringe.

4.2. WATER QUALITY UNITS AND CONCENTRATION CONSIDERATIONS

Before we delve into water quality standards and monitoring activities, it is important to understand and characterize chemical and physical measurements of complex aquatic systems. To do this, a working understanding of how environmental parameters monitored are recorded in different units is essential. The concepts presented below provide a basic review of common units of weights and measure, and discussion of how concentrations of trace elements are typically expressed. Such knowledge will strengthen concepts presented in this chapter and provide guidance for students tackling material presented in subsequent chapters.

4.2.1. SI Units of Measure and Conversions

The **metric system** was first proposed in 1791 during the French Revolution and has since evolved into a global understanding of common units of weights and measure. The modern metric system, termed the **International System of Units (SI units)**, precisely defines fundamental units for each

TABLE 4.1 The Seven Base SI Units

Physical quantity	Unit	Symbol
Length	meter	m
Mass	kilogram	kg
Time	second	s
Amount of substance	mole	mol
Temperature	kelvin	K
Electric current	ampere	A
Luminous intensity	candela	cd

quantity of measure. SI units are built upon seven base units, one for each of the standard physical quantities (Table 4.1). Additional units are derived from combinations of the base units. It is often advantageous, especially to beginning environmental chemistry students, to form decimal fractions and decimal multiples of the base and derived units. This limits the use of exponential notation and is used to express both small and large measured quantities in simple terms. Appendix XI lists the multiplying prefixes. Notice how the units are multiplied by simple powers of 10, mostly based on the unit of 10^3. The SI unit for an amount-of-substance is the mole: e.g., 1 molar $= 1.0 \times 10^{-6}$ or (0.000001) micromolar. Note that many trace elements measured in natural waters will be factors lower than this. For example, selenium species (among many trace metals) in seawater can be measured at the low picomolar (pM, 1×10^{-12} M) range by the use of stripping voltammetry (SV).

4.2.2. Chemical Concentrations

It will also become evident during your studies that concentration of solutes in solution will be expressed in a variety of ways to suit the environmental chemist's needs. For example, concentration profiles of dissolved species in natural waters (e.g., Figure 4.7) are typically expressed in molar fashion, with the number of moles of solute dissolved per liter of solution, M, given by

$$M = \frac{\text{moles of solute}}{\text{liters of soloution}}. \tag{4.1}$$

As chemists, we are often asked to find the molar concentration of a resulting solution. Knowledge of the number of grams of a given substance and calculation of its molecular mass (the sum of the atomic masses of the atoms in the molecule) will allow us to find the number of moles. From there we can calculate the molarity given the number of liters of solution.

TEXTBOX 4.3

Voltammetric Techniques

All voltammetric instrumental techniques have a commonality involving the application of a potential to an electrode resulting in a current flowing through an electrochemical cell. In **stripping voltammetry (SV)**, for example, preconcentration (ratio of the concentration of microcomponents increases) of analyte at the electrode surface in order to lower the detection limit for an analyte in solution occurs. This is shown in Figure 4.6 as an example metal species, deposited Cd^{2+}, forms a Cd(Hg) amalgam:

Cd^{2+} (solution) $+ 2e^- + Hg \rightarrow Cd(Hg)$.

The second step is a short resting period (typically a few seconds) during which the potential applied to the electrode is maintained at the same value as that used for deposition. Finally, the preconcentrated material is redissolved (stripped) from the electrode by scanning the electrode potential in the anodic direction. In the case shown in Figure 4.6, the following reaction represents the stripping step:

Cd(Hg) (at the electrode) $- 2e^- \rightarrow Cd^{2+}$

As a result, well-defined voltammograms can be acquired.

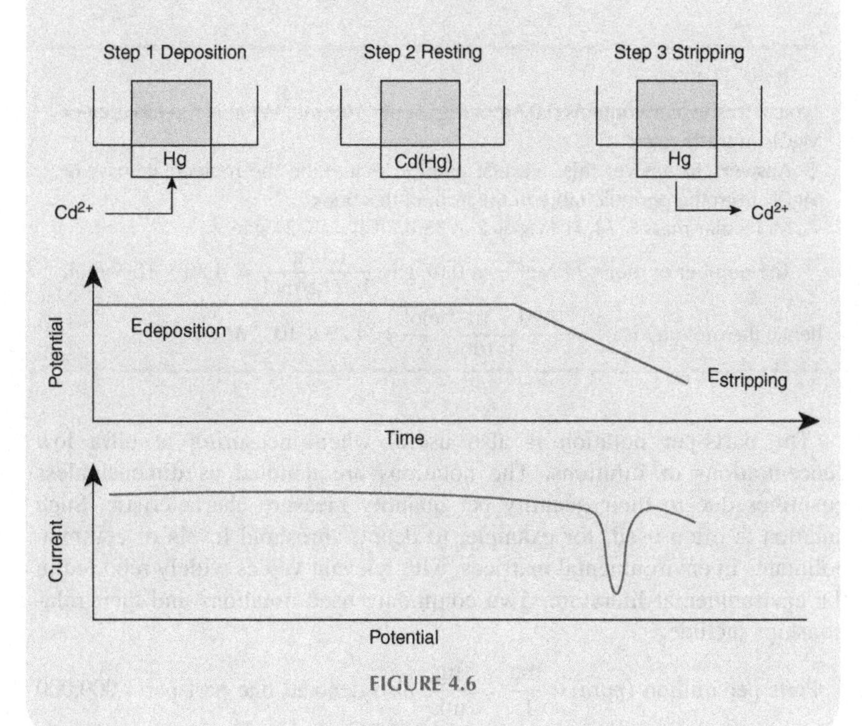

FIGURE 4.6

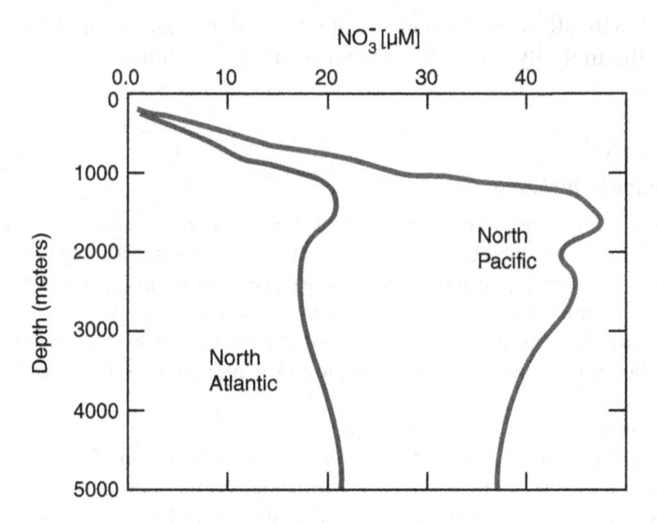

FIGURE 4.7 Vertical profile illustrating changing nitrogen as nitrate (NO_3^-) concentrations with depth in North Atlantic and North Pacific Oceans waters. Notice that concentrations are lower in surface waters, but generally increase as depth increases.

Example Problem 4.1

Typical freshwater contains 0.07 g of $MgCl_2$ per 100 mL. What is the molarity of $MgCl_2$ in freshwater?

Answer: To answer this, we first need to determine the molecular mass of $MgCl_2$ from the periodic table in the front of this book:

Molecular mass = 24.31 (Mg) + 2 × 35.45 (Cl) = 95.21 g/mol.

The number of moles of $MgCl_2$ in 0.07 g is $\frac{0.07\text{g}}{95.21\text{g/mol}} = 4.20 \times 10^{-4}$ mol,

hence the molarity is $M = \frac{4.20 \times 10^{-4}\text{ mol}}{0.100\text{L}} = 4.20 \times 10^{-3}\ M$.

The parts-per notation is also useful when measuring at ultra low concentrations or dilutions. The notations are denoted as dimensionless quantities due to their quantity per quantity measure characteristic. Such notation is often used, for example, to denote threshold levels of common pollutants in environmental matrices, with relevant values widely reported in the environmental literature. Two commonly used notations and their relationships include:

1. Parts per million (ppm) $= \frac{\text{mg}}{\text{L}} = \frac{\mu\text{g}}{\text{mL}}$, also denoted one part per 1,000,000 parts, one part in 10^6, and a value of 1×10^{-6}.

2. Parts per billion (ppb) $= \frac{\mu g}{L} = \frac{ng}{mL}$, also denoted one part per 1,000,000,000, one part in 10^9, and a value of 1×10^{-9}.

Note that one percent (1%) = 10,000 ppm. Consequently, a water sample that has a concentration of 200 ppm for species X could also be said to have a concentration of

$$\frac{(200 \text{ ppm})}{(10,000 \text{ ppm}/\%)} = 0.02\% \text{ (by mass)}.$$

Parts-per notation can easily be converted into molarity as evident in Example Problem 4.2.

Example Problem 4.2

Hydrogen sulfide (H_2S) is one of the most problematic compounds to be dealt with in potable water supplies. It is formed by the reduction of sulfides by organic matter (or bacteria) under anaerobic conditions, ultimately resulting in the "rotten egg" odor associated with septic wastewater. The concentration of H_2S in potable water has been reported to reach the 75 ppb level. Find the molarity of this compound at this level.

Answer: A concentration of 75 ppb $= 75 \times 10^{-9}$ g or 75 ng of H_2S per gram of potable water. We next need to convert to moles per liter as follows:

$$75 \times 10^{-9} \frac{g}{mL} \times \frac{1000 \text{ mL}}{L} = 75 \times 10^{-6} \frac{g}{L}$$

Finally, find the molarity with the use of the molecular mass of H_2S (34.08 g/mol):

$$\text{molarity of } H_2S = \frac{75 \times 10^{-6} \text{ g/L}}{34.08 \text{ g/mol}} = 2.2 \times 10^{-6} M.$$

Chemists also express concentration as the percentage of a component in a given mixture or solution. Three ways of expression include:

1. $$\text{Weight (mass) percent(w/w)} = \frac{\text{weight of solute}}{\text{weight of total solution}} \times 100 \quad (4.2)$$

2. $$\text{Volume percent(v/v)} = \frac{\text{volume of solute}}{\text{volume of total soloution}} \times 100 \quad (4.3)$$

3. $$\text{Weight/volume percent(w/v)} = \frac{\text{weight of solute (g)}}{\text{volume of solution (mL)}} \times 100 \quad (4.4)$$

Do be aware of a few key items. Both weight and volume percent are relative values that do not depend on the units of weight or volume used in the calculation (when both the numerator and denominator have the same units). When considering weight/volume, the percentage is expressed in % g/mL, not straight % units. Chemists are routinely faced with converting one unit to another, with numerous assumptions being made during the process. For example, a useful heuristic when converting weight percent (w/w) to molarity is to assume 100 g of solution. Conversely, when converting molarity to (w/w), one can logically assume 1 L of solution.

Example Problem 4.3

A solution is prepared by mixing 13.0 g $NaNO_3$ and 325 g of H_20. Calculate the weight percent of the solution.

Answer:

Weight of solute = 13.0 g $NaNO_3$

Weight of total solution = 13.0 g $NaNO_3$ + 325 g H_20 = 326. g

$$\text{Weight percent} = \frac{13.0\ \text{g}}{326\ \text{g}} \times 100 = 3.99\%\ NaNO_3$$

4.3. WATER QUALITY PARAMETERS AND CRITERIA GUIDELINES

There is a never-ending list of biological and chemical substances that can impact the quality of surface and groundwater, and consequently potential drinking water supplies. The impacts range from an unpleasant taste to imminent health hazards. As will be evident below, water quality is determined by the quantity and diversity of substances residing in it and the various factors that control their transport and behavior.

4.3.1. Drinking Water Standards

Drinking water contaminants are assessed worldwide by established regulations concerning the quality of the water intended for human consumption and typically fall into distinct groups including inorganic compounds, organic and synthetic compounds, such as pesticides, and microbial and other contaminants. Table 4.2 highlights common contaminants and suggested levels from the World Health Organization's (WHO) drinking water standards, the European Union's Council Directive 98/83/EC, and the U.S. EPA's National Drinking Water Standards (based on maximum contaminant level (MCL)). More specific detail on selected contaminants found in groundwater is provided below.

TABLE 4.2 Selected U.S. EPA/WHO/EU Drinking Water Standards Comparative Table[a]

Contaminant	U.S. EPA Standards (1996)	EU Standards (1998)	WHO Standards (1993)
Acrylamide	Zero	0.0001 mg L^{-1}	0.0005 mg L^{-1}
Arsenic	0.01 mg L^{-1}	0.01 mg L^{-1}	0.01 mg L^{-1}
Barium	2 mg L^{-1}	2 mg L^{-1}	Not listed
Benzene	0.005 mg L^{-1}	0.001 mg L^{-1}	0.001 mg L^{-1}
Benzo(a)pyrene	0.0002 mg L^{-1}	0.0001 mg L^{-1}	Not listed
Cadmium	0.005 mg L^{-1}	0.005 mg L^{-1}	0.003 mg L^{-1}
Color	15 color units	Not listed	Not listed
Copper	1 mg L^{-1}	2 mg L^{-1}	2 mg L^{-1}
1,2-Dichloroethane	0.005 mg L^{-1}	0.003 mg L^{-1}	0.003 mg L^{-1}
Iron	0.3 mg L^{-1}	0.2 mg L^{-1}	Not listed
Lead	0.015 mg L^{-1}	0.01 mg L^{-1}	0.01 mg L^{-1}
Mercury (inorganic)	0.002 mg L^{-1}	0.001 mg L^{-1}	0.001 mg L^{-1}
Nitrate	10 mg L^{-1}	50 mg L^{-1}	50 mg L^{-1}
Odor	3 threshold odor	Not listed	Not listed
pH	6.5–8.5	Not listed	Not listed
Selenium	0.05 mg L^{-1}	0.01 mg L^{-1}	0.01 mg L^{-1}
Sulfate	250 mg L^{-1}	250 mg L^{-1}	500 mg L^{-1}
Total dissolved solids	500 mg L^{-1}	Not listed	Not listed
Zinc	5 mg L^{-1}	Not listed	3 mg L^{-1}

[a](U.S. EPA, 1996); (The Council of the European Union, 1998); (WHO, 1993).

4.3.2. Inorganic and Organometallic Compounds

Nitrogen and Phosphorus

Chapter 1 provided detailed coverage of the nitrogen and phosphorus cycles and the behavior, transport, and fate of associated species within and between environmental compartments. It also exposed environmental equilibria concepts, details on the transformation of nitrogen and phosphorus species, and assessed the influence of and perturbation by anthropogenic activities. Recall that biological processes executed by microorganisms in the soil transform organic nitrogen to inorganic forms, where organic nitrogen is ultimately converted to inorganic ammonium. Nitrate is the form of nitrogen that is most used by plants for growth and development and the primary form of nitrogen at the surface/subsurface interface. Nitrate is supplied by inorganic fertilizer, animal manure, septic systems, and airborne nitrogen compounds given off by industry and automobiles that are deposited on the land in precipitation and dry particles. In contrast to ammonium, nitrate does not readily attach to soil particles and, as a result, is easily influenced by the movement of water. Its solubility (see Chapter 8 for extended coverage) in water allows it to be effortlessly passed through soil to the groundwater table.

As expected, nitrate concentrations in groundwater generally increase with higher nitrogen input and higher aquifer vulnerability (dependent upon underlying soil and bedrock conditions, as well as distance or depth to groundwater) (Hubbard and Sheridan, 1989). For example, if the distance to the groundwater is relatively shallow and the underlying soil is, e.g., sandy, the propensity for nitrates to invade the groundwater is elevated. Although nitrate generally is not a threat for adults, ingestion in drinking water by infants can cause low oxygen levels in the blood (Spalding and Exner, 1993). Given this information, the U.S. EPA has established a drinking water standard of 10 mg L^{-1} nitrate as nitrogen (see Table 4.2). Other than isolated instances, groundwater nitrate concentrations are typically shown to be at levels <2 mg L^{-1}.

Although the impact of phosphorus species on surface waters has been demonstrated (e.g., eutrophication), its role in groundwater contamination is less understood and/or accepted. Like nitrogen, phosphorus is a common constituent of agricultural fertilizers, manure, and organic wastes in sewage and industrial effluent. It is its mobility and migration that have been under scrutiny. To address such concerns, scientists have employed plume studies as a method of assessing the short- and long-term migration of phosphorus in the subsurface. For example, U.S. Geological Survey scientists recently studied the long-term migration of phosphorus in a subsurface plume of treated sewage at the Toxic Substances Hydrology Program's research site located in Cape Cod, Massachusetts. The observed extent of the phosphorus plume and the interaction of the plume with an associated pond challenged these scientists to re-evaluate their understanding of the mobility of phosphorus in groundwater and the interactions between groundwater and surface water. The potential contribution of groundwater phosphorus to surface waters was also recently examined by the Scottish & Northern Ireland Forum for Environmental Research (SNIFFER). Significant correlations were found between measured groundwater phosphorus concentrations and land use, which provides a quantity of evidence that anthropogenic activities may have had some impact on groundwater phosphorus, at least in England and Wales.

Heavy Metals and Metalloids

Groundwater scientists estimate that over 60% of the world's contaminated sites have elevated levels of toxic metals such as arsenic, cadmium, chromium, copper, lead, mercury, nickel, and zinc. What is the significance of this occurrence and how do such metals infiltrate the groundwater? In the study of contaminated land, chromium is of particular importance given its widespread release from industrial sources and its two stable oxidation states in natural waters: Cr(VI) and Cr(III), each varying considerably in their toxic effects and transport mechanisms. We will use this as a model metal of study for our discussion. The trivalent form, typically present in the cationic form as Cr^{3+}, is an essential nutrient, while hexavalent (either chromate (CrO_4^{2-}) or dichromate ($Cr_2O_7^{2-}$)) is extremely toxic (Calder, 1998). The investigation of a groundwater source influenced by Cr(VI) requires analysis of groundwater for both Cr(VI) and total chromium.

Recall from Chapter 3 that metal speciation, especially in a groundwater environment, is influenced by Eh (in the presence of either oxidizing or reducing conditions) and pH. For chromium, Cr(VI) exists in moderately oxidizing and reduced environments. Hexavalent chromium exists in alkaline, strongly oxidizing environments. At constant pH, Cr(VI) species predominate in solution under high Eh (oxidizing) conditions. The form, concentration, and transport of chromium from sources are also influenced by the type of chromium utilized in the industrial process and the presence of other inorganic and organic constituents present in solution. For example, Cr(VI) species can be readily reduced to Cr(III) in the presence of organic matter, ferrous (Fe(II)) iron, and sulfide. The amount of metal (in this case Cr(VI)) adsorbed can be described by the traditional partition coefficient, K_d, which is defined as the ratio of the concentration of metal bound on the surface of the solid (C_s) to the total concentration of metal dissolved in the liquid phase at equilibrium (C_l), as in

$$K_d = \frac{C_S(\text{mg/kg})}{C_l(\text{mg/L})}. \tag{4.5}$$

This process can be expressed as the surface adsorption reaction (Allison, et al., 1990)

$$\text{SOH} + \text{M} \rightarrow \text{SOH}\bullet\text{M}, \tag{4.6}$$

where SOH = unreacted surface site, M = dissolved metal M, and SOH•M = adsorption site occupied by a component or surface-bound metal M (detailed sorption characteristics covered in Chapter 8). During transport of metals in groundwater systems, metal sorption to the solid matrix results in a reduction in the dissolved concentration of metal, which ultimately affects the overall rate of metal transport.

Research Application III: Arsenic Groundwater Contamination—the Cambodian Experience

Cambodia has one of the lowest percentages of population with access to safe drinking water. In fact, a majority of the 1.2 million people residing along the Mekong River floodplain (Figure 4.8) are being exposed to elevated levels of arsenic (As) due to naturally contaminated groundwater sources. In this region, average As levels as high as 163 $\mu g\ L^{-1}$ have been reported, more than ten times the World Health Organization's (WHO) drinking water guideline recommendation of 10 $\mu g\ L^{-1}$. High levels of As in drinking water have been associated with adverse health outcomes, primarily cancers.

FIGURE 4.8

A joint study by the Swiss Federal Institute of Aquatic Science and Technology and Development International-Cambodia established the first comprehensive database of groundwater composition in the Mekong River floodplain. As part of this study, 131 groundwater samples taken from private wells in three provinces were used to i) study the magnitude and distribution of arsenic (as well as manganese) contamination, ii) identify hydrogeological features of areas at risk, iii) elucidate the mechanisms of arsenic release, and iv) study the temporal trends of groundwater contamination. Arsenic levels ranged from 1–1,340 $\mu g\ L^{-1}$ (average = 163 $\mu g\ L^{-1}$), with 48% of the samples collected exceeding the recommendation level of 10 $\mu g\ L^{-1}$. The authors attribute the release of As from Holocene sediments present to the reductive dissolution of metal oxides. In regards to health considerations, Figure 4.9 shows the cumulative frequency distribution (a set of data showing the number of items less than or equal to the upper class limit of each class) for As in all three regions. The highest As concentrations were found in the Kandal Province (empty circles). The health threat posed by these levels justifies intensive study and international acknowledgment. Lowering of the ingested inorganic arsenic level and introduction of newer treatment options are thus of urgent need to safeguard the mass As poisoning and internal As related health problems this population faces.

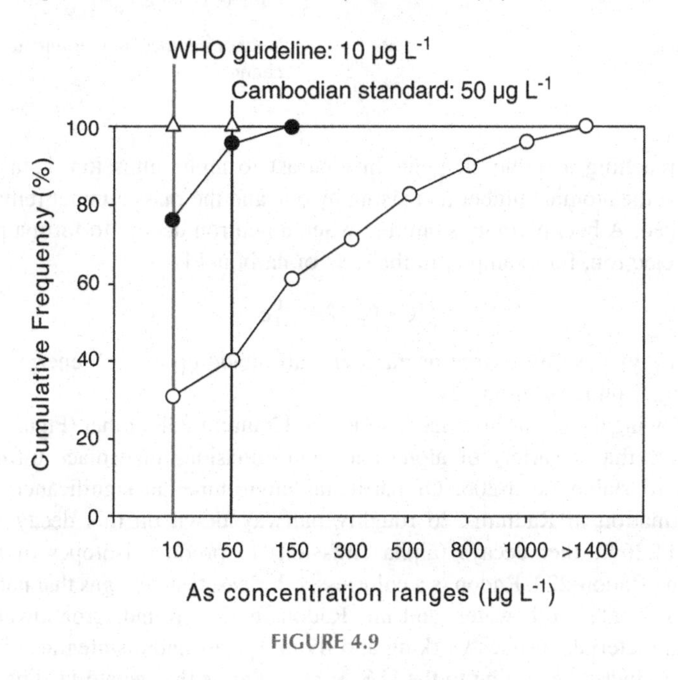

FIGURE 4.9

Radionuclides

There are a variety of naturally occurring and manmade radionuclides that are regulated by adopted drinking water standards. Radionuclides are species of atoms that emit radiation as they undergo radioactive decay through the emission of **alpha particles** (α), **beta particles** (β), or **gamma rays** (γ). Table 4.3 summarizes the characteristics of these defined decay processes. When an alpha or beta particle is emitted the particle changes identity. For example, the following nuclear equation shows the alpha decay of uranium-238 to thorium-234:

$$^{238}_{92}U \rightarrow ^{234}_{90}Th + ^{4}_{2}He.$$

Notice that the loss of an alpha particle (helium) results in the mass number of the product to be four lower. As we will learn shortly, uranium undergoes further radioactive decay; a chain of products is formed as a result of one by-product itself decaying to another element, which in turn decays further until

TABLE 4.3 Three Types of Radiation and Associated Characteristics

Type of Radiation	Symbol	Charge	Transformations
Alpha	$^{4}_{2}He$	+2	Mass number decreases by 4; atomic number decreases by 2
Beta	$^{0}_{-1}e$	−1	Atomic number increases by 1; mass number is unchanged
Gamma	$^{0}_{0}\gamma$	0	Neither the mass nor atomic number changes

finally reaching a stable element. In contrast to alpha emission, beta decay results in the atomic number increasing by one and the mass number remaining unchanged. A beta particle is emitted when a neutron decays to form a proton and an electron, for example, in the case of carbon-14:

$$^{14}_{6}C \rightarrow {}^{14}_{4}N + {}^{0}_{-1}e.$$

Gamma rays have no charge or mass and are made up of high-energy, short-wavelength photons of energy.

Following the radioactive decay series for Uranium-238 further (Figure 4.10), we notice that a variety of alpha and beta emissions take place ultimately leading to stable Lead-206. Of particular environmental significance is the transformation to Radium-226 roughly halfway down on this decay series. Radium-226 further decays (alpha emission) to produce isotopes of radon, including Radon-222. Radon is a colorless, odorless, tasteless gas that naturally occurs in rocks, soil, water, and air. Radon in the ground, groundwater, or building materials invades working and living spaces and disintegrates into its decay products. According to the U.S. EPA, radon is the second leading cause of lung cancer and has been implicated in as many as 20,000 premature deaths.

The decay of radionuclides follows first-order kinetics. Using the **half-life** (the amount of time it takes for an element undergoing decay to decrease by half) of a first-order reaction, we can institute the rate constant, k, with the decay rate constant, λ, and obtain a half-life equation specifically for radioactive decay rates:

$$t_{1/2} = \frac{0.693}{\lambda}. \quad (4.7)$$

Note that 0.693 is derived from the natural logarithm 2. Radon-222, with a half-life of 3.85 days, is long-lasting compared with its first four decay products, as shown in Figure 4.10. The longest half-life of the short-lived decay products of Radon-222 is Lead-214 at 26.8 minutes, and hence the decay products of Radon-222 deposited in the bronchial tree will largely decay in the lungs before biological removal mechanisms counteract their deleterious effects.

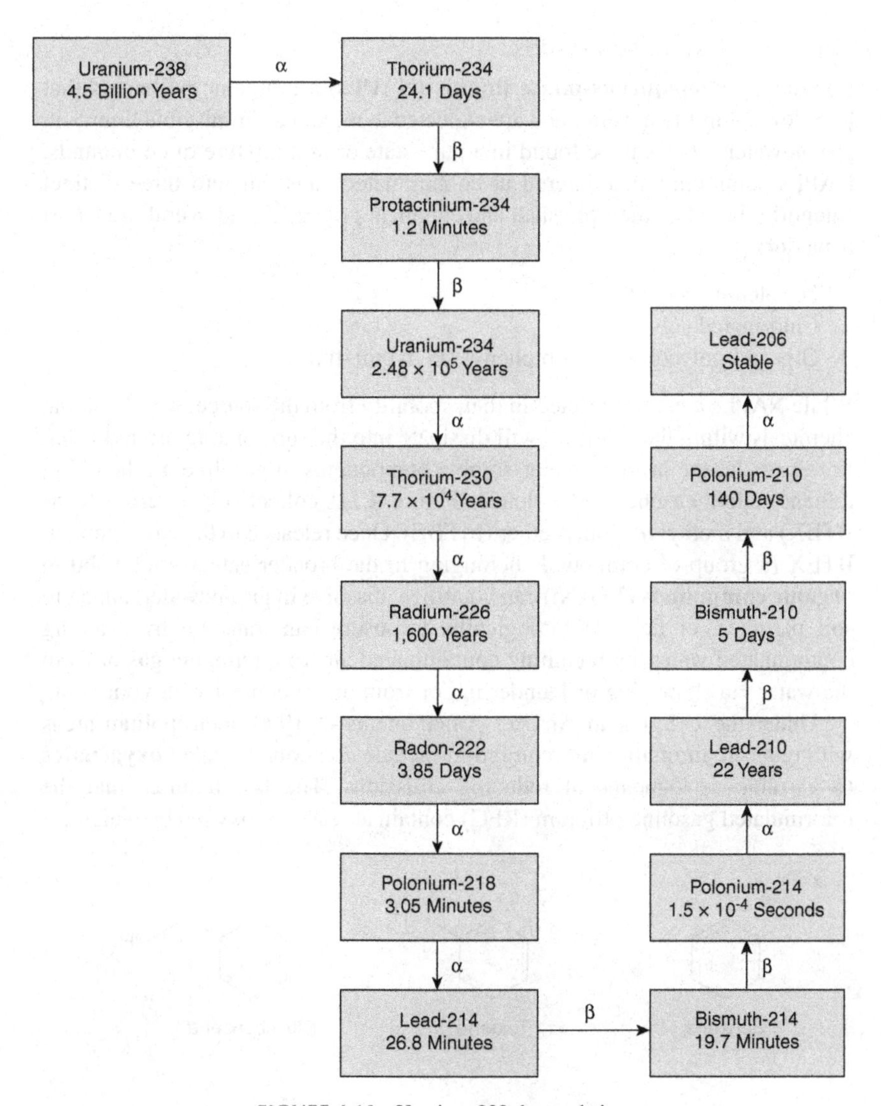

FIGURE 4.10 Uranium-238 decay chain.

Example Problem 4.4

Determine the decay rate constant for Carbon-14, a naturally occurring isotope produced in the upper atmosphere by cosmic ray bombardment, if it has a half-life of 5,730 years.

Answer: Using Equation 4.7, set $t_{1/2} = 3.85$ days, and solve for λ.

$\lambda = 1.209 \times 10^{-4}$ years^{-1}

Nonaqueous-phase Liquids

Chemically, **nonaqueous-phase liquids (NAPLs)** are organic pollutants that have low solubility in water and are expected to remain as immiscible liquids in groundwater. They can be found in a pure state or as a mixture of compounds. NAPLs commonly encountered at contaminated sites fall into three distinct categories based on their physical and chemical properties and overall transport behavior:

1. Petroleum products;
2. Chlorinated solvents;
3. Oils and polychlorinated biphenyl (PCB) mixtures.

While NAPLs can be restricted in their mobility from the source, water-soluble chemicals within these liquids will dissipate into the surrounding groundwater. For example, the primary water-soluble components in gasoline are benzene, toluene, ethylbenzene, and xylenes (Figure 4.11; collectively referred to as BTEX) and methyl *tert*-butyl ether (MTBE). Once released to the environment, BTEX (a group of compounds belonging to the broader category of **volatile organic compounds (VOCs)**) can volatilize, dissolve in groundwater, attach to soil particles, or degrade biologically. Exposure can transpire by drinking contaminated water, by breathing contaminated air from pumping gas or from the water via showering or laundering, or from direct contact with your skin.

Under the U.S. Clean Air Act Amendments of 1990, metropolitan areas with reduced air quality are required to include compounds called **oxygenates** to gasoline as a means of reducing emissions. The law requires that the reformulated gasoline program (RFG) contain at least 2% oxygen by weight. It

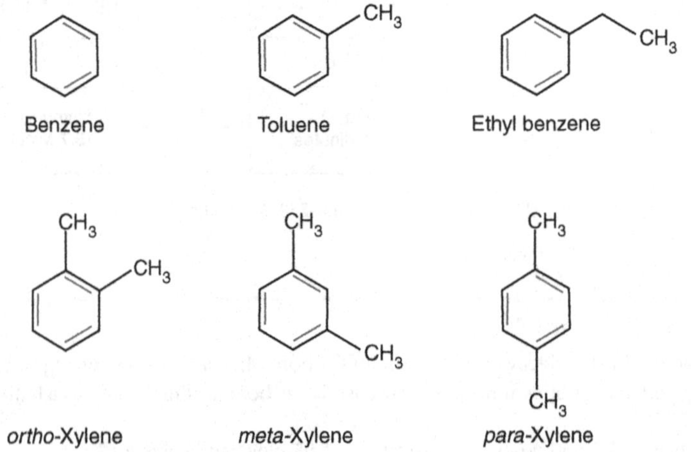

Benzene

Toluene

Ethyl benzene

ortho-Xylene

meta-Xylene

para-Xylene

FIGURE 4.11 The BTEX group of contaminants: benzene, ethyl benzene, toluene, and three isomers of xylene.

was met by adding a number of ethers or alcohols, including fuel ethanol, MTBE, ethyl *tert*-butyl ether (ETBE), and *tert*-amyl methyl ether (TAME). We will concentrate on MTBE as a model oxygenate of study. Although MTBE has provided air quality benefits, the addition of MTBE in gasoline has been a growing concern for several reasons: i) it is much more soluble in water, ii) it demonstrates a lower taste and odor threshold, and iii) it has a higher overall transport rate (U.S. Environmental Protection Agency, 1998). In regards to solubility, MTBE can move quickly through soil and into groundwater more rapidly than related compounds.

MTBE has been shown to be transformed under aerobic conditions by bacteria called aerobes, which use oxygen as their electron acceptor. For example, in the presence of oxygen, the most likely pathway, MTBE to *tert*-butyl alcohol (TBA) (Figure 4.12), starts with an oxidation of the methoxy group by mono-oxygenases (Steffan, et al., 1997). The resulting *tert*-butoxy methanol either reacts directly to form TBA and formaldehyde or is further oxidized to *tert*-butyl formate, which then hydrolyses to form TBA. The potential for anaerobic degradation of MTBE has been studied but not fully elucidated as of yet. One possibility is a nucleophilic substitution reaction as depicted, either via S_N1 or via S_N2 mechanisms or (nucleophilic substitutions). Recall from your organic chemistry training that a nucleophilic substitution is the reaction of an electron pair donor (the nucleophile, Nu) with an electron

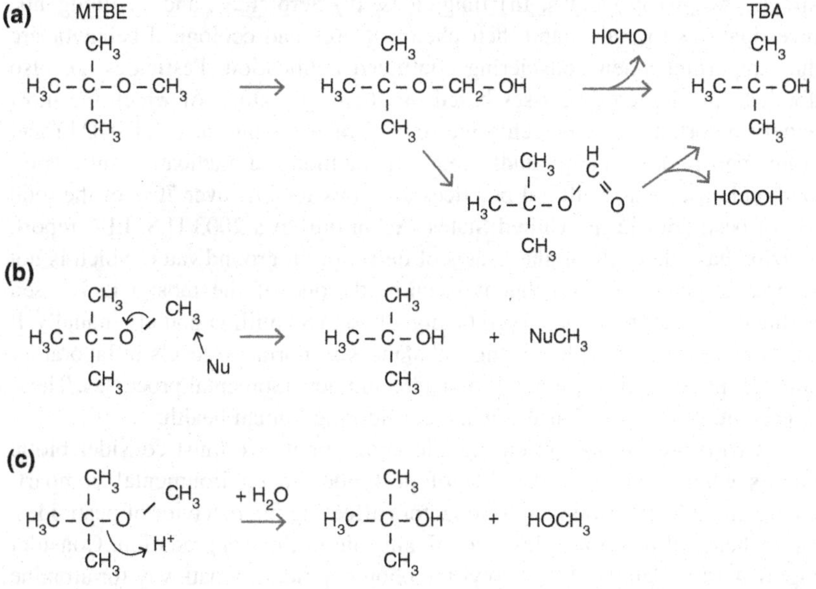

FIGURE 4.12 Proposed aerobic and anaerobic pathways for the enzymatic degradation of MTBE. *Based partially on a scheme by (Wang, et al., 2005).*

pair acceptor (the electrophile). In both reactions, the nucleophile competes with the leaving group.

Microbial degradation of **chlorinated aliphatic hydrocarbons (CAHs)** in groundwater, in particular perchloroethylene (PCE) and trichloroethylene (TCE), is also well documented. Redox conditions often determine the pathways for transformations of CAHs. Moreover, CAH concentrations vary significantly with depth and associated sulfate and methane concentrations. Some anaerobic bacteria use nitrate (denitrification), Fe(III) (iron reduction), and Mn(IV) (manganese reduction) as their electron acceptors, and break down organic constituents into smaller compounds, often producing CO_2 and CH_4 as the final products (Shim, et al., 2001). Consider the sequential anaerobic reduction schemes of PCE and TCE in Figure 4.13, which have been observed under methanogenic conditions. Partial transformation to -1,2-*cis*-dichrlorotheylene (*c*-DCE), 1,1-dichloroethylene (1,1-DCE), and 1,2-*trans*-dichloroetyheylene (*t*-DCE) can occur. Of added importance is the possible transformation to vinyl chloride (VC), a known carcinogen, in the subsurface. VC can be further reduced biologically to ethylene (E).

Pesticides

The widespread use of synthetic organic **pesticides** in agricultural and residential practices over the last few decades has led to their increased association with groundwater contamination. Pesticides fall into five general categories: i) insecticides, ii) larvicides, iii) fungicides, iv) herbicides, and v) fumigants. Investigations to understand their chemical uses and ecological behavior are thus important when considering continued application. Pesticides are also divided into numerous classes based on chemical nature, of which the most widely reported are organochlorine (e.g., atrazine, dieldrine, 4,4'-DDT) and organophosphorus compounds (e.g., malathion, dimethoate, sulfotepp). Herbicide use in agricultural practices accounts for just over 70% of the total use of pesticides in the United States. According to a 2003 U.S. EPA report, atrazine has one of the highest rates of detection in groundwater, which is not surprising given the fact that it is currently one of the most widely used herbicides, with an estimated production of 76 to 88 million pounds annually. It has been associated with causing imbalances in hormone levels in laboratory animals, possibly disrupting reproductive and developmental processes. These effects are of obvious concern when considering human health.

As with our discussion on organic compounds, we must consider biotic factors when evaluating the fate of pesticides in environmental compartments. In addition, among abiotic factors affecting the behavior of pesticides, photochemical reactions play a dominant role in their degradation. Consider Figure 4.14 in detail, which shows the photodegradation pathway for atrazine in natural waters (Evgenidou and Fytianos, 2002). Formation of product (1) occurs after dechlorination and hydroxylation. A second proposed route

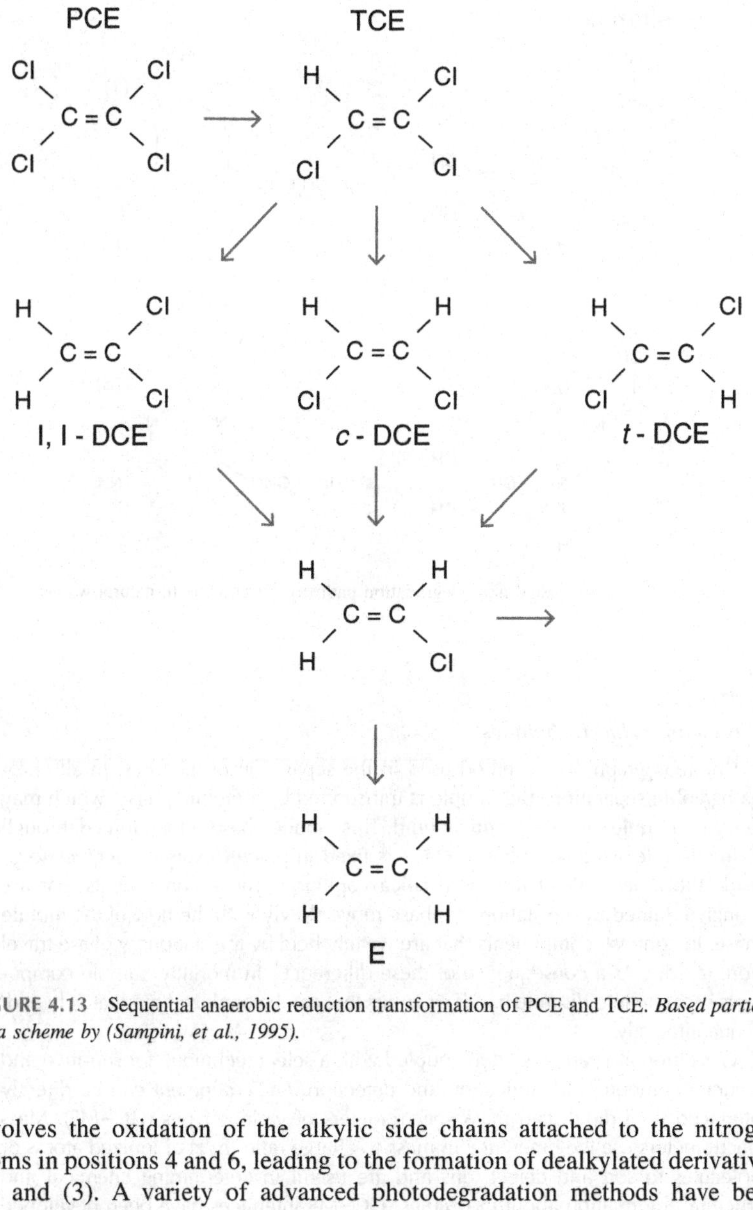

PCE

TCE

I, I - DCE

c - DCE

t - DCE

E

FIGURE 4.13 Sequential anaerobic reduction transformation of PCE and TCE. *Based partially on a scheme by (Sampini, et al., 1995).*

involves the oxidation of the alkylic side chains attached to the nitrogen atoms in positions 4 and 6, leading to the formation of dealkylated derivatives (2) and (3). A variety of advanced photodegradation methods have been tested including exposure to hypochlorous acid (HOCl), hydrogen peroxide (H_2O_2), Fe-tetra-amido macrocyclic ligands (Fe-TAML), and ozone (O_3). Main by-products are routinely determined by gas chromatography-mass spectrometry (GC-MS) in order to estimate the photolysis mechanisms that take place. Greater discussion of pesticide use and degradation is contained in Chapter 8.

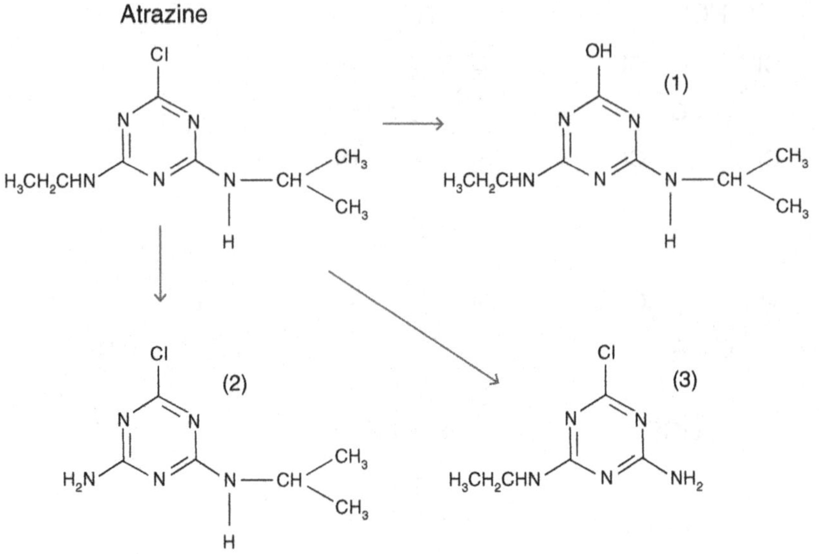

FIGURE 4.14 Suggested photodegradation pathway for atrazine in natural waters.

TEXTBOX 4.4
Chromatographic Techniques

Chromatography is a method used in the separation of mixtures. In all chromatographic separations the sample is transported by a mobile phase, which may be a gas, a liquid, or a supercritical fluid. This mobile phase is then forced through an immiscible stationary phase, which is fixed in place to varying degrees (e.g., inside the inner wall of the fused silica capillary). Those components that are strongly retained by the stationary phase move slowly with the flow of the mobile phase. In contrast, components that are weakly held by the stationary phase travel more rapidly. As a consequence of these differences in mobility, sample components separate into discrete bands or zones that can be analyzed qualitatively and/or quantitatively.

Gas chromatography is often coupled with a select technique for sensitive and accurate compound identification and detection. GC equipment can be directly interfaced with rapid-scan mass spectrometers of various types (GC–MS). Mass spectrometers use the difference in mass-to-charge ratio (m/z) of ionized atoms or molecules to sort and detect ions and are useful in determining chemical and structural information about molecules. GC–MS interfaces have been developed for packed column systems that allow for analyte molecules to be dynamically extracted from the carrier gas stream at the end of a packed column, and thereby selectively "sucked" into the MS for analysis. Typical components of the GC–MS include a capillary GC system, an ionization source, a mass separator, and an ion detector.

4.4. PHYSICO-CHEMICAL WATER QUALITY INDICATORS

The Federal Water Pollution Control Act Amendments of 1972, commonly referred to as the Clean Water Act (CWA), is one of the most important and far-reaching environmental statutes ever passed by the U.S. Congress. Similar legislation has recently been adopted by the European Union (EU) under the Water Framework Directive 2000/60/EC of the European Parliament. Both establish criteria for water protection and management and prepare and direct comprehensive programs for eliminating or reducing the contamination of surface and underground waters. Section 304(a)(1) of the CWA requires the development of criteria for water quality based solely on data and scientific judgments on pollutant concentrations and environmental or human health effects. Criteria are developed for the protection of aquatic life as well as for human health (Figure 4.15). In general terms, aquatic life criteria list chemical concentration goals, biological criteria, and nutrient considerations when assessing and protecting surface water for aquatic life use. Human health includes technical information and guidance on surface water, drinking water and microbials, as well as recreational considerations to protect people that leisure in, for example, coastal recreational waters. Pollutants regulated under the CWA, for example, include **"priority" pollutants**, including various toxic pollutants; **"non-priority" pollutants**, including biochemical oxygen demand (BOD), total suspended solids (TSS), oil and grease, and pH; and those of **organoleptic** effects (e.g., taste, odor).

TEXTBOX 4.5

In 2002, the U.S. EPA published revisions to many of the ambient water quality criteria for human health as the National Recommended Water Quality Criteria: 2002 (EPA-822-R-02-047). In 2003, they published an additional 15 revised Human Health criteria. The documents distributed under this action include a poster, brochure, and an electronic table reflecting the compilation of the updated information already published by the Agency in 2002 and 2003. This information is combined into one place for easy reference by interested students and faculty. Visit the following link for details: http://www.epa.gov/waterscience/criteria/wqctable/.

Threshold values exist for a more specific range of physico-chemical indicators: key indicators that affect water quality and can be monitored to measure the physico-chemical condition of aquatic systems. Table 4.4 provides a list of key indicators with brief descriptions of each. Let's again consider the concept of pH. In Chapter 3, we learned that natural water composition is controlled by a combination of geochemical and biological processes, which are largely affected by pH. For example, low pH values can adversely affect freshwater macrophytes and other aquatic biota. Possible causes of lowered pH values include the release of strongly acidic waste into streams, bacterial

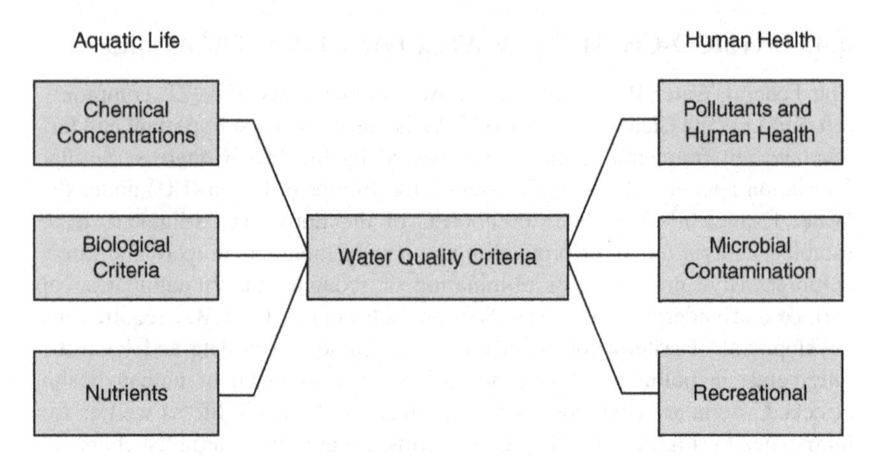

FIGURE 4.15 Water quality criteria developed for the protection of aquatic life as well as for human health.

nitrification, or sulfate reduction. Nutrients (nitrogen and phosphorus compounds) in excess can result in increased primary productivity in aquatic systems. On a related note, turbidity and suspended particulate matter can also influence primary production.

4.4.1. Electrical Conductivity (EC)

The concept of **electrical conductivity (EC)** may be used as a means of classifying the **potability** of water. Potable water is water that is fit for consumption by humans and other animals. Table 4.5 shows potability classification based on a measure range of conductivity ($\mu S \cdot cm^{-1}$). Until the late 1970s the units of EC were micromhos per centimeter (μmhos/cm), after which they were changed to microSiemens/cm (1 μS/cm = 1 μmho/cm). Although you may encounter both in the scientific literature, their numerical values are the same.

The ability of water to conduct a current is temperature dependent. Hence, all conductivity data are temperature adjusted to 25°C (usually called specific EC). In other words, investigators reference all EC readings to 25°C to eliminate temperature differences associated with seasons and depth. **Salinity**, another physico-chemical indicator, has a direct relationship with EC, and may be determined accurately from conductivity measurements. The **Practical Salinity Scale** is set relative to the conductivity of a potassium chloride (KCL) solution of known concentration. For example, seawater with Practical Salinity = 35 has the same conductivity as 32.4356 g kg^{-1} KCl solution at 15°C. Moreover, EC can be used to estimate the **total dissolved solids (TDS)** concentration (ppm) in freshwaters. Total dissolved solids include inorganic salts (e.g., calcium, magnesium, potassium, sodium, bicarbonates, sulfates, and chlorides) and minor amounts of organic matter

TABLE 4.4 Key Physico-Chemical Water Quality Indicators

Indicator	Description
Chlorophyll *a*	Indicator of phytoplankton abundance and biomass in coastal and estuarine waters.
Color	Change in color is indicative of dissolved organic material, inadequate treatment, high disinfectant demand, and the production of surplus disinfectant by-products.
Dissolved oxygen (DO)	Measure of the quantity of oxygen present in water reflecting atmospheric dissolution and biological processes that produce and consume oxygen.
Electrical conductivity (EC)	Measure of the electric current carrying ability of water with direct relation to the concentration of dissolved ions present in solution.
Flow	The amount of water and the rate (stream velocity and discharge) at which it moves affecting both ambient water quality and downstream processes.
Nutrients	e.g., nitrogen and phosphorus compounds, which in excess, can result in increased primary productivity.
pH	A measure of the acidity or basicity of solution.
Salinity	Dissolved salt content of a waterbody.
Turbidity and total dissolved solids (TDS)	Opacity caused by the amount of suspended particles (clay, silt, sand, algae, and other substances).
Temperature	The temperature of a waterbody directly affects many physical, biological, and chemical characteristics.

TABLE 4.5 Water Potability Classification Based on a Measured Range of Conductivity

Classification	Conductivity Range ($\mu S \cdot cm^{-1}$)	Designation
Fresh	<325	Potable water
Marginal	>325 but <975	Suitable for livestock watering and irrigation
Brackish	>975 but <3,250	Selective irrigation and livestock watering
Saline	>3,250	Suitable for coarse industrial processes

that are dissolved in water. Modern field instruments will automatically convert between the two while correcting for any temperature changes. To manually convert the electric conductivity of a water sample into the approximate concentration of TDS, the μS/cm is multiplied by a conversion factor, which depends on the chemical composition of the TDS and shown to vary between 0.54–0.96. A value of 0.67 is frequently used as an approximation if the actual factor is nonexistent:

$$TDS_{ppm} = \text{Conductivity } \mu S/cm \times 0.67. \tag{4.8}$$

As stated earlier, adjustments will need to be made based on relevant temperature changes often encountered in field situations.

4.4.2. Dissolved Oxygen (DO)

The analysis of **dissolved oxygen (DO)** is also of extreme importance in water quality monitoring programs. DO is essential for the metabolism of aerobic organisms and also directly influences inorganic chemical reactions. Knowledge of the solubility and dynamics of oxygen is therefore necessary for proper study and interpretation of both biological and chemical processes within aquatic systems. As we learned in Chapter 3, the amount of any gas (oxygen for this discussion) that can dissolve in pure water is inversely proportional to water temperature. The colder the water, for example, the more dissolved oxygen present. Atmospheric pressure and depth also have an effect on dissolved oxygen levels. In stratified lakes, for instance, DO concentration levels show the general trend of decreasing with depth, where the bottom hypolimnetic waters become oxygen-depleted through bacterial respiration.

DO is measured in standard solution units such as mL of O_2 per liter (mL/L) or mg of O_2 per liter (mg/L). As a general indicator of water quality, DO levels that fall below 5.0 mg/L are indicative of biological stress and will likely result in large fish kills. This is often the result of nutrient-rich waterbodies due to increased photosynthesis by the large quantities of algae present. Given that DO concentrations vary over a range of temperature, pressure, and salinity values, correction factors are often needed (unless these features are built in automated meters) when comparing multiple bodies of water. Expressing the results in % saturation, given by

$$\text{DO}(\% \text{ saturation}) = \frac{\text{measured DO (mg/L)}}{\text{DO (mg/L at 100\% saturation)}}, \tag{4.9}$$

is also recommended and taken as a more useful calculation of the **saturated concentration** of DO. Most commercial DO probes have this option for data readout. Alternatively, one can measure DO concentration via the **Winkler method**, a technique that uses titration. Oxygen in the collected water sample oxidizes iodide ion (I^-) to iodine (I_2) quantitatively. The amount of iodine generated is then determined by titration with a standard thiosulfate ($S_2O_3^{2-}$)

solution. The endpoint is typically determined by using soluble starch as an indicator. Note that at the time of sampling the DO is fixed by the addition of Mn(II) under basic conditions, resulting in a brown precipitate ($MnO(OH)_2$). The following five equations are the basis behind the Winkler method:

$$4Mn^{2+} + O_2 + 8OH^- + 2H_2O \rightleftharpoons 4Mn(OH)_{3(s)}$$

$$2Mn(OH)_{3(s)} + 3H_2SO_4 \rightleftharpoons 2Mn^{3+} + 3\,SO_4^{2-} + 3H_2O$$

$$2Mn^{3+} + 2I^- \rightleftharpoons 2Mn^{2+} + I_2$$

$$I_2 + I^- \rightleftharpoons I_3^-$$

$$I_3^- + 2S_2O_3^{2-} \rightleftharpoons 3I^- + S_4O_6^{2-}.$$

The Winkler method is routinely performed in undergraduate laboratories as part of a lab-based Environmental Chemistry course.

A related concept, **biochemical oxygen demand (BOD)**, measures the amount of oxygen that microorganisms consume while decomposing organic matter:

$$\text{Organic Matter} + O_2 \rightarrow H_2O + CO_2.$$

BOD itself directly affects the amount of DO in waterbodies. For example, the greater the BOD, the more rapidly oxygen is depleted. BOD levels can be as low as <5 mg/L (unpolluted, natural waters) to as high as 300 mg/L in raw sewage. BOD is affected by the same factors that affect DO, with additional considerations given to chlorine content as this can inhibit or kill microorganisms that decompose organic matter. Also related is the term **chemical oxygen demand (COD)**, which refers to the measure of the capacity of water to consume oxygen during the decomposition of organic matter and the oxidation of inorganic chemicals (e.g., ammonia and nitrate).

Example Problem 4.5

Would you expect purified water (e.g., distilled and/or deionized) to have a high electrical conductance?

Answer: No. Purified water is lacking excess dissolved ions with EC values reported in the range of 22–60 $\mu S \cdot cm^{-1}$.

4.5. DEVELOPMENTAL TOXICANTS

The presence and effect of **developmental toxicants** in water are of increasing concern. Such substances include natural hormones, synthetic steroids, and alkyl phenolic surfactants in domestic wastewaters, phytoestrogens in effluent from papermaking, perchlorate from jet fuel and lubricating oils, and phthalates from plastic industries. These substances are believed to interfere with normal bodily hormone function and reported to affect the reproduction process and increase individual susceptibility to cancer and other diseases, even at low ppb levels. Perchlorates, in particular, have received a lot of press, especially in the Southern California region. Perchlorates are the salts derived from perchloric acid known to interfere with iodine uptake into the thyroid gland, thus changing thyroid hormone levels that are suspected to cause thyroid gland tumors and improper development and regulation of metabolism in children.

In a 2003 memorandum released by the Los Angeles Regional Quality Control Board, perchlorate was shown to be present in 38 water supply systems within the Los Angeles Region. The State of California has recognized the severity of the problem and has proposed an action level of 4 ppb. The U.S. EPA is currently in the process of performing toxicity assessments that reflect the science and health effects of perchlorate due to the severity of the California occurrences and the additional releases of perchlorate reported in 20 other states across the U.S. Some form of regulation is expected after the toxicity assessments are completed.

4.6. BIOLOGICAL ASSESSEMENTS

In sharp contrast to analyzing physico-chemical parameters, **biological assessments** are increasingly used to evaluate the condition of a waterbody using direct measurements of the resident biota in surface waters. Here, the cumulative impacts of chemical, physical, and biological **stressors** on aquatic life are integrated into one complete study. In fact, many countries are now adopting national biological monitoring programs that concentrate on key biological indicators:

1. Species richness;
2. Species composition;
3. Primary production;
4. Ecosystem function.

Although specific information on which species to target or how the measurements should be taken is limited, biomonitoring programs do focus on detecting changes in patterns of key indicators: species richness, abundance, and composition. Numerous biological communities are studied and include algae, fish, plants, and macroinvertebrates. Additional measures of ecosystem processes (e.g., gross primary production) and functions are performed considering that changes in pattern do not always equate to changes in **biological integrity**—the ability to support and maintain a balanced, integrated, and adaptive community of organisms having a species composition, diversity, and functional organization comparable to those of natural habitats within a region (Karr and Dudley, 1981).

Case Study I: Environmental Cost of China's Rural Industrial Activities

Global aquatic environments are among the world's most valuable assets. This is particularly true in rural China where large-scale industrial expansion is transpiring to support their bourgeoning economic growth. Consider the Yangtze River (Figure 4.16), Asia's longest river (6,300 km) and fourth in terms of total water discharge (900 km^3/year) (Chen, et al., 2002). The main Yangtze channel is joined by a large number of tributaries, each housing significant industrial enterprises, with many lacking appropriate waste treatment technologies. In fact, by the end of 2003, China reported nearly 22 million rural enterprises (including those on the Yangtze Plain), accounting for over 30% of the country's gross domestic product (GDP) (Wang, et al., 2008).

FIGURE 4.16 The Yangtze River, China. More than 500 drinking water pumping stations are reported to be located on the Yangtze's banks. This is particularly disturbing considering that nearly 80% of waste entering the river is only partially treated or left untreated completely.

What are the consequences of such developments and how do they relate to our discussion of water quality? These questions can only be answered by considering the amount and nature of established environmental regulations and how widely implemented they are. Contrary to popular belief, China has some of the world's most stringent water quality policies. However, many of these policies have not yet reached the rural areas, and those that have are not fully implemented, nor are they enforced as stringently as more urbanized locations. Many believe this is a direct result of poorly designed policy instruments, an unsupportive work environment for regulators, and legislative shortcomings (Ross, 1992). More recent work by Wang et al. (1998) sheds light on this complicated process and provides a current view of underlying issues. They believe that the most fundamental factors of water pollution are the same factors that have underpinned the economic success of rural enterprises. These enterprises have close relationships with local government entities, thus limiting the enforcement of environmental regulations. Their recommendation: raising environmental awareness, providing media with more freedom to report on environmental conditions, and giving citizens more voice in local governmental affairs.

4.7. WATER QUALITY MONITORING

Water quality monitoring activities are carried out to provide information to answer questions relating to the management of waterbodies and their watersheds. They may be a single activity to examine a particular issue, i.e., a specified priority pollutant, or they may be ongoing monitoring programs to ensure that water quality standards are upheld. Both approaches require careful planning and defined goals, documentation, should be cost-effective, and employ the correct analytical method(s) for comprehensive environmental monitoring. Water quality monitoring programs require the development and implementation of a well-designed protocol (Figure 4.17) for proper collection of physical, chemical, and biological data, as well as interpretation of resulting measurements.

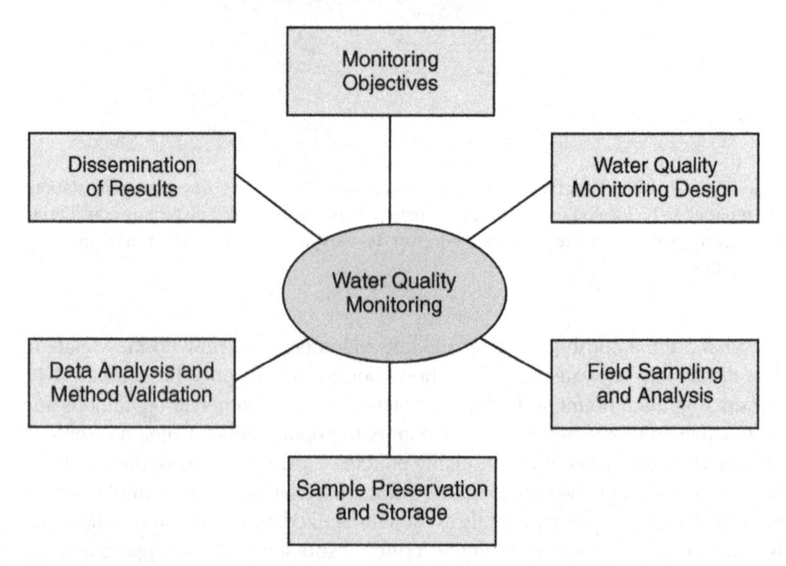

FIGURE 4.17 Recommended protocol for the development and implementation of effective water quality monitoring programs.

TEXTBOX 4.6
Water on the Web

Water on the Web (WOW) is an internet-based curriculum effort that contains historical data from numerous lakes and rivers in the United States, extensive primers on instrumental techniques and water chemistry, and data interpretation tools. It is a quick and reliable source of information for students to gain a better understanding of real-world environmental issues using advanced technology.

This site contains an extensive database of environmental data collected from sampling campaigns and numerous monitoring stations located in key areas of selected streams and rivers from 1997–2005. Students are encouraged to visit this site (http://waterontheweb.org/index.html) and explore the wide range of tutorials and graphical data to supplement their learning.

4.7.1. Water Quality Monitoring Design

The monitoring design phase is vital for guaranteeing a cost-effective sampling and analysis program. Based on the monitoring program objectives, information related to the following key areas satisfies the design process:

1. Spatial boundaries;
2. Study duration and sampling frequency;
3. Measurement parameters and scales;
4. Study type and site selection.

Investigators must have some understanding of the spatial boundaries of the aquatic system under study and the measurement scale considered. The term *scale* refers to the unit of space or time at which the system is under investigation. For example, it is vital to avoid boundary areas, e.g., at the confluence of streams or rivers and below sewage treatment plants, unless their impact on the system is being investigated. Other waterbodies pose additional complications, and these must be considered. In lakes and reservoirs, for example, representative sampling is often difficult due to environmental heterogeneity, both spatial and temporal (e.g., thermal stratification). Location and frequency must also be considered when designing a monitoring protocol. Site selection will ultimately depend on the problem to be addressed with safety and accessibility of paramount importance. The frequency of sampling (e.g., seasonal, continuous) will depend on the monitoring objectives and weather patterns. For example, the highest nutrient loadings in rivers and streams are generally correlated with intense, short-term flows during autumn and winter months, while the lowest loadings occur in the summer months when flow is low and biological activity is high. In water, processes that affect nutrient concentrations that must also be considered include plant, algal, and bacterial turnover, anthropogenic inputs (e.g., sewage effluent), matrix considerations (e.g., water hardness), and resuspension of bottom sediments from increasing river flow.

4.7.2. Sample Preservation and Storage for Subsequent Laboratory Analysis

Proper sample preservation and storage of aquatic samples before laboratory-based analyses are crucial to maintain sample integrity and prevent unwanted contamination. A suitable sample preservation protocol that minimizes the physical, chemical, and biological processes that can alter the physico-chemical forms of target analytes during storage must be used. The effectiveness of such protocols depends on a variety of factors including:

1. Sampling procedure;
2. Filtration;
3. Storage container;
4. Physical and chemical treatment;
5. Sample matrix;
6. Location and frequency;
7. Cleaning procedure;
8. Appropriate quality control measures.

Nutrients are especially prone to physico-chemical effects including calcium concentration, salinity, biological uptake, and various matrix considerations. Prior to any sampling campaign, it is essential to adopt an efficient cleaning protocol for all sampling equipment and storage bottles and continue this throughout the study. The walls of sample containers, for example, are excellent substrates for bacterial growth and therefore rigorous cleaning of all laboratory ware is necessary. In addition, sampling blanks should be taken to monitor and control the sampling process. For a representative list of parameters commonly evaluated, refer to Table 4.6 for recommended preservation and storage procedures.

4.7.3. Field Sampling and Analysis

Sampling (covered in detail in Chapter 2) or field-based analysis will follow the water quality design process. The advantages of field measurements are that they avert the need to collect, transport, and store samples of interest for subsequent laboratory analysis. For this reason, there is increase use of on-site or *in situ* instrumental techniques for monitoring of common water quality parameters. Regardless of whether or not field-based methods are used, sample preparation and QA/QC methodologies are needed for simplification of the sample matrix and enrichment of the target analyte(s) and for ensuring that collected data are of adequate scientific credibility to permit statistical interpretations. Common sample preparation techniques utilized in the analysis of aqueous samples include:

1. Filtration;
2. Dilution;
3. Centrifugation;
4. Extraction.

TABLE 4.6 Recommended Sample Preservation and Storage Procedures for Selected Parameters

Parameter	Preservation Method	Container Type	Holding Time
pH	None specific	None specific	Measured at the time of sampling
Nitrate, phosphate	Cool to 4°C. Chloroform or other chemical additives are often used to prevent biological growth	HDPE[a]	Within 24–48 hours ideally
Total metals	Acidify to pH 2 with HNO_3	Teflon	6 months
Polychlorinated biphenols (PCBs)	Cool to 4°C	Glass or Teflon	7 days to extraction, 30–40 days after
Chlorine	None specific	Plastic or glass	Within 24 hours

[a]HDPE = High-density polyethylene

Preliminary treatment often involves filtration that differentiates between the dissolved phase (operationally defined as that fraction which passes through a 0.45 or 0.2 μm membrane filter) and suspended matter collected on the filter. Filtration with a 0.2 μm filter is preferred, as it removes the majority of bacteria and plankton that would otherwise alter species concentrations during storage. For example, the determination of individual trace metal species is often compulsory, as total concentrations are poor indicators to understanding their availability and toxicity to organisms. In many cases only a relatively small fraction of the total dissolved metal concentration is present in the free hydrated state or complexed with inorganic ligands, while a large fraction is complexed with organic ligands. As mentioned in Chapter 3, factors influencing the speciation of inorganic metals include pH, temperature, major ion composition (e.g., metal-carbonate complexes), and ionic strength.

A variety of post-extraction procedures is performed including concentration of sample extracts and clean-up. Often, target analytes are diluted in the presence of a large volume of solvents used in the extraction procedure. An additional concentration step is therefore necessary to increase the concentration of the extract. If the analyte is non-volatile (e.g., non-volatile pesticides), for example, the solvent can be vaporized by a stream of nitrogen gas flowing across the surface. Larger volumes may need to be exposed to a rotary vacuum

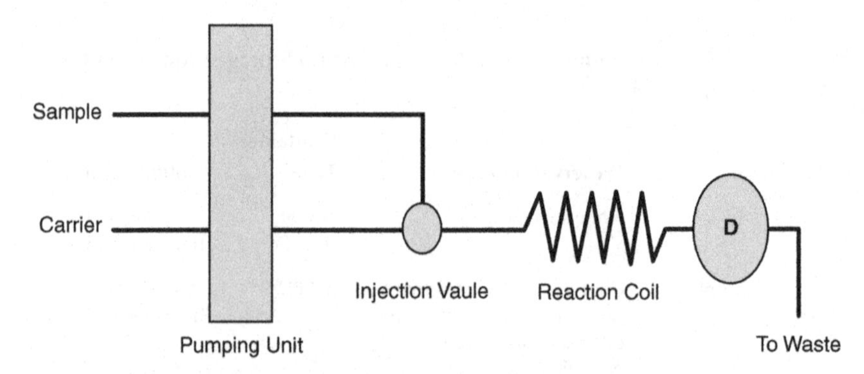

FIGURE 4.18 The basic FIA setup comprises a pumping unit, which propels the carrier stream through the narrow tube; an injection port, by which a well-defined amount of sample is injection into the carrier stream; a reaction coil where the sample zone reacts with components of the sample stream; and a flow-through detector.

evaporator. Sample clean-up is vital for the removal of interfering species, especially prior to instrumental analysis.

Finally, continuous aquatic environmental monitoring efforts have been further realized by the use of **flow injection analysis (FIA)** instrumental technology. A basic FIA setup is shown in Figure 4.18. In its most common form, FIA involves the injection of a liquid sample into a flowing carrier stream. En route, the injected sample zone can merge with one or more reagents, or be subjected to other physico-chemical processes (e.g., dialysis, thermal digestion, photo-oxidation, or enzymatic reaction) that generate species that are detected in one or more flows through detection devices (Worsfold, et al., 2008). Photometric detection in the visible range is the most common form of detection used for environmental applications, but other detection techniques such as UV spectrophotometry, fluorescence, chemiluminescence, atomic and mass spectrometry, potentiometry, and voltammetry have all been used. Attractive features of FIA for on-site or *in situ* environmental monitoring include:

1. High temporal and spatial resolution;
2. Portable, automated instrumentation;
3. Contamination-free, enclosed environment;
4. Sensitive and selective detection;
5. In-line filtration, treatment, and removal of matrix interferences;
6. On-board calibration and quality control.

Early FIA field systems tended to consist of slightly repackaged laboratory systems that were not necessarily suitable for continuous use for extended periods, and consequently it was difficult to replicate the high precision and reliability achieved in the laboratory setting (Gardolinski, et al., 2002).

However, advancements in technology and miniaturization have led to the development and deployment of numerous portable and automated systems for simultaneous, high temporal and spatial resolution monitoring of, for example, total dissolved or total phosphorus, nitrogen (Figure 4.19), and carbon species in freshwater systems, NO_x and NH_4^+ in activated wastewater sludge, pH, H_2O_2, pCO_2, Fe(II)/Fe(III), and a variety of metal species in natural waters.

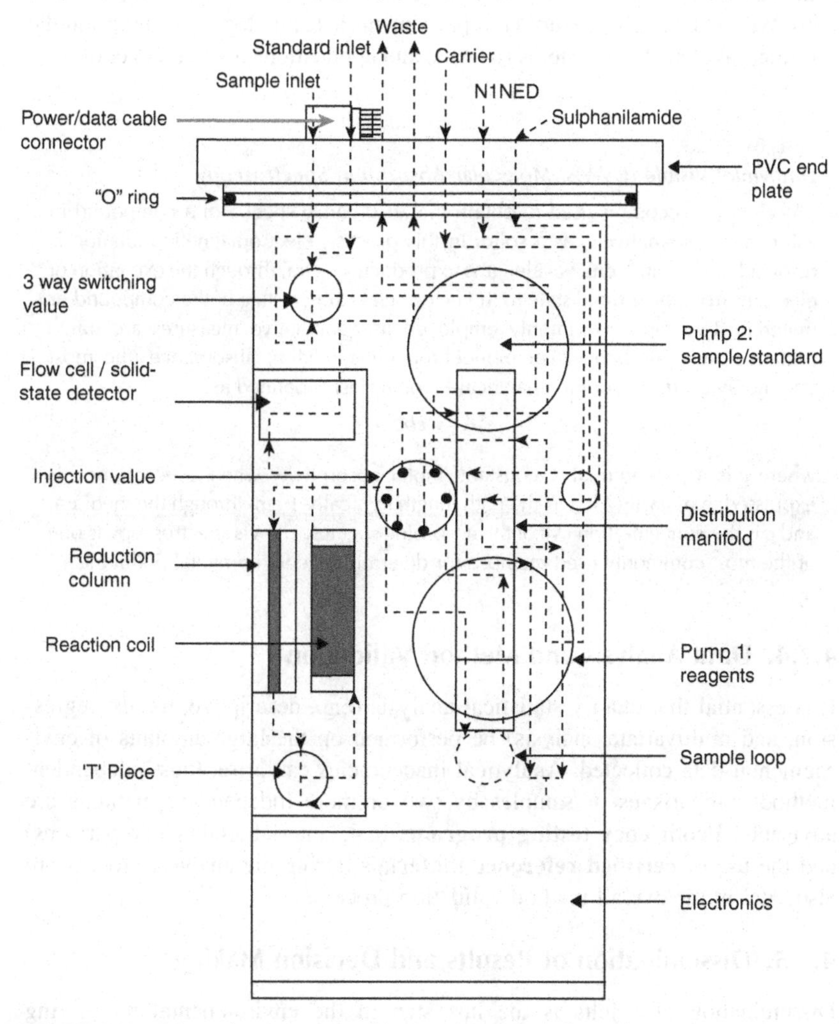

FIGURE 4.19 Schematic of a submersible flow injection analyzer for the determination of nitrate in estuarine and coastal waters. Such a system uses cadmium reduction followed by diazotization with solid-state spectrophotometric detection (modified from Gardolinski, et al., 2002 with permission from Elsevier). The versatility of the instrument was demonstrated by results of laboratory, shipboard, and *in situ* deployments in the North Sea and Tamar Estuary, UK.

Regardless of the technique employed, the application of on-site and *in situ* field instrumentation is useful in obtaining high-quality chemical data for the purpose of environmental modeling (see section below) and improved understanding and protection of environmental systems. These data are often of high temporal and spatial resolution, which have and will continue to benefit the study of the rapid dynamics of many environmental processes, such as in-stream dynamics of phosphorus exchange in freshwater systems, and spatial variations in concentrations, such as oceanic-scale iron and watershed-scale nutrient distributions. We will no doubt see further improvements in technology (e.g., microfluidic and micrototal analysis systems (μTAS)) and applications for years to come.

TEXTBOX 4.7

Ultraviolet-Visible (UV-Vis) Molecular Absorption Spectroscopy

UV-Vis spectroscopy is used to obtain the absorbance spectra of a compound in solution, or alternatively, as a solid. In this process, electromagnetic radiation is resolved into component wavelengths to produce spectra, through the excitation of electrons from the ground state to the first singlet excited state of the compound or material. Two terms commonly employed as quantitative measures are transmittance and absorbance. For monochromatic radiation, absorbance (the most commonly used measure in analytical laboratories) is defined as:

$$A = \varepsilon bc,$$

where ε is a proportionality constant (molar absorptivity when concentration is expressed in molarity M), b is the path length (typically 1 cm) through the medium, and c is the concentration (M) of the absorbing species. UV-Vis spectroscopy is one of the most commonly used methods for determining environmental pollutants.

4.7.4. Data Analysis and Method Validation

It is essential that quality statistical analysis (e.g., descriptive, trends, regression, and multivariate analysis) be performed on the large amounts of environmental data collected. Analytical inaccuracies can arise, thus independent method comparisons of samples by two or more independent methods are advisable. **Proficiency testing programs** (e.g., interlaboratory comparisons) and the use of **certified reference materials** during the analysis process are also vital in the overall method validation process.

4.7.5. Dissemination of Results and Decision Making

Dissemination of results is the last step in the environmental monitoring sequence, but not the least important. Recall that environmental analysis deals with sizeable data sets from a wide range of matrices. The ability to successfully archive, search, and distribute such data sets are compounded by both the heterogeneity of the data and the heterogeneous nature of environmental science inquiry. Therefore, data must be archived in a systematic and easily accessible manner, as timely decision making often calls upon such information. A large number of commercial and governmental databases (e.g., EIMS, SOFIA, ACCESS, FoxPro) are currently available in which investigators can incorporate such information as performing analyst, techniques used, validation of entered data, evaluation and interpretation of results, and end users.

4.8. INTRODUCTION TO ENVIRONMENTAL MODELING: A WATER QUALITY PERSPECTIVE

A wide array of scientific and engineering disciplines develop environmental models to advance fundamental understanding of natural systems under study. Such models allow investigators to perform numerical experiments in lieu of real experiments. This is especially relevant in environmental analyses where issues of control, scale, complexity, and reproducibility often make actual experiments nearly impossible. Models reveal and quantify uncertainties, provide enhanced information about a given environmental process, and allow simulation and/or prediction of events that may be either confirmed or disproven by future observational data. Like any scientific discipline, good practice is needed to efficiently construct and employ mathematical models for given environmental systems. The process outlined in Figure 4.20

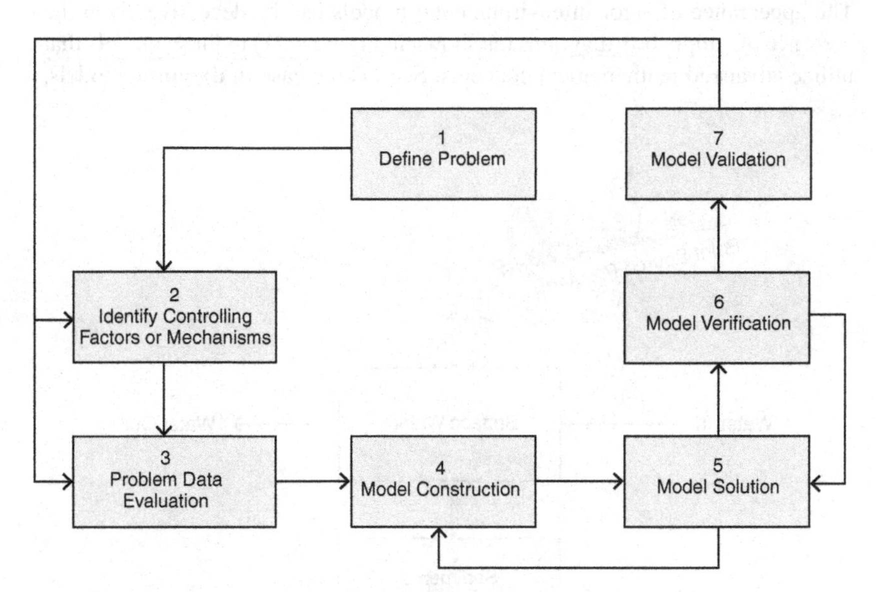

FIGURE 4.20 A systematic approach to the efficient construction and implementation of mathematical models. Mathematical modeling can be applied to a wide number of natural systems with formulations taking on different meanings depending on the type of system, the mathematical approaches used, the nature of the variables, and the behavior of the system under study.

represents a systematic approach to mathematical model development and implementation. Modeling is not a perfect science, with many steps being iterative in nature and the repetition of steps possible when unwanted problems arise.

The overriding goal of mathematical modeling in the context of biological and natural systems is to gain a better understanding of how real systems work. For example, examining and simulating pollutant behavior and transport enhance our understanding of environmental systems and contribute significantly to decision-making processes. Moreover, mathematical models can be used to demonstrate a better understanding of the relationship between contaminants and human health. How do such models apply to aquatic systems? The answer to this question is complicated by the numerous hydrological and water quality models of different scales available for use in water resource management. Let's consider static or **steady-state** models that describe a system where its inputs and outputs do not vary with time. Consider Figure 4.21 below, which depicts a single point source discharge affecting an aquatic environment. Assuming that sedimentation will remove only a very small portion of material, and if the inflow and outflow of water remain constant, the aquatic system can be modeled by a static model. The total mass balance equation can thus be written as

$$\text{Input}(I) = \text{Output}(O). \tag{4.10}$$

The appearance of rigor in environmental models can be deceptive given the presence of simple box diagrams (as shown in Figure 4.21) to those models that utilize advanced mathematical concepts. Such is the case in **dynamic** models,

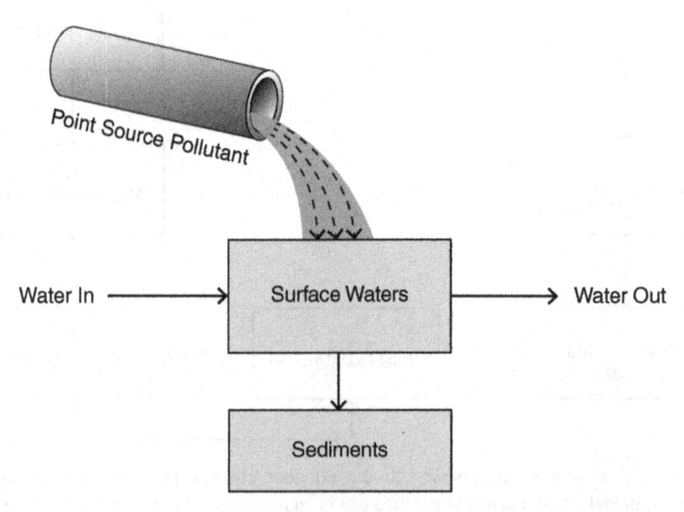

FIGURE 4.21 A simple box diagram showing the input of a point source pollutant into surface waters. Such a system can be modeled by a static model if the inflow and outflow remain constant.

which are built in algorithmic terms with characteristic functions and model the change of given parameters as a function of time.

Students should understand that models are a simplification of reality and therefore there is always some degree of associated risk or uncertainty. Applied models may also have narrow scopes and defined objectives, but when used judiciously, such models have been proven to be effective in advancing the fundamental understanding of complex natural systems. This is especially true for their use as environmental management and decision-making tools. Our modeling discussion throughout will center on the mathematical approach, although empirical and physical concepts will be integrated. Chapter sections will stress the development, application, and validation of mathematical models used to investigate environmental systems, with special emphasis on complexity, variability, scale, purpose, prediction, uncertainty, model robustness, and sensitivity analysis. The use of complex mathematical concepts is avoided as much as possible, although some rigor is needed to complement learning.

4.9. END OF CHAPTER PROBLEMS

4.1 Based on the physical units covered in this chapter, express ppm concentrations in the formats below. Assume that density = 1 g/mL.
   a. weight/weight (w/w)
   b. volume/volume (v/v)
   c. weight/volume (w/v)

4.2 Complete the following table of conversions:

ppm	ppb	%
0.001		0.0000001
	10	
0.1		
		0.0001

4.3 From Example Problem 4.1 in the main text, convert the molarity of H_2S calculated to millimoles per liter (mM).

4.4 Water is considered to be hard if it has a hardness of 100 mg/L or more as calcium carbonate ($CaCO_3$). What is the molarity of Ca^{2+} in a 100 ppm solution of $CaCO_3$?

4.5 Calculate the molar concentration of a 1.98 ppm solution of Ca^{2+}. (Molecular mass = 40.078 g mol^{-1}).

4.6 In 1991, the U.S. EPA lowered the federal standard for the allowable level of lead (Pb) in drinking water from 50 ppb to 15 ppb. Showing all necessary calculations, express the value (15 ppb) in percent (%) by mass.

4.7 Derive a mass action expression for the surface reaction of metals, as depicted in Equation 4.6.

4.8 The first-order rate constant for Plutonium-240 = $1.06 \times 10^{-4}\ yr^{-1}$. Calculate the half-life for this radionuclide.

4.9 Rhodium-106 is reported to have a half-life of approximately 30 seconds. If one started with a 50 g sample of this material, how many grams would be remaining after one minute?

4.10 Describe in detail how changes in redox conditions might affect the development and spread of inorganic and organic contaminant plumes.

4.11 State the names of three types of pollutants regulated under the Clean Water Act (CWA). Correctly name one contaminant in each category.

4.12 a) What is meant by the term "water quality indicator"? b) How are they incorporated into the establishment of water quality guidelines?

4.13 a) Provide chemical justification for the following statement: "electrical conductivity will vary with water source." b) How does salinity factor into the concept of electrical conductivity?

4.14 Explain how dissolved oxygen (DO) levels can indicate an increased load of organic matter has entered a body of water.

4.15 How does the concept of "representativeness" relate to proper sampling and analysis of environmental contaminants?

4.16 The use of dissolved oxygen (DO) field probes are now the preferred method (compared to the Winkler method) when determining DO in natural waters. However, field probes do suffer a number of drawbacks. Based on discussion in section 4.4, provide specific examples to support this last statement.

4.17 Refer to Figure 4.4, which illustrates how nitrate (NO_3^-) in oceans is lower in surface waters, but generally increases as depth increases. Provide reasoning for this trend. Hint: Refer back to Chapter 1 concepts.

4.18 Refer to Textbox 4.6. Visit the Water on the Web site and click on the Data tab. Choose a location on the map and click on the Real-Time Data tab associated with this lake or river (water quality data). There will be various options to choose from including weekly, monthly, and in-stream/in-lake trends. a) Choose one and list five water quality indicators that were routinely measured during associated sampling/monitoring campaigns. b) Comment on any noticeable trends in the data including seasonal changes. Hint: The associated Data Visualization Tools tab will help with part b.

4.10 REFERENCES

Allison, J. D., Brown, D. S., & Novo-Gradac, K. J. (1990). *MINTEQA2/PRODEFA2, A geochemical assessment model for environmental systems: Version 3.0 user's manual*. EPA/600/3e91/021. Athens, GA: U.S. Environmental Protection Agency.

Berkowitz, B., Dror, I., & Yaron, B. (2008). *Contaminant geochemistry: Interactions and transport in the subsurface environment*. Berlin: Springer-Verlag.

Calder, L. M. (1998). Chromium contamination in groundwater. In J. O. Nridgu, & E. Bieboer (Eds.), *Chromium in the natural and human environments*. New York: Wiley-Interscience.

Chen, J., Wang, F., Xia, X., & Zhang, L. (2002). Major element chemistry of the Changjiang (Yangtze River). *Chemical Geology, 187*, 231–255.

Evgenidou, E., & Fytianos, K. (2002). Photodegradation of triazine herbicides in aqueous solutions and natural waters. *Journal of Agricultural and Food Chemistry, 50*, 6423–6427.

Fitts, C. R. (2002). *Groundwater science*. Burlington, MA: Academic Press.

Freeze, R. A., & Cherry, J. A. (1979). *Groundwater*. Englewood Cliffs, N.Y: Prentice-Hall.

Gardolinski, P. C. F. C., David, A. R. J., & Worsfold, P. J. (2002). Miniature flow injection analyser for laboratory, shipboard and in situ monitoring of nitrate in estuarine and coastal waters. *Talanta, 58*, 1015–1027.

Hubbard, R. K., & Sheridan, J. M. (1989). Nitrate movement to groundwater in the Southeastern Coastal Plain. *Journal of Soil and Water Conservation, 44*, 20–27.

Karr, J. R., & Dudley, D. R. (1981). Ecological perspectives on water quality goals. *Environmental Management, 5*, 55–68.

Nielsen, P., & Perrochet, P. (2000). Watertable dynamics under capillary fringes: Experiments and modeling. *Advances in Water Resources, 23*, 503–515.

Rees, H. C., Oswald, S. E., Banwart, S. A., Pickup, R. W., & Lerner, D. N. (2007). Biodegradation processes in a laboratory-scale groundwater contaminant plume assessed by fluorescence imaging and microbial analysis. *Applied and Environmental Microbiology, 73*, 3865–3876.

Ross, L. (1992). The politics of environmental policy in the People's Republic of China. *Policy Studies Journal, 20*, 628–642.

Sampini, L., Kitanidis, P. K., Kampbell, D. H., & Wilson, J. T. (1995). Anaerobic transformation of chlorinated aliphatic hydrocarbons in a sand aquifer based on spatial chemical distributions. *Water Resources Research, 31*, 1051–1062.

Shim, H., Ryoo, D., Barbieri, P., & Wood, T. K. (2001). Aerobic degradation of mixtures of tetrachloroethylene, trichloroethylene, dichloroethylenes, and vinyl chloride by toluene- *o*-xylene monooxygenase of Pseudomonas stutzeri OX1. *Applied Microbiology and Biotechnology, 56*, 265–269.

Spalding, R. F., & Exner, M. E. (1993). Occurrence of nitrate in groundwater: A review. *Journal of Environmental Quality, 22*, 392–402.

Steffan, R. J., McClay, K., Vainberg, S., Condee, C. W., & Zhang, D. (1997). Biodegradation of the gasoline oxygenates methyl tert-butyl ether, ethyl tert-butyl ether, and tert-amyl methyl ether by propane-oxidizing bacteria. *Applied and Environmental Microbiology, 63*, 4216–4222.

The Council of the European Union. *Council directive 98/83/EC of 3 november 1998 on the quality of water intended for human consumption*.

Tóth, J. (2009). *Gravitational systems of groundwater flow: Theory, evaluation, and utilization*. Cambridge: Cambridge University Press.

U.S. Environmental Protection Agency. *Safe drinking water act amendments of 1996*, General Guide to Provisions, Washington, D.C.

U.S. Environmental Protection Agency. (1998). *Oxygenates in water: Critical information and research needs, Office of Research and Development*. EPA/600/R-98/048.

Wang, M., Webber, M., Finlayson, B., & Barnett, J. (2008). Rural industries and water pollution in China. *Journal of Environmental Management, 86*, 648–659.

World Health Organization. *Guidelines for drinking water quality, Volume I-Recommendations*, Geneva, 1993.

Worsfold, P. J., McKelvie, I. D., & Hanrahan, G. (2008). Environmental applications of flow injection analysis: Waters, sediments and soils. In S. D. Kolev, & I. D. McKelvie (Eds.), *Advances in flow injection analysis and related techniques*. Amsterdam: Elsevier.

Zwank, L., Elberg, M., Elsner, M., Schmidt, T., Schwarzenback, R. P., & Haderlein, S. B. (2005). New evaluation scheme for two-dimensional isotope analysis to decipher biodegradation processes: Application to groundwater contamination by MTBE. *Environmental Science & Technology, 39*, 1018–1029.

Chapter 5

Abstracts and Keywords

Abstract

This chapter provides a solid overview of water treatment processes and related technologies. It begins with discussion on municipal wastewater treatment, specifically primary, secondary, and tertiary methods of treating municipal sewage and water. Next, detailed information on wastewater disinfection is presented. Such concepts as chlorination, ozonation, and ultraviolet radiation treatment will be covered. The chapter ends with detailed information on the use and disposal of biosolids and on point-of-entry and point-of-use technologies, including activated carbon, reverse osmosis, and ion exchange processes as applied to real-world situations.

Keywords

Water treatment, Municipal wastewater, Disinfection, Chlorination, Ozonation, Biosolids, Point-of-entry technologies, Point-of-use-technologies

Chapter 5

Water Treatment and Related Technologies

Most sewage produced in the more developed countries is treated, primarily at large-scale facilities owned by municipalities such as that shown in Figure 5.1. According to industry reports, an estimated 21,600 publicly-owned treatment plants were in operation in the United States in 2004, treating nearly 35 billion gallons of wastewater daily and serving nearly 230 million people (U.S. Environmental Protection Agency, 2009). An additional 1 billion gallons per day of treated wastewater is reclaimed to meet non-potable water needs, including numerous public and private irrigation activities. This has all been made possible through technological advances in treatment processes: developments that have led to improved effluent water at municipal water treatment plants and other point sources of discharge.

FIGURE 5.1 Aerial view of the Hill Canyon Wastewater Treatment plant, a 14,000,000 gallon a day capacity, tertiary wastewater treatment facility located in Thousand Oaks, California. This facility serves domestic, commercial, and industrial customers. Photo courtesy of Chuck Roberts.

5.1. MUNICIPAL WASTEWATER TREATMENT

In Chapter 4 we learned that water used in residential properties and industrial facilities originates either from a groundwater source, such as a well, or from a surface water source, such as a river, lake, or reservoir. Perhaps less obvious, or at least less thought about, is where the water goes after its final use. Fortunately, dedicated wastewater treatment and collection facilities are in operation thanks largely to the Clean Water Act (CWA) previously discussed in Chapter 4. The CWA established a control program ensuring that communities have clean water by regulating the release of contaminants into dedicated waterways. In the United States, permits that limit the amount of pollutants discharged are required of all municipal and industrial wastewater dischargers under the National Pollutant Discharge Elimination System (NPDES) permit program. In addition, a construction grants program was set up to assist publicly-owned wastewater treatment works to build the improvements required to meet these new limits.

In the United States, over 75% of the nation's population is served by centralized wastewater collection and treatment systems, with approximately 16,000 municipal wastewater treatment facilities in operation nationwide (U.S. Environmental Protection Agency, 2004). Schematic representation of a general municipal water treatment plant is shown in Figure 5.2. As will be discussed in section 5.2, a combination of biological, chemical, and physical treatment processes that allow treatment plants the ability to make available a sufficient supply of clean water for domestic, agricultural, and industrial use is employed. Such processes also help keep surrounding surface waterways uncontaminated as the result of regulated water discharge.

5.2. THE TREATMENT OF MUNICIPAL SEWAGE AND WATER

Although municipal sewage is mainly comprised of water (99.9%), it does house potentially damaging solids that contain organic matter, suspended

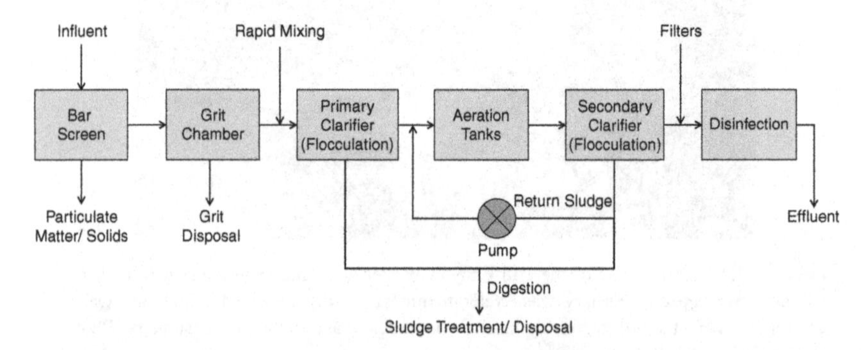

FIGURE 5.2 The multi-stage process of a wastewater treatment plant in schematic form.

sediments, living organisms derived from feces, viruses, protozoa, detergents, oil, grease, fats, soaps, pesticides, salts, nutrients, metals, paper products, and refractory organic compounds. Wastewater characteristics such as turbidity, dissolved oxygen (DO), pH, chemical oxygen demand (COD), biological oxygen demand (BOD), and total organic carbon (TOC) are used to describe sewage. Given this potentially threatening cocktail, appropriate treatment measures are needed to ensure removal of as much of this material as possible to protect human health and surface waterways and their inhabitants. The three main categories of wastewater treatment are described below. They each contain some form of treatment operation with methods classified as:

1. **Physical unit operations**: treatment brought about through the application of physical forces (e.g., screening, mixing, sedimentation, and filtration);
2. **Chemical unit operations**: removal or treatment of contaminants is brought about by the addition of chemicals, or by chemical reactions (e.g., chemical precipitation and disinfection);
3. **Biological unit operations**: removal of contaminants is brought about by biological processes (e.g., nitrification and denitrification).

More detailed coverage of these processes, along with categorized treatment stages, are provided below.

5.2.1. Primary Treatment

The initial stage in the treatment of domestic wastewater is termed **primary treatment**. During this stage, physical processes such as screening and shredding in the **bar rack** remove debris (e.g., sticks and rags) and a portion of the contaminants that settle or float. Grit, having specific gravities greater than those of the organic solids, is removed in the **grit chamber**. Heavier organic solids are then allowed to settle to the bottom of the **primary clarifier** and are removed by submerged pumps. During advanced primary treatment, **coagulation** agents are introduced, followed by the formation of aggregations and clumps of suspended material. Consideration of the process by which a solid phase is formed in solution is crucial in understanding coagulation, and thus aids in the design and operation of water treatment facilities. Consider Figure 5.3 as a general example. The particles of a **colloid** selectively absorb ions and acquire an electric charge. All of the particles of a given colloid take on the same charge (either positive or negative) and are thus repelled by one another. Chemical coagulants are used to neutralize the charges on the suspended particles to prevent electrostatic repulsion. If positively charged, they will attract to the negative particles in the water, resulting in electrical neutrality. As a result, van der Waal's forces take over and allow attraction to occur. When enough particles have "joined," they become **floc** and will settle out of solution.

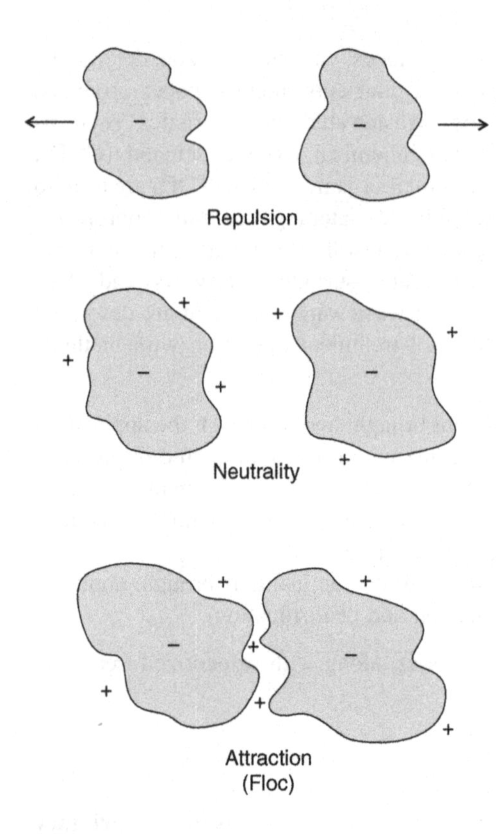

FIGURE 5.3 Charged particles and chemical coagulants joined together in a solid phase to form floc and settle out of solution. As shown, chemical coagulants act to neutralize the charges on the suspended particles to prevent electrostatic repulsion.

Aluminium sulphate (also termed alum) is a commonly used water treatment coagulant that, when utilized with the associated alkalinity, produces the reaction below. The alum reacts with bicarbonate to form aluminium hydroxide, a precipitate, as follows:

$$Al_2(SO_4)_3 \cdot xH_2O + 6HCO_3^- \rightleftharpoons 2Al(OH)_3 + 3SO_4^{2-} + 6CO_2 + xH_2O.$$

Note that the number of water molecules bound (x) to the central molecule can vary substantially from 2–24. While this has no effect on the behavior of the coagulant, it is necessary to know how much bound water is present in order to compute the number of moles or equivalents of alum. See the discussion on the tertiary wastewater treatment process below for an extended review of coagulation and its role in, for example, phosphorus removal. The formation of resultant clumps, termed **flocculation**, involves the trapping or attraction of suspended solids into clusters (or floc). Stirring is required for complete flocculation with settling of the formed floc termed **sedimentation**. Primary sedimentation tanks remove up to 50–70% of the suspended solids and 25–40% of the BOD. Lighter solids, such as grease and scum, float to the surface and are skimmed off and placed in a container called a scum box. Chemical additives are routinely employed to reduce odors, neutralize acids or bases, reduce corrosion, reduce BOD, and improve solids removal.

Example Problem 5.1

Note that in the above reaction, for every mole of alum consumed, six moles of CO_2 are produced. What effect will this have on pH? Name one consequence of this change.

Answer: The overall reaction produces CO_2, which will then react with water to produce carbonic acid (H_2CO_3). Upon dissociation of this weak acid, a slight reduction in pH occurs. Lowered pH values mean that there are more positively-charged particles loose in the water to react with the negatively-charged colloids. For example, the decrease in pH will likely affect the promotion of, for example, the dissociation of hydrous metal oxides.

5.2.2. Secondary Treatment

Secondary treatment begins with biological oxidation through **activated sludge, trickling filters**, or **stabilization pond** processes (or a combination of all of these). In each of these, the organic matter of wastewater provides an energy source for the production of new cells for a mixed population of microorganisms in an aerobic aquatic environment. The microbes convert carbon into cell tissue and oxidized end products that include CO_2 and water. The activated sludge process, the most common approach, involves blending settled primary effluent wastewater with a culture of microorganisms into a fluid called "**mixed liquor**." This mixed liquor is passed through an aeration tank that provides an adequate oxygen source for the type of activated sludge process(s) selected and to mix the mixed liquor. Common processes include: conventional activated sludge, step-feed activated sludge, contact stabilization activated sludge, extended aeration activated sludge, and selector activated sludge. The microbes present consume the organic waste in the water. This conversion of BOD into a microbial culture necessitates the removal of microbes from the process in what is termed **waste activated sludge (WAS)**. The solids and microbes are separated from the wastewater being treated in a secondary clarifier. A large portion of the microbes are returned to the aeration basin(s) as **return activated sludge (RAS)** to further extend this much needed cycle. In order to meet the U.S. EPA definition of secondary treatment, a wastewater treatment plant must produce an effluent with suspended solids and BOD levels below 20 mg L^{-1}.

A process termed **Enhanced Biological Phosphorus Removal (EBPR)** is a modification of the activated sludge process that will eliminate phosphorus from wastewater. Recall from our discussions in Chapter 1 that too much

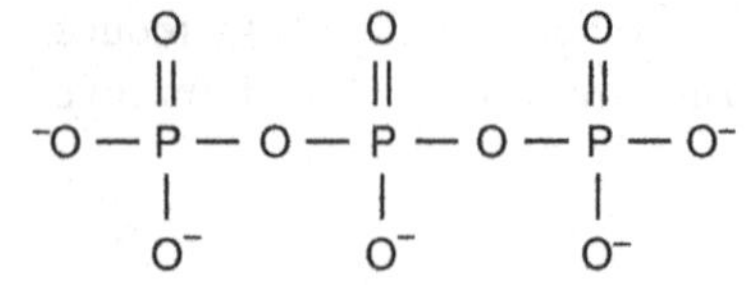

```
      O         O         O
      ||        ||        ||
-O —  P — O —  P — O —  P — O-
      |         |         |
      O-        O-        O-
```

FIGURE 5.4 Tripolyphosphate. A class of pentavalent phosphorus compounds in which various numbers of tetrahedral PO_4 groups are linked together by oxygen bridges.

phosphorus causes excessive growth of nuisance algae. The respiration and decomposition of algae in eutrophic waters deplete dissolved oxygen, where aquatic organisms have difficulty surviving. Excess algae growth may also create unpleasant taste and odors in drinking water supplies. Thus, developing novel and efficient technologies to control measures to reduce phosphorus discharges from point and non-point sources to the surface waters is of great importance.

During the anaerobic stage of EBPR, no electron acceptors (e.g., O_2 or NO_x) are available. Organic substrates (matter) are supplied from influent wastewater into the anaerobic stage, and the return sludge comes into contact with the carbon source (but only in the anaerobic stage). The anaerobic stage is said to function as a biological selector to facilitate the proliferation of polyphosphate-accumulating bacteria (PAB) with high phosphorus content (Mino, 2000). Structurally, polyphosphates are found to be polymers of orthophosphate, with phosphoanhydride linkages (Figure 5.4) thermodynamically equivalent to the "energy-rich" phosphate of adenosine triphosphate (ATP) of which we are so familiar with in biological systems. Without electron acceptors, aerobic bacteria and denitrifying bacteria are unable to obtain the energy required for the utilization of organic substrates, and they are thus unable to compete with PAB. Therefore, the introduction of the anaerobic stage leads to the superiority of PAB and to a rise in phosphorus content of the sludge. By withdrawing the phosphorus-rich sludge from the system as excess sludge, high phosphate removal efficiency can be realized. In terms of removal efficiency, EBPR systems have been shown to remove 70–80% of the free phosphorus (e.g., PO_4^{3-}) from the wastewater influent (Mullan, et al., 2006). In fact, if operating efficiently, it is possible to achieve effluent phosphate concentrations of <0.1 mg/L (most facilities have target effluent concentrations of <1.0 mg/L). Factors known to influence the removal efficiency of phosphorus include a high influent BOD/total phosphorus (TP) ratio, exclusion of oxidants (e.g., nitrate and oxygen) from the anaerobic zone, moderate temperature, the presence of glycogen accumulating organisms (GAOs) that are known to compete with PABs, and the addition of an external carbon source (Majed and Gu, 2010).

5.2.3. Tertiary Treatment

Following secondary treatment, the clarified effluent may require additional aeration and/or other chemical treatment (**tertiary treatment**) to destroy bacteria remaining from the secondary treating stage and to increase the content of DO needed for oxidation of the residual BOD. Chemical tertiary treatment can also be used to remove nitrogen and phosphorus compounds. For example, chemical precipitation is used to remove the inorganic forms of phosphate by the addition of an appropriate coagulant and thorough mixing. The multivalent metal ions most commonly used are calcium, aluminum, and iron. There are two general steps for the removal of phosphate by chemical means:

1. The adsorption of soluble phosphate (PO_4^{3-}) into metal hydroxide floc;
2. The subsequent separation of floc from the liquid phase by clarification and filtration.

Hydrated aluminium sulphate is widely used to precipitate phosphate and aluminium phosphates ($AlPO_4$). The basic reaction is

$$Al^{3+} + H_nPO_4^{3-n} \rightleftharpoons AlPO_4 + nH^+.$$

When coupled with filtration, chemical removal can reduce final phosphorus concentrations to the order of 0.2–0.3 mg L^{-1}. The dosage rate required is a function of the phosphorus removal required. The efficiency of coagulation falls as the concentration of phosphorus decreases. In practice, an 80–90% removal rate is achieved at coagulant dosage rates between 50 and 200 mg L^{-1} (Mino, 2000). If iron is used as a coagulant, ferric ions in solution combine to form ferric phosphate ($FePO_4$). However, a coagulant aid (e.g., lime) is normally added to raise the pH in order to enhance coagulation.

Nitrogen is present in wastewater in several forms, predominately organic nitrogen (both soluble and particulate), ammonium, ammonia, and nitrate. In the activated sludge process several reactions may occur, which will likely alter the form of the nitrogenous matter. Two processes are involved in biological removal of nitrogen: nitrification and denitrification. These processes were briefly covered in the opening chapter. Let's consider them in more detail as they relate to nitrogen removal in wastewater streams. During nitrification, ammonium is converted into nitrate by autotrophic bacteria. Recall the two-step process:

1) oxidation of ammonia to nitrite

$$NH_4^+ + \frac{3}{2}O_2 \rightarrow NO_2^- + H_2O + 2H^+$$

and, 2) the subsequent oxidation of nitrite to nitrate

$$NO_2^- + \frac{1}{2}O_2 \rightarrow NO_3^-.$$

There are many operational factors that influence the nitrification process, in particular the nitrifier growth rates. These include concentrations of NH_4^+ and

NO_2^-, mixed liquor temperature, dissolved oxygen concentration (DO), and pH (optimum range = 7.5–8.5). The influence of temperature on the growth rate of the nitrifiers has particular significance for the activated sludge process. Moreover, a bulk DO concentration of 2 mg O_2/L has been shown to be sufficient to prevent oxygen limitation in the nitrification process (Beccari, et al., 1992). Also recall that denitrification requires organic carbon as an electron donor as shown below:

$$2NO_3^- + CH_2O \rightarrow 2NO_2^- + CO_2 + H_2O$$

$$4NO_2^- + 3CH_2O + 4H^+ \rightarrow 2N_2 + 3CO_2 + 5H_2O.$$

Many denitrifying bacteria consume organic matter while utilizing NO_3^- as their electron acceptor. The source of the organic substrate can be internal (i.e., present in the wastewater) or external (i.e., added directly to the wastewater). Thus, denitrification can readily occur in the presence of organic matter, in soils devoid of oxygen, and with the help of denitrifying bacteria. For denitrification to proceed efficiently, a few requirements are needed. These include the presence of nitrate and absence of DO (<1 mg/L) in the mixed liquor, pH (roughly neutral pH) suitable conditions for bacterial growth, and as discussed above, the presence of an electron donor (i.e., organic material). The concepts of nitrification and denitrification **removal capacities** are used in practice to describe nitrogen removal in the activated sludge process. It (the removal capacity) is an important parameter that assesses the amount of nitrate (in the case of denitrification) that can be removed from 1L of influent. Factors used to determine the removal capacity include temperature, sludge age, the concentration and composition of influent organic matter, and the anoxic sludge mass fraction (Komorowska-Kaufman, et al., 2006).

Removal of heavy metals (e.g., cadmium, chromium, lead, and mercury) also occurs during the tertiary treatment process. As metals enter the treatment process, they are typically in a stable, dissolved aqueous form and are unable to form solids easily. However, if you recall our discussion of the formation of metal hydroxides in Chapter 3, metals can be removed as solids upon insoluble precipitate formation. For example, in the electroplating industry, metal hydroxide sludge is a dried waste containing metal hydroxides $M(OH)_n$ and other salts (e.g., $CaSO_4$, $CaCO_3$, NaCl, and $NaHCO_3$). Chromium metal is of particular significance in electroplating waste owing to its distinctive chemistry. The chrome-plating process is done in very highly concentrated chromic acid (H_2CrO_4) solutions. Chromium metal exists primarily in one of two stable oxidation states: trivalent Cr(III) and hexavalent Cr(VI). Considering Cr(VI) in greater detail, it is known to exist in solution as chromate (CrO_4^{2-}), dichromate ($Cr_2O_7^{2-}$), and hydrogen chromate ($HCrO_4^-$), depending on the pH and total chromate concentration.

Example Problem 5.2

Start with chromic acid (H_2CrO_4) and show all pertinent equilibrium reactions in solution.

Answer:

$$H_2CrO_4 \rightleftharpoons H^+ + HCrO_4^- \quad (1)$$
$$HCrO_4^- \rightleftharpoons H^+ + CrO_4^{2-} \quad (2)$$
$$2HCrO_4^- \rightleftharpoons Cr_2O_7^{2-} + H_2O \quad (3)$$

Let's also extend our discussion of arsenic species from Chapter 3. For example, granular ferric hydroxide can be used to remove As(V) from wastewater streams. Consider the following dissolved As species reactions:

$$FeOH + H_2AsO_4^- \rightleftharpoons FeH_2AsO_4^- + OH^-$$
$$FeH_2AsO_4^- \rightleftharpoons FeHAsO_4^{2-} + H^+.$$

If the speciation of arsenate changes from $H_2AsO_4^-$ to $HAsO_4^{2-}$ then

$$FeOH + HAsO_4^{2-} \rightleftharpoons FeHAsO_4^{2-} + OH^-$$
$$FeOH + HAsO_4^{2-} \rightleftharpoons FeAsO_4^{3-} + H_2O.$$

In the former set of reactions, removal at higher pH is characterized by the udissociated acid donating a proton to the surface hydroxyl group to form water that can be displaced by the anion (Streat, et al., 2008). All of the above processes are dependent upon the concentration of the metal and the pH of the aqueous environment. Note that not all metals have the same minimum solubility, thus, pH is routinely adjusted to an average pH value of 9.0 when multiple metals are present in solution (Tünay and Kabdaşli, 1994).

Interestingly, pilot-scale **constructed wetlands (CW)** have also been used as tertiary treatment in the removal of heavy metal ions from wastewater. CWs consist of shallow aquatic compartments, with a natural or constructed subsurface of clay or geotechnical material to prevent seepage and a suitable substrate to support rooted emergent plants. For example, shallow ponds containing hyacinth plants have been used to remove any number of heavy metals from common industrial applications. Studies have shown high removal efficiencies of common metals, including, for example, up to 60–65% removal of Cd, Pb, Cu, and Zn concentrations in wastewater streams. Numerous other processes exist for removing dissolved metals from wastewater including ion exchange, reverse osmosis, and activated carbon. The basic principles behind these processes are described in detail in section 5.5.

5.3. DISINFECTION

Wastewater that remains after the tertiary phase is disinfected to kill harmful microorganisms before being released into receiving waters. When determining disinfection needs for wastewater, care should be taken in using bacterial indicators such as *E. coli* or thermotolerant coliforms to design and control disinfection processes as these indicator organisms are known to be more susceptible to some disinfection processes than other pathogens. The disinfection methods generally considered for use consist of chemical methods (e.g., chlorine, chlorine dioxide, and ozone), physical methods (e.g., UV irradiation and membrane microfiltration), and biological methods (e.g., ponds).

TEXTBOX 5.1

Coliform bacteria, as typified by *E. coli*, are considered indicator organisms when assessing microbiological water quality treatment for human consumption. They can also be used as an indicator of water quality for reuse of reclaimed water, provided that biologically-treated wastewater is chemical coagulated and filtered.

5.3.1. Chlorination

The more established technique of water disinfection involves chemical treatment with chlorine, with its use in water treatment dating back to 1904 by the London Metropolitan Water Board. Chlorine is a yellow-green gas that dissolves in water according to the following reaction:

$$Cl_2 + H_2O \rightarrow HOCl + HCl.$$

The hypochlorous acid formed is readily ionized into hypochlorite ion:

$$HOCl \rightleftharpoons H^+ + OCl^-.$$

Note that the fraction of the free available chlorine, HOCl, demonstrates much stronger disinfection kinetics than OCl^-. The most applied kinetic model of the disinfection process, **Chick's Law**, states that the rate of the reaction is first order with respect to the concentration of the organisms being inactivated by the disinfectant (Chick, 1908):

$$N = N_0 e^{-kt}, \tag{5.1}$$

where N = the number of surviving microorganisms after treatment, N_0 = the initial number of organisms, k = the disinfection constant (death rate), and t = the contact time. This law has been extensively applied across disciplines with most disinfection data evaluated initially with the use of a semilog graph of survival, e.g., log (S) versus time (t), where $S = N/N_0$. Finally, the effect of disinfection is strongly dependent on pH, temperature, and the coexistence of other matter in the waste. Disinfection performance is often assessed through changes in concentrations of indicator organisms (primarily fecal coliform (FC) and total coliform (TC)) over time.

Example Problem 5.3

Determination of the disinfection rate (death rate, k) is vital in assessing disinfection performance through changes in concentrations of indicator organisms. Show the rearrangement of Equation 5.1 so that k can be calculated appropriately.

Answer: Starting with Equation 5.1, $N = N_0 e^{-kt}$, we can rearrange as follows:

$$-kt = \log(N/N_0).$$

Logarithms are an essential part of the environmental chemist's repertoire. Logarithms allow us to convert an exponential equation into an equation that can be readily solved. Practice by performing the functions necessary to calculate death rates, as asked in the End of Chapter Problem 5.9.

If ammonia is readily present in the wastewater, HOCl will react to form three chloroamine species:

(1) $NH_3 + HOCl \rightarrow NH_2Cl + H_2O$
(2) $NH_2Cl + HOCl \rightarrow NHCl_2 + H_2O$
(3) $NHCl_2 + HOCl \rightarrow NCl_3 + H_2O$.

The formation of monochloroamines (1) is favored at higher pH conditions and a low Cl_2/NH_3 ratio. Although the potential for oxidation drops, it still retains a comparable oxidizing capacity as free chlorine. Dichloroamines (2) are relatively unstable species. The formation of trichloroamines (3) is favored at low pH conditions and a high Cl_2/NH_3 ratio. The sum of the three reaction products is referred to as combined chlorine, with the total equaling free chlorine + combined chlorine.

Despite chlorine's accepted use as a disinfectant, byproduct formation accounts for increased human health risks. For example, halogenic disinfection byproducts of Cl_2 and HOCl (e.g., trihalomethanes, chlorine hydrates, and chlorophenols) are likely to form as the dose of the disinfectant increases. Trihalomethanes (THMs) (Figure 5.5) are of particular concern given their ubiquitous presence in wastewater and link to several types of cancers from prolonged exposure. Of the four primary THMs, chloroform is the most common and detected at the greatest concentration levels. In 1998, the U.S. EPA published the State 1 Disinfectants/Disinfection Byproducts Rule to regulate total trihalomethanes (TTHMs) at a maximum allowable annual average level of 80 ppb for large surface water and groundwater systems (U.S. Environmental Protection Agency, 1998).

In regards to chloroamines, organohalogenic disinfection byproducts include haloacetonitrils, cyano chlorine, organic chloramines, chloramino acids, chlorohydrates, and haloketons. Ultimately, reaction time, temperature,

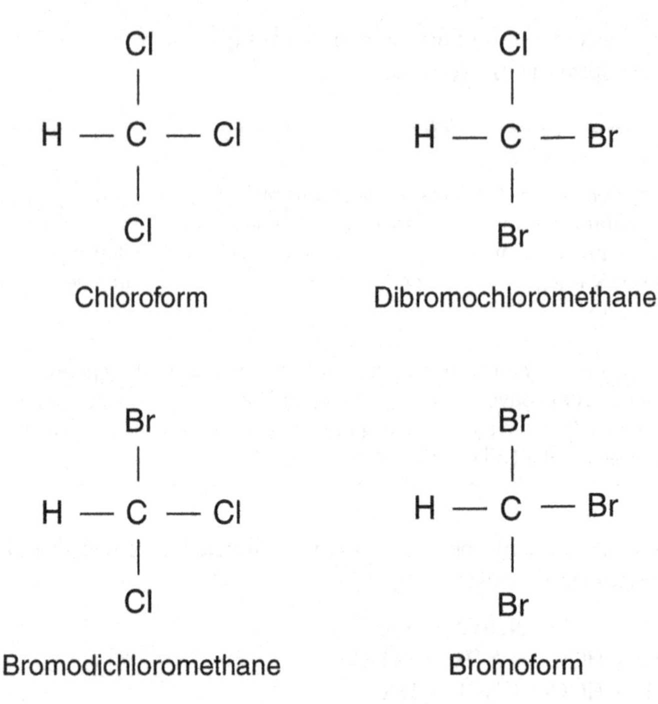

FIGURE 5.5 Four common trihalomethane byproducts of chlorine disinfection.

pH, and natural organic matter content affect the circumstances of disinfection and the formation of harmful byproducts. For example, on a shorter reaction time scale, higher concentrations of trihalomethanes and halogenic acetic acids will likely form. When the reaction time is longer, some temporary forms of disinfection byproducts may become disinfection end products, e.g., tribromine acetic acid or bromoform. At this point, haloacetonitrils and haloketons are primarily decomposed. As pH values are increased, the affectivity of chlorine disinfection dissipates as more OCl^- is formed. Moreover, higher pH levels result in increased hydrolysis reactions, thus aiding in the decomposition of haloacetonitrils and haloketons.

5.3.2. Ozonation and Ultraviolet Radiation

As a result of the complications of chlorination, the increased use of ozone (**ozonation**) and/or ultraviolet radiation has been seen in routine wastewater treatment. As with the addition of chlorine, ozonation is an oxidative disinfection process. The ozonation systems used for water treatment generate ozone at the source, and almost all of them do so by means of a crown discharge produced by the passage of oxygen or dry air between two dielectrics. Ozone is produced when oxygen (O_2) molecules are dissociated into two highly reactive oxygen singlets (O•) by the electrical discharge (spark) produced. These unstable atoms combine with other oxygen molecules to form the gas, ozone (O_3). Ozone's extensive oxidation potential enables it to eliminate organic compounds that give water a disagreeable color, taste, or smell. This ability can, however, lead to potential byproducts. For example, exposing organic molecules containing points of unsaturation to ozone yields many transitional compounds and decomposition products including aldehydes, ketones, acids, or alcohols. Despite these minimal residual risks, ozonation is very effective in destroying waterborne pathogens. Inactivation kinetics with ozone has been extensively studied, with results showing that lower concentrations of ozone and shorter contact times are required (when compared with other agents (e.g., chlorine)) for disinfection against harmful organisms including bacteria, amoebic cysts, and viruses. Let's consider the mechanism of inactivation in bacteria, for example. Due to its oxidation potential, ozone oxidizes components of the bacterial cell wall, penetrates the cell, oxidizes essential components (e.g., enzymes, DNA, and RNA), and ultimately causes cell lysis (Kim, et al., 1980). In slight contrast, chlorine enters the cell through diffusion and largely affects the actions of several enzyme types.

The use of UV radiation as a disinfection tool is gaining increased popularity due to its lowered cost and limited formation of detrimental disinfection byproducts. UV radiation is defined as that portion of the electromagnetic spectrum (Figure 5.6) between 40 and 400 nm. The UV spectrum is divided into Vacuum UV (40–190 nm), Far UV (190–220 nm), UV-C (220–290 nm), UV-B (290–320), and UV-A (320–400 nm). Peak germicidal (both inactivation and cumulative dosage) is maximized in the UV-C region ($\lambda = 254$ nm). UV radiation causes microbial damage by deactivation of cells. More specifically, UV radiation is absorbed by nucleotides, the building blocks of cellular RNA and DNA. Absorbed UV promotes the formation of bonds between adjacent nucleotides, creating dimers (typically thymine-thymine and cytosine-cytosine). Formation of a sufficient number of dimers within a microbe prevents it from

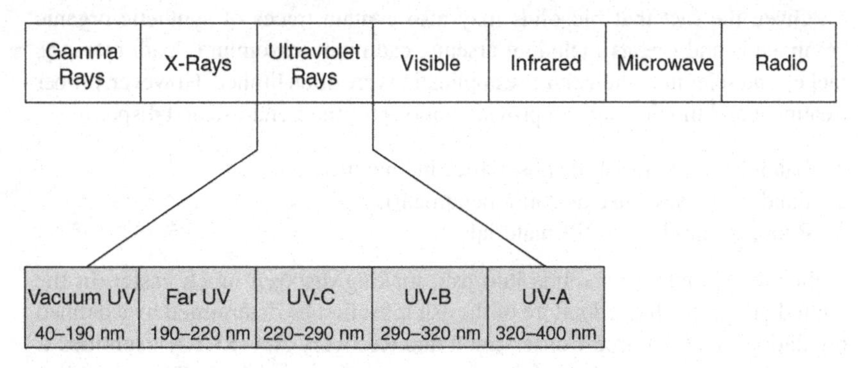

FIGURE 5.6 The generalized electromagnetic radiation spectrum with subdivision of the ultraviolet (UV) region displayed.

replicating its DNA and RNA. The degree of inactivation is a function of energy dosage; adapting Equation 5.1 as follows:

$$N = N_0 e^{-kIt} \quad (5.2)$$

allows examination of dosage (in the form of UV radiation) over the time of exposure. In this equation, I = the intensity of the applied radiation as a function of time. As expected, inactivation rates vary depending on the species, its population, and the wavelength of UV light. In general, bacteria are less resistant to UV at 254 nm than viruses, which in turn are less resistant than bacterial spores (Setlow, 2001). Note that protozoan cysts and oocysts are regarded as the most resistant pathogenic microbes to UV at 254 nm. The minimum UV dose required for pathogen reduction (and prevention of microbe reactivation) is a contentious issue and depends on numerous considerations including source water quality, level of microbial contamination, epidemiological/health correlations, and risk levels defined by regulatory agencies.

5.4. USE AND DISPOSAL OF BIOSOLIDS

Biosolids are composed of nutrient-rich organic byproducts resulting from the sewage treatment process. Only after treatment does sewage sludge become biosolids, which can lead to acceptable disposal under regulated conditions. In our discussion on secondary treatment, we briefly covered aerobic and anaerobic digestion processes and their role in sludge digestion. Expanding upon this, primary solids from settling tanks and secondary solids from clarifiers are forced into a **digester**. It is in the digester where microorganisms use the organic material present as a food source and convert it to byproducts (e.g., methane gas and water). Typically, digestion results in the production of biosolids that routinely contain up to 97% water. Installed pressure filters or centrifuges are employed to rid excess water from the biosolids. Onsite air drying of biosolids (Figure 5.7) is a common alternative in the industry.

Given the fact that biosolids may also contain traces of synthetic organic compounds and metals including arsenic, cadmium, chromium, lead, mercury, nickel, and selenium, their end-use applications are often limited. However, proper treatment and monitoring can provide options for their end-use and disposal:

1. Landfilling (inert ash disposal after incineration);
2. Land disposal (reuse as soil amendment);
3. Reuse in building or fill materials.

Incineration transforms solids into ash, making disposal much easier. In the United States, the hazard nature of the ash must first be determined by a defined regulation: the Resource Conservation and Recovery Act (RCRA), subtitles C and D, for hazardous and non-hazardous materials, respectively. The ash is then placed in the appropriate type of landfill. In regards to land disposal, biosolids

FIGURE 5.7 Onsite air drying of biosolids at a wastewater treatment facility. Photo courtesy of Chuck Roberts.

add organic bulk to the soil, thus improving water retention and strengthening erosion defense. Although reuse as building or fill material is not routine, it does have site specific impacts.

Case Study II—Characterizing Organic Wastewater Contaminants in Biosolids

Researchers at Eastern Washington University and the U.S. Geological Survey have recently conducted an extensive survey of organic wastewater contaminants (OWCs) in biosolids destined for land application (Kinney, et al., 2006). OWCs are organic compounds produced in industrial, medical, and household products and applications, including pharmaceuticals, hormones, detergent metabolites, fragrances, plasticizers, and pesticides. In this survey, nine different biosolid products, produced by municipal wastewater treatment plants, were analyzed for 87 different OWCs. It was designed to determine ranges of compositions and concentrations of OWCs present in representative biosolid products produced under a wide range of treatments. Ultimately, estimation of the possible impacts of OWCs on land application was assessed.

In total, 55 of the 87 OWCs were detected in one or more of the biosolids studied. The types and concentrations of OWCs varied, but were most notably characterized as pharmaceuticals (e.g., fluoxetine, Figure 5.8), detergent metabolites (e.g., 4-*tert*-octylphenol), steroids (e.g., cholesterol), PAHs (e.g., phenanthrene), synthetic fragrances (e.g., indole), and disinfectants (e.g., triclosan). The composition and total OWC content did not vary greatly between biosolids. Moreover, biosolids were highly enriched with OWCs compared to effluents or effluent-impacted water. Ultimately, these studies provided evidence for a need to further study and characterize OWCs in biosolids, especially those being applied to land since these may contribute to non-point source emission into the environment.

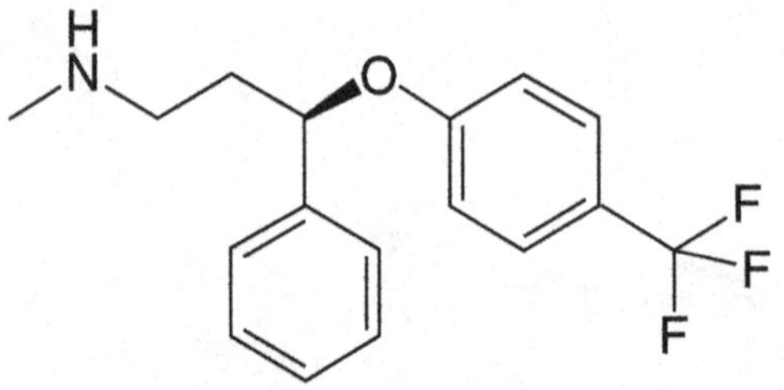

FIGURE 5.8 Chemical structure of fluoxetine, a serotonin reuptake inhibitor for the treatment of depression and various obsessive-compulsive and panic disorders. Potential environmental risks are associated with fluoxetine given its persistence and acute toxicity to non-target organisms.

5.5. POINT-OF-ENTRY AND POINT-OF-USE TECHNOLOGIES

As we learned in section 5.2, the main objective of municipal water treatment is to provide a potable supply of water for human consumption. A quick review of the various drinking water regulations presented in Chapter 4 demonstrates the need and level at which potable water must be chemically and microbiologically safe. Many public water systems employ the use of **point-of-entry (POE)** and **point-of-use (POU)** treatment devices for compliance with maximum contaminant levels. In addition, individual homeowners have the option of installing such technologies to enhance the aesthetic quality of the water, or if they feel that a specific contaminant's (or group of contaminants') presence in water is causing concern. The most widely used methods include activated carbon, reverse osmosis, and ion exchange.

TEXTBOX 5.2

Point-of-entry (POE) systems are installed units (e.g., water softeners) that modify the water quality of potable water entering a building or residence. **Point-of-use (POU)** treatment systems are units installed on individual water faucets or bubblers that change the water quality.

5.5.1. Activated Carbon

Conventional biological treatment is efficient in removing biodegradable organics but limited in its ability to rid residual organic compounds including lignins, humic substances, pesticides, and color- and odor-producing organics (Yen, 1999). **Activated carbon** has proven effective in removing these organic compounds. Activated carbon filters used for home water treatment typically contain either powdered activated carbon (PAC) or granular activated carbon (GAC). Note that activated carbon treatment is also routinely employed on a larger scale during the tertiary wastewater treatment process. There are two principal mechanisms by which activated carbon remove contaminants from water: adsorption and catalytic reduction. The adsorption process was covered in detail in Chapter 3. Catalytic reduction is a process involving the attraction of negatively-charged contaminant ions to the positively-charged activated carbon. Organic compounds are removed by adsorption, and residual disinfectants such as chlorine and chloramines are removed by catalytic reduction. Activated carbon is also commonly used as pre-treatment as part of a reverse osmosis system (see below) to reduce organic contaminants, chlorine, and other materials that could foul the reverse osmosis membrane.

5.5.2. Reverse Osmosis

Reverse osmosis is a process whereby water is forced through a semipermeable membrane via pressure over a concentration gradient, as depicted in Figure 5.9. Notice in the diagram that pressure is exerted on the side with the concentrated solute solution to force the molecules across the membrane to the fresh water side. It is important to overcome the osmotic pressure equilibrium across the membrane as it is natural for the flow to move from dilute to concentrate. The semipermeable membrane allows the passage of water, but essentially no ions or

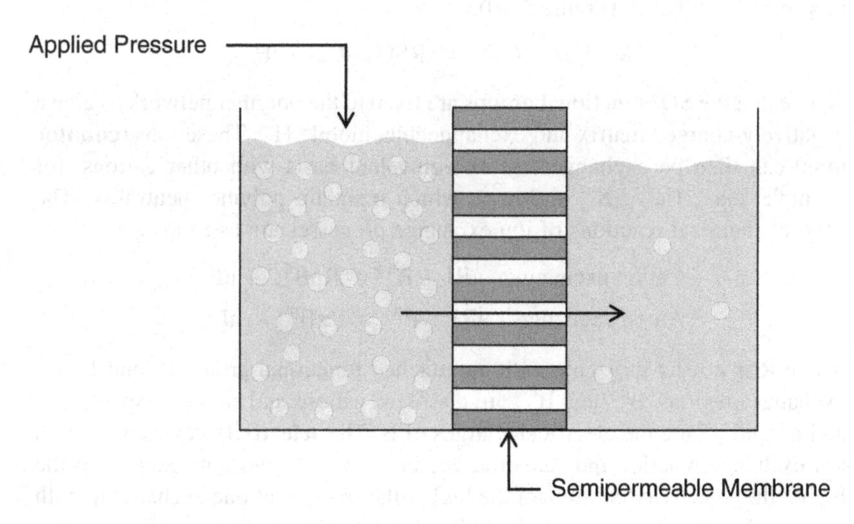

FIGURE 5.9 The reverse osmosis process. Water is forced through a semipermeable membrane via pressure across a concentration gradient.

larger molecules, although this is dependent upon the chemical make-up of the membrane and number of stages (3, 4, or 5 stage treatment) employed. Reverse osmosis has been shown to remove chlorides, sulphates, arsenic, asbestos, fluoride, lead, and when combined with an activated carbon system, has been effective in removing soluble organic compounds. Ultimately, purity is determined by the rejection ratio of the membrane and the type of salts in solution. Membranes are very good at rejecting high molecular weight compounds and multivalent ions. Monovalent ions such as Na^+ and Cl^- are not rejected well. The amount of leakage is determined by the reject ratio or rejection percentage. For example, a membrane rated at 99% would result in a 2 ppm output stream. The reject ratio changes over the life of the membrane. The potential leakage also increases. Moreover, where water contains more than one contaminant, the rejection rate for each contaminant may be reduced. In addition, a given contaminant may be reduced in preference to other contaminants.

5.5.3. Ion Exchange

Ion exchange is a reversible chemical process in which ions from solution are exchanged with exchangeable cations or anions attached to an immobile solid phase. Ions in solution are exchanged for a stoichiometrically-equivalent amount of ions of the same sign when the ion exchanger is in contact with an electrolyte solution. Solid ion exchange particles are either naturally occurring inorganic materials or, of predominant use today, synthetically-produced organic resins falling into four major classes: strong acid cation, weak acid cation, strong base anion, and weak base anion exchangers. For example, strong acid cation exchange resins are typically prepared by sulfonating benzene rings in a selected polymer (Figure 5.10):

$$2RSO_3H + Ca^{2+} \rightleftharpoons (RSO_3)_2Ca + 2H^+$$

Notice that the SO_3 functional groups are fixed to the polymer network to give a negatively-charged matrix and exchangeable, mobile H^+. These ions (**counter ions**) can then be exchanged on an equivalent basis with other cations, for example, Na^+, Ca^{2+}, K^+ or Mg^{2+}, which maintain polymer neutrality. The general chemical reactions of ion exchange processes are as follows:

$$\text{Cation exchange} \quad nRI + B^{n+} \rightleftharpoons RnB^{n+} + nI^+$$

$$\text{Anion exchange} \quad nRI + B^{n-} \rightleftharpoons RnB^{n-} + nI^-,$$

where R is a resin molecule with an attached functional group, I^+ and I^- are exchangeable ions, B^{n+} and B^{n-} are dissolved cations and anions, respectively, and n^+ and n^- are the electrical charges of B. The **selectivity coefficient** of an ion exchange reaction indicates the degree by which two ions exchange; the higher the selectivity coefficient the higher the fraction of one exchanging with another. The selectivity coefficient also determines a preference of one ion to another ion; ions with higher selectivity will replace the ions with lower

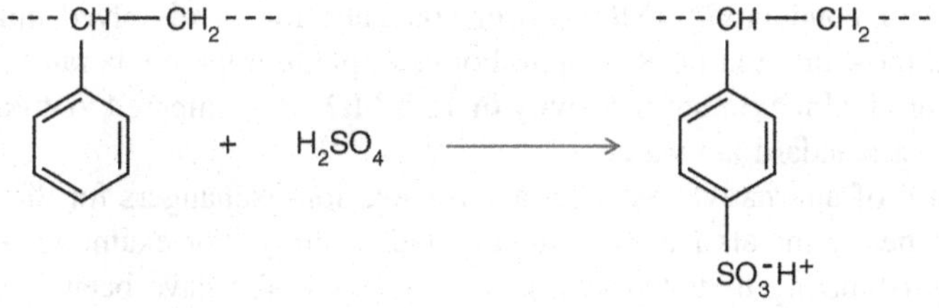

FIGURE 5.10 An example preparation of a strong acid cation exchange resin by sulfonating benzene rings.

selectivity in the body of the resin molecule. The **exchange capacity** of a particular resin is defined as the total number of chemical equivalents available for exchange per some unit weight or unit volume of resin. The capacity may be expressed in terms of meq/g of dry resin or in terms of meq/mL of wet resin. This value is typically determined by titration.

Ion exchange resins are used primarily for two types of water treatment processes: **water softening** and **demineralization (deionization)**. For example, a typical household water softening system primarily removes calcium and magnesium ions from hard water and replaces them with sodium ions. Calcium and magnesium ions interfere with the action of household soaps and detergents. During the ion exchange water softening process, water is passed through a bed of sulfonated styrene-divynalbenzene resin beads. The sulfonation process allows the formation of exchange sites on the beads that are saturated with sodium ions by introducing a brine solution to the resin. The Ca^{2+} and Mg^{2+} ions then attach themselves to the resin beads, causing Na^+ ions to be released into the water:

$$Ca^{2+} + Na_2R(\text{solid}) \rightleftharpoons CaR(\text{solid}) + 2Na^+,$$

where R = an ion exchange resin site. This release of sodium maintains a balance of electrical charge on the resin. Upon saturation with Ca^{2+} and Mg^{2+} ions, the resin is regenerated by passing a NaCl solution through the resin. This causes the above reaction to be reversed with Na^+ ions, replacing Ca^{2+} and Mg^{2+} ions.

During the deionization process both cations and anions are removed from solution. Typically, this process is employed for the production of high purity water for process/utility applications. Two resin types are characteristically used: strong acid cation resins and strong base anion resins. They can be used separately or in combination in what is termed a **mixed-bed deionizer**. If used separately, water first passes through the acid cation resin where the dissolved cations are bound by the resin and replaced with the equivalent amount of H^+. The water then passes through the base anion resin where the dissolved anions are replaced with OH^-. In a mixed-bed deionizer, cation and anion resins are thoroughly mixed in a single tank. The mixed resins act like a series of alternating cation and anion exchange tanks to produce very high quality water. The effectiveness of the deionization process is determined by measuring the resistivity or conductivity (MΩ•cm) or concentration of dissolved minerals, measured most often in units of ppm. For example, ultrapure laboratory water has a theoretical maximum resistivity of 18.3 MΩ•cm compared to around 15 MΩ•cm for standard tap water.

The use of alternative materials as probable ion-exchangers for the elimination of heavy metals has been highlighted recently. For example, **zeolites** (naturally-occurring hydrated aluminosilicate minerals) have been evaluated with respect to their removal performance for the treatment of effluents contaminated with mixed heavy metals, namely Pb, Cd, Cu, Zn, Cr, Ni, and Co (Ouki and Kavannagh, 1997). The structures of zeolites consist of three-dimensional frameworks of SiO_4 and AlO_4 tetrahedra, which demonstrate the following general empirical formula (Erdem, et al., 2004):

$$M_{2/n}O \cdot Al_2O_3 \cdot xSiO_2 \cdot yH_2O,$$

where M is any alkali or alkaline earth atom, n is the charge on that atom, and x and y are numbers. For instance, the natural zeolite clinoptilolite has the following formula:

$$(Na_3K_3)(Al_6Si_{40})O_{96} \cdot 24H_2O.$$

In regards to structural characteristics and functionality, the isomorphous replacement of Si_4^+ by Al_3^+ produces a negative charge in the lattice (Erdem, et al., 2004). The net negative charge is balanced by the placement of exchangeable cations such as Na^+, K^+, Ca^{2+}, and Mg^{2+} (as demonstrated in the example of clinoptilolite). The innocuous nature of the exchangeable cations makes them particularly suitable for use in treating heavy metal ions from industrial effluent waters. During the ion-exchange process, metal ions move through the pores of the zeolite mass and channels of the lattice to replace the exchangeable cations. The adsorption of metal ions on natural zeolite is primarily a function of pH and metal concentrations in solution.

5.6. END OF CHAPTER PROBLEMS

5.1 Discuss the importance of environmental regulations and related standards and how they have impacted the development and operation of water treatment facilities.

5.2 Describe the differences between physical, chemical, and biological unit wastewater treatment processes and provide an example of each.

5.3 In section 5.1.1 we discussed the use of alum or hydrated aluminium sulphate to precipitate phosphates and aluminium phosphates ($AlPO_4$) out of solution. In addition, calcium, typically in the form of lime $(CaOH)_2$, is used to help remove phosphate from wastewater. It reacts with the natural alkalinity in the wastewater to produce $CaCO_3$:

$$Ca(HCO_3)_2 + Ca(OH)_2 \rightleftharpoons 2CaCO_3 + 2H_2O.$$

When the pH value of the wastewater increases to >10, excess calcium ions will then react with the phosphate and precipitate as hydroxylapatite. Show the complete, balanced reaction for this process.

5.4 Provide a basic mechanism for the use of ferric chloride for phosphorus removal (through coagulation) in wastewater streams.

5.5 Explain why the amount of Ca^{2+} ion needed to drive the reaction in problem 5.3 is almost completely independent of the amount of phosphate present in solution.

5.6 What factors are known to affect/influence the coagulation process?

5.7 Describe how a charge on a given ion will affect its ability to coagulate.

5.8 Provide reasoning as to why it is important to provide an initial rapid mixing of the coagulant (and other chemicals) during the flocculation process.

5.9 Initial total chloroform (TC) count (numbers) in a wastestream was 10,003 per 100 mL of sample collected. Upon disinfection via chlorine treatment, TC numbers were 6779, 3993, and 582/100 mL, after treatment for 12 hr, 24 hr, and 48 hr, respectively. Calculate the death rate (k) achieved (in days) after each time treatment.

5.10 As mentioned, disinfection data (Chick's Law concepts) can be evaluated with the use of a semilogarithmic plot of survival. Plot the data generated from problem 5.9 in Excel and construct a straight line.

5.11 Discuss the potential impacts of DO levels on the performance of activated sludge bioreactors.

5.12 Describe two important factors affecting the efficiency of activated carbon filtration.

5.13 Explain why the reverse osmosis processes would be particularly affected by high total dissolved solid (TDS) content during the water treatment process.

5.14 Extending our discussion on membrane treatment processes, speculate on various physical/chemical processes that would help distinguish which species (contaminants) could pass through a given membrane.

5.15 What is the role of alkalinity in the wastewater treatment process?

5.16 List common ions that are associated with alkalinity.

5.17 Name two disadvantages of the chlorination process used in wastewater disinfection.

5.18 What is meant by water softening? What water treatment technique is routinely used in this process?

5.19 What are the two main causes of hardness in water?

5.20 As discussed, when an ion-exchange water softening system becomes saturated with Ca^{2+} and Mg^{2+} ions, it is regenerated by passing a NaCl solution through the resin. Write a balanced chemical reaction for the replacement of Ca^{2+} during this process.

5.21 Ion exchange can be used to remove excess salt and other minerals from water (e.g., seawater) during the desalination process. What types of resins would be used to individually remove Cl^{-} and Na^{+} ions?

5.22 List possible water treatment processes/techniques that could be used to treat the following contaminants:
a. Pathogens
b. Organic carbon
c. Phosphate

5.7. REFERENCES

Beccari, M., Di Pinto, A. C., Ramadori, R., & Tomei, M. C. (1992). Effects of dissolved oxygen and diffusion resistances on nitrification kinetics. *Water Research, 26*, 1099–1104.

Chick, H. (1908). An investigation of the laws of disinfection. *Journal of Hygiene, 8*, 92–158.

Erdem, E., Karapinar, N., & Donat, R. (2004). The removal of heavy metal cations by natural zeolites. *Journal of Colloid and Interface Science, 280*, 309–314.

Kim, C. K., Gentile, D. M., & Sproul, O. J. (1980). Mechanism of ozone inactivation of Bacteriophage f2. *Applied and Environmental Microbiology, 39*, 210–218.

Kinney, C. A., Furlong, E. T., Zaugg, S. D., Burkhardt, M. R., Werner, S. L., Cahill, J. D., & Jorgensen, G. R. (2006). Survey of organic wastewater contaminants in biosolids destined for land application. *Environmental Science & Technology, 40*, 7207–7215.

Komorowska-Kaufman, M., Majcherek, H., & Klaczynski, E. (2006). Factors affecting the biological nitrogen removal from wastewater. *Process Biochemistry, 41*, 1015–1021.

Majed, N., & Gu, A. Z. (2010). Application of raman microscopy for simultaneous and quantitative evaluation of multiple polymers dynamics functionally relevant to enhanced biological phosphorus removal processes. *Environmental Science & Technology, 44*, 8601–8608.

Mino, T. (2000). Microbial selection of polyphosphate-accumulating bacteria in activated sludge wastewater treatment processes for enhanced biological phosphate removal. *Biochemistry, 65*, 341–348.

Mullan, A., McGrath, J. W., Adamson, T., Irwin, S., & Quinn, J. P. (2006). Pilot-scale evaluation of the application of low pH-inducible polyphosphate accumulation to the biological removal of phosphate from wastewaters. *Environmental Science & Technology, 40*, 296–301.

Ouki, S. K., & Kavannagh, M. (1997). Performance of natural zeolites for the treatment of mixed metal-contaminated effluents. *Waste Management Research, 15*, 4383–4394.

Setlow, P. (2001). Resistance of spores of Bacillus species to ultraviolet light. *Environmental and Molecular Mutagenesis, 38*, 97–104.

Streat, M., Hellgardt, K., & Newton, N. L. R. (2008). Hydrous ferric oxide as an adsorbant in water treatment. Part 2. Adsorption studies. *Process Safety and Environmental Protection, 86*, 11–20.

Tünay, O., & Kabdaşli, N. I. (1994). Hydroxide precipitation of complexed metals. *Water Research, 28*, 2117–2124.

U.S. Environmental Protection Agency. (1998). *National primary drinking water regulations: Disinfectants and disinfection byproducts;* Final Rule. 40. CFR, Parts 9, 141, and 142.

U.S. Environmental Protection Agency. (2004). *Primer for municipal wastewater treatment systems, EPA 832–R–04–001*. Washington, D.C.: Office of Wastewater Management.

U.S. Environmental Protection Agency. (2009). *National water quality inventory 2004 report.* Washington, D.C.

Yen, T. F. (1999). *Environmental chemistry: Chemical principles for environmental processes.* Upper Saddle River, NJ: Prentice-Hall, Inc.

Chapter 6

Abstracts and Keywords

Abstract

The purpose of this chapter is to promote basic understanding of atmospheric chemical and physical processes and the impact of natural and anthropogenic pollutant releases. It begins with defining the four atmospheric layers and describing their distinguishing characteristics. Next, a review of basic photochemistry and its role in the formation of ground-level ozone and stratospheric ozone destruction are given. Pressure and temperature considerations in relation to the atmospheric layers will then be presented. Subsequently, information on example exchange processes that occur between the biosphere and the atmosphere is discussed. Next, a brief introduction to example transport, transformation, and deposition processes is presented. Lastly, a short primer on global climate change including example greenhouse gases and a concept termed global warming potential (GWP) is offered.

Keywords

Atmospheric layers, Solar radiation, Photochemistry, Ozone layer, Ground-level ozone, Biosphere, Greenhouse gases, Global warming potential (GWP), Climate change

Chapter 6

The Atmosphere and Associated Processes

As will be covered in this chapter, air pollution takes many forms and affects numerous natural processes and human health. One of the most pervasive air pollutants, tropospheric ozone, is a serious threat not only to human health, but also to the conservation and sustainability of our world's forests. Tropospheric ozone is known to damage foliage of certain species of pines and conifers (Figure 6.1), thus reducing species diversity, population numbers, and growth. In 1994, the Forest Inventory and Analysis (FIA: http://fia.fs.fed.us) and Forest Health Monitoring (FHM: http://fhm.fs.fed.us) programs of the USDA Forest Service implemented a national ozone biomonitoring program to demonstrate their commitment to sustainable forest management. This effort includes a variety of field methods, sampling procedures, and analytical techniques to assess the impacts of tropospheric ozone on our nation's forests.

FIGURE 6.1 Evidence of air pollution disease (primarily by tropospheric ozone) in Ponderosa Pines of the Los Padres National Forest, California.

6.1. INTRODUCTION

Theories on the atmospheric composition of early Earth have evolved considerably over the last few decades. Scientists now generally agree that the Earth's early atmosphere contained little or no free oxygen and that oxygen concentrations increased noticeably just over two billion years ago (Kasting, 1993). The latter occurrence is likely the result of cyanobacteria, which used energy from the sun for photosynthesis with the release of oxygen as a byproduct. This net gain of oxygen likely first occurred in the oceans with subsequent transfer to the atmosphere. Ultimately, with sufficient oxygen in the atmosphere, respiration naturally balanced the photosynthetic effect. These processes led to the current atmospheric composition of 78.1% nitrogen, 21.0% oxygen, and 0.9% other trace gases. Although the present-day atmosphere was created largely as a result of biological activity over two billion years ago, its composition, while supporting life as we know it, is being modified extensively by trace gas production and destruction processes. Thus, our understanding of atmospheric chemical and physical processes, and the impact of anthropogenic releases and climate change, is of vital importance.

6.2. THE LAYERS OF THE ATMOSPHERE

In most general terms, the atmosphere is divided into lower and upper regions. However, **atmospheric layers** in these regions are more specifically characterized by variations in temperature and pressure with altitude (Figure 6.2). As depicted, four major layers are distinguished and briefly described below:

1. **Troposphere**: The lowest layer, extending from the surface of the Earth up to the tropopause (ranging from ≈10–15 km in altitude). As shown in Figure 6.2, the troposphere is characterized by decreasing temperature with increased altitude. Such a trend causes rapid vertical mixing.
2. **Stratosphere**: The next layer of the atmosphere that extends from the tropopause to the stratopause (ranging from ≈46–54 km in altitude). In contrast to the troposphere, temperature increases with increasing altitude. As a result, slower vertical mixing is demonstrated.
3. **Mesosphere**: The layer extending from the stratopause to the mesopause (ranging from ≈80–85 km in altitude). As shown, temperature decreases in the mesosphere with increased altitude to the ultimate lowest point (−80°C) occurring at the boundary between the mesosphere and the thermosphere (mesopause).
4. **Thermosphere**: The layer above the mesosphere extending ≈85–640 km. It is characterized by higher temperatures (as a result of the absorption of short-wavelength radiation by N_2 and O_2). The lower thermosphere contains the ionosphere. Although not a separate region of the atmosphere, it is an area of great photoionization.

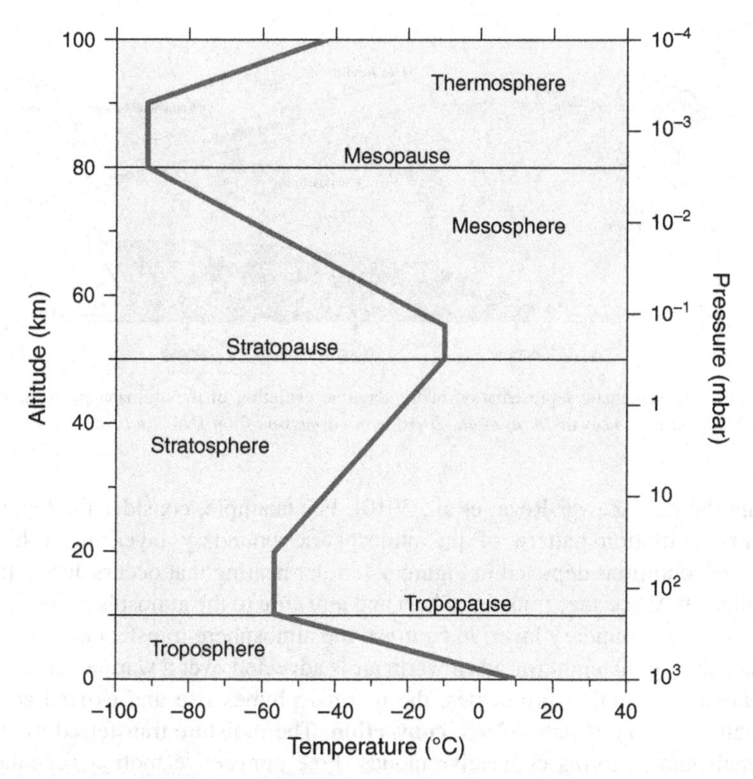

FIGURE 6.2 The layers of the atmosphere distinguished based on temperature and pressure. Note: The blue line represents the temperature change with height.

A fifth layer, termed the **exosphere**, is the outermost layer of Earth's atmosphere extending to the edge of interplanetary space. It is this region where atmospheric gas molecules can, to any appreciable extent, escape from the Earth's gravitational pull. Although air is fairly well mixed throughout the atmosphere, the atmosphere itself is not physically uniform but has significant variations in temperature and pressure with altitude. As we will see below, such variations help define atmospheric layers and the interactions of chemical and physical processes contained within them.

6.2.1. The Boundary Layer

The **boundary layer**, also referred to as the atmospheric boundary layer, is a portion of the troposphere that is directly influenced by the Earth's surface and responds to surface forcing on a short time scale (typically on the order of <1 hour) (Stull, 1988). This layer is an important player in pollutant dispersion and chemistry due to rapid mixing depth and temperature fluctuations. It also varies greatly in space, ranging from tens of meters at night to 1 km or more

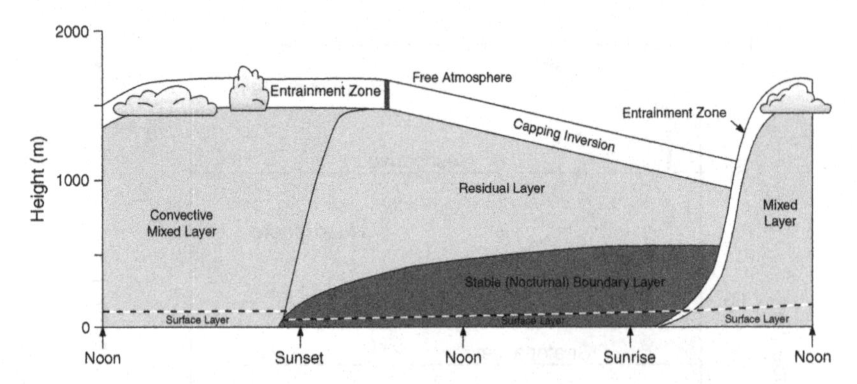

FIGURE 6.3 Schematic representation of the daytime evolution of the atmospheric boundary layer. *Modified from (Zawar-Reza, et al., 2010) with permission from ILM Publications.*

during the day (Zawar-Reza, et al., 2010). For example, consider the typical daytime evolution pattern of the atmospheric boundary layer under high pressure conditions depicted in Figure 6.3. Solar heating that occurs during the day dictates the surface transfer of heat and moisture to the atmosphere forming an "unstable" boundary layer. In contrast, the atmosphere transfers heat to the surface during the night (or when warm air is advected over a warmer surface). As daytime evolution progresses, the thermal plumes rise and expand adiabatically, creating a state of free **convection**. The moisture transferred by the thermal plumes forms convective clouds. Free convective motion generates **turbulent mixing.** Turbulent events affect physical processes that lead to cloud formation, aerosol formation, and enhancement or limitation of the ability of aerosols and reactive chemical species to engage in common physical/chemical processes (e.g., chemical transformation, dispersion and mixing of species, and deposition) (Elperin, et al., 2000). For example, inefficient mixing can disrupt the chemical equilibrium and thus slow or limit associated chemical transformations. Studies have also shown that an altitude-dependent chemistry in the stable (nocturnal) boundary layer leads to zones where different chemical mechanisms deplete ozone and other pollutants of environmental interest. For example, nearer to the Earth's surface, atmospheric chemistry is impacted by the upward transport of freshly emitted species, while the top of the nocturnal layer is influenced by the **entrainment** of compounds from the residual layer (Stutz, et al., 2004). A zone of reactivity then develops between these two borders where active radical chemistry takes place on a continuous basis.

6.3. ATMOSPHERIC RADIATION AND PHOTOCHEMICAL REACTIONS

Nearly all of the energy that drives the Earth's biological and physical cycles originates from the sun. A significant portion of this **solar radiation** consists of high energy photons capable of dissociating molecules. The discussion of atmospheric layers above should be an indication that the atmosphere, as a whole, is not a transparent entity, but rather regions containing trace gases and suspended particles that modify the solar radiation passing through. This process, termed **scattering**, occurs when particles and gas molecules diffuse a portion of the incoming solar radiation without any modification to the wavelength λ of the electromagnetic energy. Recall that radiant energy is arranged in order of its wavelength in the electromagnetic spectrum. Of particular interest in atmospheric chemistry is the optical region extending over the UV, the visible, and the near-infrared, although our discussion will center mainly on the UV region. A portion of the scattered shortwave solar radiation is redirected back to interplanetary space. Luckily, a portion of the solar radiation is retained or absorbed by gases and particles and converted into heat energy. Yet more incoming solar radiation is reflected, a process where radiation is redirected by 180° after it strikes atmospheric particles. This overall exchange process is tantamount to the Earth's energy balance.

6.3.1. Energy of a Photon: Understanding Basic Photochemistry

Of importance in atmospheric chemical processes is the extent to which photon energies are necessary to break chemical bonds. The energy, E, of a photon is given by

$$E = h\nu = \frac{hc}{\lambda}, \tag{6.1}$$

where h = Planck's constant (6.626×10^{-34} J·s), ν = the frequency of light, and c = the speed of light in vacuum. Notice that the frequency is inversely proportional to its wavelength, λ. Recall from our discussion in Chapter 5 that UV radiation can be defined as that portion of the electromagnetic spectrum between 40 and 400 nm, with further division as UV-C (220–290 nm), UV-B (290–320), and UV-A (320–400 nm). UV radiation is very energetic and more likely to break chemical bonds than visible light.

A generalized photochemical reaction may be written (Seinfeld and Pandis, 2006) as

$$A + h\nu \rightarrow A^*, \tag{6.2}$$

where A^* = an electronically excited state of A. This excited molecule may then partake in the following reactions:

$$\text{Dissociation}: A^* \rightarrow B_1 + B_2$$

$$\text{Direct reaction}: A^* + B \rightarrow C_1 + C_2$$

$$\text{Fluorescence}: A^* \rightarrow A + h\nu$$

$$\text{Collisional deactivation}: A^* + M \rightarrow A + M$$

$$\text{Ionization}: A^* \rightarrow A^+ + e$$

To ascertain the **photochemical reaction rate**, we must consider the absorption spectrum of A, the available light at the point in the atmosphere where A exists and the **quantum yield** for dissociation of A once it has become electronically excited by absorption of radiation. The quantum yield for a given process is defined as the ratio of the number of molecules of A* undergoing that process to the number of photons absorbed (Seinfeld and Pandis, 2006). A photodissociation rate constant J_i can be defined just as we define a rate constant for a chemical reaction in General Chemistry. The first-order rate constant is given the symbol $J(\sec^{-1})$ by atmospheric scientists, hence the name J-value. However, unlike regular chemical rate constants, J is a function that must be computed for each compound under photochemical study.

Example: Stratospheric Ozone

Some of the most important reactions in the atmosphere are **termolecular** in nature, described on the simplest level as

$$A + B + C \rightarrow \text{Product}, \tag{6.3}$$

with the rate expression equal to $k\ [A][B][C]$. Absorption of UV radiation by molecular oxygen, for example, leads to the formation of **stratospheric ozone**, as described by the four steps of the **Chapman reaction mechanism** (Zawar-Reza, et al., 2010):

(1) $$O_2 + h\nu\ (\lambda < 240\ \text{nm}) \rightarrow O + O \tag{6.4}$$

(2) $$O_2 + O + M \rightarrow O_3 + M \tag{6.5}$$

(3) $$O_3 + h\nu\ (240\ \text{nm} < \lambda < 320\ \text{nm}) \rightarrow O_2 + O \tag{6.6}$$

(4) $$O_3 + O \rightarrow O_2 + O_2 \tag{6.7}$$

Recall that a reaction mechanism is a step-by-step sequence of reactions by which an overall chemical transformation transpires. Also notice that in the above steps, the photon that is a reactant in a chemical reaction is written as $h\nu$. Reactions (1) and (2) are **elementary reactions**; in other words, they cannot be subdivided into two or more simpler reactions. As shown in (1), molecular oxygen photodissociates into two reactive oxygen atoms by solar radiation with a $\lambda<240$ nm. The rate of reaction for (1) = J_{O_2} $[O_2]$, where J_{O_2} denotes the photolysis rate constant of Reaction (1). This is representative of an effective first-order rate constant. Greater examination of the photon energy/wavelength relationship explains why O_2 cannot be photodissociated at a $\lambda>240$ nm. First, photon energy is expressed per mole of a substance by multiplying $h\nu$ by Avogadro's number (6.022×10^{23} molecules mol^{-1}):

$$E = 6.022 \times 10^{23} \frac{hc}{\lambda}. \tag{6.8}$$

The energy associated with a given wavelength λ is as follows:

$$E = \frac{1.19625 \times 10^5}{\lambda} \text{ kJ mol}^{-1}. \quad (6.9)$$

The enthalpy change for Reaction (1) above is $\Delta H = 498.4$ kJ mol^{-1}. Using Equation 6.9 to calculate λ_{max} gives us

$$E = \frac{1.19625 \times 10^5}{498.4} \text{ nm} = 240 \text{ nm}.$$

Example Problem 6.1

How much energy is required to break the O-O bond in O_2 in Reaction (1) of the Chapman mechanism?

Answer: Recall that the enthalpy change for Reaction (1) is $\Delta H = 498.4$ kJ mol^{-1}. To express energies associated with molecules in kcal mol^{-1} (a typical expression used by chemists), we multiply kJ mol^{-1} × 0.2390. In the case of the O-O bond, (498.4 kJ mol^{-1}) × 0.2390 = 119.1 kcal mol^{-1}.

In Reaction (2), the reactive oxygen species O reacts rapidly with O_2 in the presence of M (a non-reactive species), typically an additional O_2 or N_2. This termolecular reaction follows third-order kinetics (a concept introduced in Chapter 1). In Reaction (3), O_3 absorbs UV radiation (240 nm $< \lambda <$ 320 nm) and decomposes into $O_2 + O$; a reverse of the spontaneous reaction in (2). Finally, in Reaction (4), O_3 will likely react with O to regenerate two molecules of O_2. One criticism of Chapman's mechanism is the over-prediction of atmospheric O_3 levels in the stratosphere. Referring to this region as the **ozone layer** is something of a misnomer, given the fact that O_3 does not occur in a defined "layer" in the atmosphere. More appropriately, it is a region of the atmosphere, approximately 10–50 km in altitude (with maximum concentration between 20–35 km, depending on season, latitude, and related factors), where the intensity of short-wavelength UV radiation is suitably high to convert O_2 to O_3. In general, atmospheric measurements reveal that the concentration of O_3 increases with altitude; the concentration is higher in the stratosphere (90% of all atmospheric ozone) and lower in the troposphere (Finlayson-Pitts and Pitts, 2000). Like all gaseous atmospheric components, ozone measure is typically expressed in ppbv (parts per billion by volume), the unit of **mixing ratio**. The mixing ratio in ppbv relates the fractional concentration of ozone as the number of ozone molecules per billion molecules of air. Typical concentrations are 10–40 ppbv and 10 ppbv for the clean troposphere and the maximum stratospheric region (20–35 km), respectively.

TEXTBOX 6.1

The U.S. National Weather Service calculates the predicted UV Index each day to inform citizens of the level of solar radiation in which they are exposed and the potential level of risk they face as a result. The scale, categorized as a value with associated risk level, is as follows:

2 or less: Low
3–5: Moderate
6–7: High
8–10: Very High
11+: Extreme

The UV Index is calculated using a computer model that relates the ground-level strength of solar UV radiation to forecasted stratospheric O_3 concentration, forecasted cloud amounts, and ground elevation. This information is published in mid-afternoon (Eastern time zone) at the U.S. EPA website (http://www.epa.gov/sunwise/uvindex.html). If the level of solar UV radiation is predicted to be unusually high (risk of overexposure is high), the forecast includes a UV Alert. Citizens can use this index to take suitable protective behaviors and avoid overexposure to potentially harmful UV radiation.

Stratospheric Ozone Destruction: The Role of Halogen Compounds

Scientists now know that additional destruction of O_3 occurs through common catalytic destruction cycles involving nitrogen (NO_x), hydrogen (HO_x), chlorine (ClO_x), and bromine (BrO_x) oxides, to name a few. Halogen species in particular have garnered attention as chlorofluorocarbons (CFCs), which were shown (as early as the 1970s) to contribute greatly to the overall halogen burden in the stratosphere. CFCs were used extensively as refrigerants and solvents. Efforts have been underway for years to manufacture CFC replacements. These include hydrofluorocarbons (HFCs) and hydrochlorofluorocarbons (HCFCs), both of which have already been recently detected in the atmosphere. As an example, let's consider the photochemical breakdown of CFCs and their role in stratospheric ozone destruction. Upon transport to the stratosphere, photochemical breakdown (with eventual release of free Cl atoms) of an example CFC proceeds as follows:

$$CFCl_3 + h\nu \rightarrow CFCl_2 + Cl \quad (6.10)$$

$$CFCl_2 + O_2 + M \rightarrow CFCl_2O_2 + M \quad (6.11)$$

$$CFCl_2O_2 + NO \rightarrow CFCl_2O + NO_2 \quad (6.12)$$

$$CFCl_2O + M \rightarrow COFCl + Cl + M \quad (6.13)$$

$$COFCl + h\nu \rightarrow FCO + Cl. \quad (6.14)$$

Note that additional breakdown of CFCs occurs as a result of reaction with O (^1D), an excited electronic state of the oxygen atom produced in the photolysis of ozone (see section 7.3 for details on the photolysis process). Free chlorine atoms can then react with O_3, resulting in the formation of ClO:

$$Cl + O_3 \rightarrow ClO + O_2. \tag{6.15}$$

Subsequently, ClO can react with O atoms:

$$ClO + O \rightarrow Cl + O_2, \tag{6.16}$$

leading to

$$\text{Net}: O + O_3 \rightarrow 2O_2. \tag{6.17}$$

Fortunately, there are many processes that regenerate the ozone lost as a result of destruction processes. For example, ClO can be converted back to Cl via numerous reactions, including the following:

$$ClO + NO \rightarrow Cl + NO_2 \tag{6.18}$$

$$NO_2 + hv \rightarrow NO + O \tag{6.19}$$

$$O + O_2 + M \rightarrow O_3 + M \tag{6.20}$$

and

$$ClO + hv \rightarrow Cl + O \tag{6.21}$$

$$O + O_2 + M \rightarrow O_3 + M. \tag{6.22}$$

One of the first and formalized environmental agreements to recognize the need for the phase out of ozone-depleting substances was the Montreal Protocol. This Protocol was initially established to implement the phase out of CFCs, but the protocol was amended in 1992 to schedule the phase out of HCFCs. Despite these efforts to curb ozone-depleting substances, ozone depletion will continue to be of concern in the coming decades. This will be especially true over the Arctic and Antarctic regions of the Earth, where the development of polar stratospheric clouds (PSCs) gives rise to a series of chemical reactions that destroy ozone far more effectively than in regions of warmer air masses. Ultimately, eliminating ozone-depleting substances from the atmosphere is expected to take between 50–100 years, given their long lifetime in the atmosphere and relatively slow breakdown.

Example Problem 6.2

The reaction of ClO and HO_2 has also been shown to lead to stratospheric O_3 destruction. Provide the multi-step reaction for this process.

Answer: Starting with free Cl:

$$Cl + O_3 \rightarrow ClO + O_2$$
$$ClO + HO_2 \rightarrow HOCl + O_2$$
$$HOCl + hv \rightarrow OH + Cl$$
$$OH + O_3 \rightarrow HO_2 + O_2.$$

TEXTBOX 6.2

Fourier Transform Infrared Spectroscopy (FTIR)

The evolution and distribution of O_3 and stratospheric species like HF, HCl, $ClONO_2$, and HNO_3 are routinely studied by the use of ground-based solar absorption **Fourier Transform Infrared Spectroscopy (FTIR)** measurements. In infrared spectroscopy, IR radiation is passed through a sample. A portion of the IR is absorbed by the sample, and a portion of it is passed through (transmitted). The resulting spectrum represents the molecular absorption and transmission, ultimately creating a "molecular fingerprint" of the sample. Because the strength of the absorption is proportional to the concentration, FTIR can be used as a quantitative analysis method. Moreover, by measuring the absorption lines from pressure-broadened transitions in the IR part of the solar spectrum, up to 25 atmospheric species can be measured by FTIR.

6.4. PRESSURE AND TEMPERATURE CONSIDERATIONS

In each of the four layers detailed above, temperature changes in response to local pressure as air moves vertically. However, in order to truly understand this relationship, we must first review pressure and temperature concepts independently. In simplest terms, the pressure of the atmosphere at any point is due to the mass of air being brought down by gravity, causing an increase in the overall weight exerted. The SI unit of pressure is $N\,m^{-2}$, or pascal (Pa). The pressure at the surface of the Earth (termed **standard atmosphere**) $= 1.01325 \times 10^5$ Pa or 1 atm. A more commonly reported unit, the millibar (mbar), can be derived, e.g., 1 mbar = 1 hPa or 100 Pa. Equation 6.23 below shows how atmospheric pressure decays exponentially from its value at the Earth's surface where the height h is equal to 0:

$$p = p_0 e^{-\left(\frac{h}{h_0}\right)}, \tag{6.23}$$

where $p =$ the atmospheric pressure (measured in bars), $h =$ height (altitude), $p_0 =$ the pressure at height 0 (Earth's surface), and $h_0 =$ the scale height (the increase in altitude by which the pressure drops by e^{-1}). When $h_0 = h$, the pressure has decreased to a value of e^{-1} times its value at the Earth's surface. The pressure on the Earth's surface $\approx$ 1 bar, and the scale height h_0 of the atmosphere $\approx$ 7 km.

Example Problem 6.3

Calculate the pressure (in bars) at an altitude of 5 km up in the Earth's troposphere.

Answer: Using Equation 6.23, we find $p = 1.0e^{-(5/7)} = 0.489542$ bars.

Notice in Figure 6.2 that the temperature in the atmosphere varies less than a factor of 2. This is in stark comparison to pressure characteristics that can display changes by six orders of magnitude. The main factors influencing atmospheric temperature gradients are energy transfer and vertical motion of air, although their interaction characteristics are still under investigation by atmospheric scientists. A majority of the energy affecting the Earth is derived from solar radiation. Release of latent heat can occur by condensation (see Chapter 1), radiational cooling of the air, and reradiation from the Earth's surface, which heats the overlying troposphere. Horizontal temperature **advection** must also be considered, with vertical motion dependent upon the type of pressure system (Barry, 1987). For example, low pressure systems are coupled with rising air that cools upon expansion and increases the vertical temperature gradient. Generally, the global temperature in the troposphere decreases with height at an average rate of 6.5°C km^1 up until the tropopause. As also demonstrated in Figure 6.2, the temperature begins to increase with altitude in the stratosphere and thermosphere due to the presence of radiation absorbing gases. Solar radiation striking the stratosphere, for example, is absorbed by the air molecules present with O_2 strongly adsorbing in the UV band.

TEXTBOX 6.3
Le Châtelier's Principle—Healthy Equilibrium at High Elevations

FIGURE 6.4 Lobuche Peal (6119 m) and Chola Che Lake, Himalayan Region, Nepal.

At higher elevations, the air pressure is reduced and thus more difficult for one to obtain the oxygen needed to feed the body's cells and tissues. In accordance with

Le Châtlelier's principle, there is a shift in equilibrium to the left, away from the oxygenated hemoglobin (Hb):

$$Hb_{(aq)} + 4O_{2(g)} \leftrightarrow Hb(O_2)_{4\ (aq)}.$$

It is this interaction that allows the transportation of oxygen to the cells. Without sufficient oxygen, a person at high altitude can feel lightheaded. Pressurized oxygen is often used by climbers at extremely high elevations such as those found in the Himalayan region of Nepal (Figure 6.4). This additional oxygen will shift the equilibrium to the right. For those that are born or train in such conditions for extended periods, their bodies perform this shift naturally by producing more hemoglobin.

6.5. BIOSPHERE-ATMOSPHERE INTERACTIONS

The exchanges that occur between the biosphere and the atmosphere affect all of Earth's living organisms, including mankind. From the development and control of the physical climate system to the interacting effects of anthropogenic emissions and nitrogen deposition on forest carbon budgets, the biosphere and atmosphere are perpetually linked. Thanks in large part to a new integrated field of study, global change biology, we are beginning to better understand the relationship between anthropogenic changes in atmospheric chemistry and the disciplines of ecosystem ecology, plant physiology, Earth system science, and climate change (Figure 6.5). For example, the performance and distribution of plant species are dependent upon climate change, which is ultimately linked to anthropogenic emission patterns, e.g., increased atmospheric CO_2 concentrations. It should thus be evident that the interactions between the biosphere and atmosphere are inexorably linked. The following two subsections present introductory material on the types of emissions and example compounds emitted from identified sources. Advanced coverage of atmospheric pollutants and their formation mechanisms is provided in Chapter 7.

6.5.1. Biogenic Emissions

Biogenic emissions emanate from natural sources such as natural vegetation areas, crops, and urban vegetation, as well as from microbial activity in soil. For example, emissions and oxidation processes of biogenic volatile organic compounds (BVOC) in the troposphere have been shown to be an important contributor to air quality. In fact, it is estimated that biogenic VOC emissions are an order of magnitude higher than anthropogenic emissions (Benkovitz, et al., 2004). Common VOCs emitted by biogenic sources include methanol, acetaldehyde, isoprene, methyl butenol, monoterpenes, and pentenone. Consider isoprene emissions in greater detail. Isoprene emission is quite obvious to us as a result of its characteristic pine and lemon scent smells. Despite these pleasantries, isoprene emission is the prevalent biogenic source of

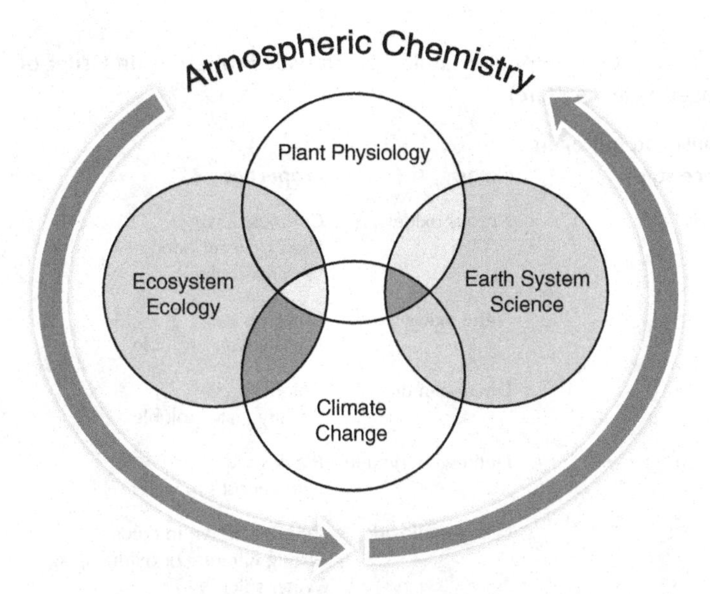

FIGURE 6.5 Anthropogenic changes in atmospheric chemistry have direct connections to the study of ecosystem ecology, plant physiology, Earth system science, and climate change.

hydrocarbons in the atmosphere, roughly equal to global emission of methane (CH_4) from all sources (Sharkey, et al., 2008). This compound (Figure 6.6) has a significant effect on atmospheric chemistry given its deep-seated connection with NO_x. The family of NO_x compounds and their properties are listed in Table 6.1. Of these, N_2O, NO, and NO_2 are the most abundant nitrogen oxides in the troposphere. Except for NO from soils, lightning, and natural fires, NO is largely anthropogenic in origin. In fact, biogenic sources account for less than 10% of total NO emissions. Further characterizing this family of oxides, N_2O_5 is considered the most highly ionized form of nitrogen oxide. It is therefore highly reactive and forms HNO_3 when it decomposes in water.

As we will learn in detail in Chapter 7, the total NO_x level is important in determining the amount of ground-level O_3 that can be formed, where sunlight photolyses NO_2 leading to O_3 formation. However, when isoprene is oxidized by atmospheric hydroxylradicals, hydroperoxides (RO_2) that convert NO to NO_2 (allowing extended ozone production) are formed (Sharkey, et al., 2008).

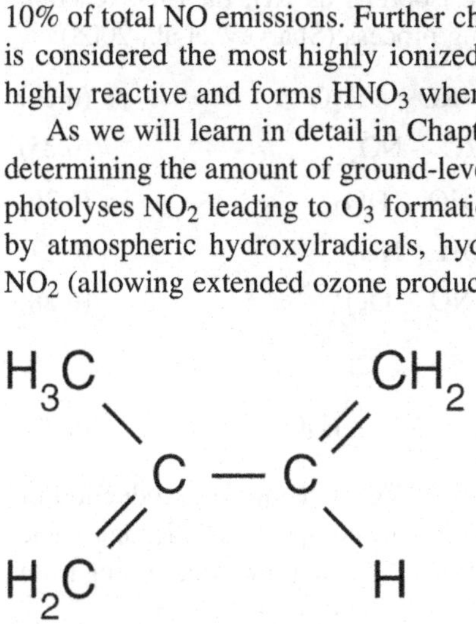

H_3C, CH_2, C — C, H_2C, H

FIGURE 6.6 The chemical structure of isoprene.

TABLE 6.1 NO_x Compounds and Characteristic Properties (in Order of Nitrogen Valence State)

Formula and (nitrogen valence state)	Name	Properties
N_2O (1)	Nitrous oxide	Colorless gas Slightly sweet odor Water soluble
NO (2)	Nitric oxide	Colorless gas Slightly water soluble
N_2O_2 (2)	Dinitrogen dioxide	Colorless gas Slightly water soluble
N_2O_3 (3)	Dinitrogen trioxide	Black solid Decomposes in water
NO_2 (4)	Nitrogen dioxide	Reddish brown in color Strong nitrating or oxidizing agent Water soluble
N_2O_4 (4)	Dinitrogen tetroxide	Reddish brown with sharp, unpleasant odor Water soluble
N_2O_5 (5)	Dinitrogen pentoxide	White solid Decomposes in water

Additional reactions can form HO_2 that can also convert NO to NO_2 and generate the OH radical. If we designate isoprene as RH, then the reaction below can help describe the ozone-forming process (Sharkey, et al., 2008):

$$RH + OH + O_2 \rightarrow RO_2 + H_2O \tag{6.24}$$

$$RO_2 + NO \rightarrow RO + NO_2 \tag{6.25}$$

$$RO + O_2 \rightarrow R'CHO + HO_2 \tag{6.26}$$

$$HO_2 + NO \rightarrow OH + NO_2 \tag{6.27}$$

$$2(NO_2 + O_2 \rightarrow NO + O_3). \tag{6.28}$$

The net reaction is then

$$RH + 4O_2 \rightarrow R'CHO + 2O_3 + H_2O. \tag{6.29}$$

Biogenic gas emissions are not limited to VOCs. Biogenic production of CH_4 occurs by the microbial breakdown of organic compounds under anaerobic conditions, e.g., in landfills and anoxic sediment layers. Emissions arising from biomass burning, although largely resulting from the activities of humans, do account for a sizeable portion of the total biogenic CH_4 budget. In addition, CO_2 emissions from biomass burning are present, although this combustion is not considered to contribute to a net addition of CO_2 to the atmosphere. Vegetation and biomass burning are also terrestrial surface emission sources of carbon monoxide (CO). It has also been shown that the ozonolysis of unsaturated hydrocarbons like isoprene and ethene also produces CO. In addition, CO can be transported over large distances, and for this reason the fraction of CO from isoprene and ethane to the total CO burden in the atmosphere depends very slightly on altitude (Pfister, et al., 2008). The annual cycle of CO at higher latitudes is strongly driven by the seasonality in OH. Finally, changes in isoprene emissions, for example, have been shown to have a significant impact on tropospheric O_3 production. By evaluating the sensitivity of O_3 production in different regions to changes in concentrations of NO_x and VOC, atmospheric scientists can better assess the changes in atmospheric chemistry that result from changing sources and identify more efficient air pollution control strategies (Wiedinmyer, et al., 2006).

Recall our discussion of the nitrogen cycle in Chapter 1. Unlike oxidized nitrogen and sulfur compounds (NO_x and SO_x), which are predominately emitted from industrial sources, ammonia (NH_3) production is primarily the result of soil and plant processes. Animal sources (e.g., waste houses) from agricultural/animal production processes and biomass burning also contribute to the NH_3 budget. Production of nitrous oxide (N_2O), an important **greenhouse gas**, is largely the result of microbial nitrification and denitrification processes in soil and aquatic systems. N_2O emissions are also the result of the decomposition of animal waste, NH_3 oxidation, and biomass burning. Nitric oxide (NO), although only having a lifetime of approximately one day in the atmosphere, does play a role in the formation of ground-level O_3. Biogenic sources of NO include soils, lightning, and biomass burning. And briefly considering the sulfur family, dimethylsulfide (DMS) and carbonyl sulfide (COS) are emitted into the atmosphere, largely from microbial and outgassing ocean processes, volcanic eruptions, and biomass burning. Both compounds contribute to the tropospheric sulfur burden and particle formation and growth in the atmosphere. Ultimately, biogenic gas exchange processes between the Earth's surface and the atmosphere are dependent upon transport through soils, sediments, and water, and on the production and consumption of gases by microbial and plant processes. Gas-specific emissions are driven largely in part by temperature and solar radiation, thus it is expected that they will be susceptible to changes in climate.

6.5.2. Anthropogenic Sources of Emission

In addition to the biogenic sources cited above, the atmosphere has been significantly impacted by human-induced, anthropogenic sources of pollutant emissions. The largest supply of anthropogenic emissions results from combustion of fossil fuels from both **mobile sources** (e.g., vehicular, airplanes, tractors, and lawn mowers) and **stationary sources** (refineries, chemical plants, natural gas processing plants, factories, and utilities). Such emissions have resulted in rapid changes in atmospheric composition observed since the start of the Industrial Revolution. For example, scientists at the National Oceanic & Atmospheric Administration (NOAA) estimate that the globally monthly mean concentration of CO_2 from fossil fuel combustion has increased from ≈280 ppm (pre-industrial concentration) to just under 390 ppm (post-industrial concentration (2010 estimates)). A recent glimpse at global monthly CO_2 levels averaged over marine surface sites from 2006–2010 (Figure 6.7) reveals the level of increase that is occurring on a relatively short time scale.

As discussed in Chapter 1, CO_2 concentrations are naturally regulated by the carbon cycle, with carbon flux (movement) between the Earth's surface and the atmosphere subjugated by natural processes (e.g., photosynthesis). However, an estimated 3.5 billion metric tons of CO_2 are added to the atmosphere annually from anthropogenic sources, thus creating an imbalance between emissions and absorption. In addition to CO_2, combustion processes emit a number of other pollutants including CO, SO_2, CH_4, NO_x, particulate matter (PM), trace metals, and hydrocarbons. For example, motor vehicle traffic and heavy truck/equipment use are major sources of CO and NO_x. Emissions from electric utilities and

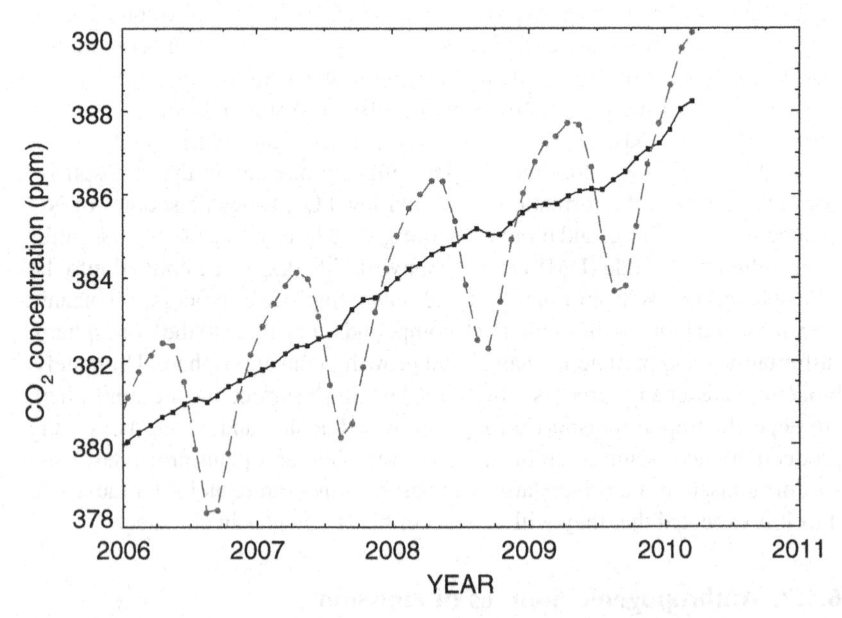

FIGURE 6.7 Recent monthly mean CO_2 levels globally averaged over marine surface sites. The blue line represents the monthly mean values, centered on the middle of each month. The black line represents the same, after correction for the average seasonal cycle. *Data source: (National Oceanographic & Atmospheric Administration, 2010).*

gas-fired boilers constitute a growing contributor of CO, NO_X, and SO_X emissions. As will be discussed below, NO_X and SO_X emissions contribute greatly to the formation of acid precipitation, the formation of aerosols, and play a role in numerous photochemical reactions in both the troposphere and stratosphere. We will also briefly discuss the significance of atmospheric pollution by heavy metals, organic pollutants, and radioactive species.

6.6. TRANSPORT, TRANSFORMATION, AND DEPOSITION PROCESSES: A BRIEF INTRODUCTION

Atmospheric transport of pollutants, their pathways into the biosphere, and their atmospheric residence times depend upon meteorological conditions, topography, and numerous chemical, biological, and physical processes. In addition, such characteristics as physical state, chemical reactivity, and particle size affect their exchange processes and forms deposited. In the troposphere, a wide range of reactions takes place, contributing to the **transformation** of atmospheric pollutants. These pollutants, or their relevant precursors, are typically emitted as gases and interact largely in the gaseous phase. However, significant atmospheric processes such as **deposition** and the formation of aerosols involve the interaction of pollutants in the gas, liquid, and solid phases. For example, aqueous particles may take up soluble gases as well as intermediates from the adjacent gas phase or dissolve material from particles directly (Carter, 1990). Compounds classed as semivolatile compounds (e.g., organochlorine pesticides and polycyclic aromatic hydrocarbons (PAHs)) tend to distribute themselves to various degrees between the gas and particle phases. Aqueous phase reactions may lead to the production of less-soluble products that could be released back to the gas phase and hence change gas-phase composition and oxidation potential (Hartmut, 2003).

TEXTBOX 6.4

Gas Exchange

The exchange of chemical constituents between the gas phase and the Earth's surfaces (e.g., soil, vegetation, and water) is largely controlled by the tendency of the constituents to be absorbed/adsorbed by these surfaces. The ratio of vapor pressure to water solubility can be used an indicator between constituent levels in the troposphere and water surfaces.

The distance of pollutant transport depends primarily on the chemical form of the emitted pollutant, the height of the emission source, chemical and physical processes, and the atmospheric conditions (e.g., wind speeds and composition of oxidizing/reducing species). When studying the global transport of pollutants, scientists rely heavily on Eulerian-type "box" transport

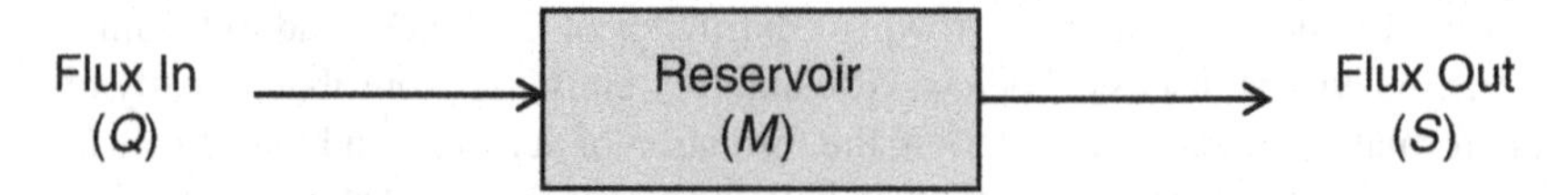

FIGURE 6.8 Generalized Eulerian-type "box" transport model illustrating a representative reservoir and the fluxes that exist between it.

models as initially described in Chapter 4. The principal cycling is illustrated in terms of **reservoirs** (e.g., the atmosphere, the oceans) and the fluxes that exist between them (Figure 6.8). A reservoir is the mass (typically grams) of material (*M*) contained (inventory) in a defined region (e.g., the atmosphere, the oceans, or the lithosphere). The flux (mass/time (g/yr)) of a material into a reservoir is termed a source (*Q*). The flux of material out of a reservoir is termed a sink (*S*). In such models, the overall mass balance is given in simple terms by

$$\frac{dM}{dt} = \sum Q - \sum S, \tag{6.30}$$

where $M =$ the global mass of a given pollutant in the region (e.g., atmosphere). This is the underlying principle for addressing the chemical fate of pollutants in each environmental compartment and for correctly accommodating the inclusion of the key physico-chemical process parameters into any global fate model (Valsaraj and Thibodeauz, 2010).

Note that the change in time in the abundance of a given pollutant inside the simple box must be equal to the difference between sources and sinks (e.g., adsorption onto land and surface waters). Also note that the use of the steady-state assumption (when the sources and sinks are in balance and do not change over time) is common in atmospheric chemistry and reduces differential equations to algebraic equations. One can assume steady state for a given pollutant as long as its source production rate and its **lifetime** (τ) (also termed **residence time**) have both remained relatively constant (Prather, 1994). For example, in the absence of anthropogenic activities, the atmospheric carbon reservoir would be in steady state given the balance between biological respiration/decomposition (flux in) and photosynthesis (flux out). How long will it take to significantly affect the carbon inventory if the flux in or flux out is altered? This question can be answered by examining the residence time in greater detail.

Under steady-state conditions, the residence time (year) in the atmosphere is calculated based on a mass balance equation, defined by

$$\tau = \frac{M}{R}, \tag{6.31}$$

where τ is the residence time, M is total mass (g) of the substance in the atmosphere, and $R =$ the removal rate. Ostensibly, τ can be thought of as the average time spent in a reservoir by an individual atom or molecule. Or, in the case of the lifetime of greenhouse gases, it can be thought of as the approximate time it would take for the anthropogenic augmentation to an atmospheric pollutant concentration to return to its natural level as a result of chemical conversion or being taken out of the atmosphere via a given sink (Finlayson-Pitts and Pitts, 2000). Lifetimes are also calculated on the basis of the substance's reactivity with sink chemicals including the hydroxyl ($OH^{\cdot}$) radical. In this instance, the lifetime can be calculated by dividing the concentration of the OH radical multiplied by the rate constant of the reaction (k) by the concentration of the substance (Godish, 2003). Notice that this calculation is based on chemical kinetics, first touched upon in Chapter 1.

Since air moves rapidly, atmospheric pollutants have the ability to travel long distances quickly and be deposited on distant land and waterbodies. However, pollutants that exhibit short residence times typically only affect local environments. For example, alkyl phenols, commonly observed above receiving wastewater effluents, adhere efficiently to atmospheric aerosols and are quickly removed by rainfall. Those with longer residence times affect air quality at scales from local to continental. Other pollutants (e.g., persistent organic pollutants (POPs)) tend to affect air quality on a global scale. In fact, POP contamination (e.g., PCBs) has been found in relatively pristine Arctic regions—thousands of miles from any known source. Table 6.2 lists residence times and travel distances of selected air pollutants.

TABLE 6.2 Residence Times and Travel Distances of Common Air Pollutants[a]

Pollutant/Category of Pollutants	Residence Time (days)	Travel Distance (km)[b,c]
Carbon monoxide	Min = 1, Max = 4	25,920 (3 months)
Nitrogen oxides	Min = 0.5, Max = 4	216 (3 days)
Organic compounds	Min = 0.5, Max = 14	216
O_3 (tropospheric)	Min = 2, Max = 21	864 (weeks to months)
$PM_{2.5}$	Min = 2, Max = 8	864
PM_{10}	Min = 2, Max = 6	864
Methane	Min = 2400, Max = 3600	1,036,800 (7–10 years)
Sulfur dioxide	Min = 1, Max = 4	432 (3 days)

[a] *Source: Organisation for Economic Co-operation and Development (Online service). Statistical link: http://dx.org/10.1787/256612755564.*
[b] *Representative windspeed = 5 m s^{-1}.*
[c] *The timeframe of distance traveled listed where data were available.*

Example Problem 6.4

The atmospheric reservoir contains roughly 750 Gton carbon (C) in the form of CO_2. If the overall removal of CO_2 at steady state was 75 Gton/yr, calculate the residence time τ of CO_2.

Answer: Recall that at the steady state, flux in = flux out. Therefore,

$$\tau = \frac{M}{P} = \frac{750\ \text{Gton}}{75\ \text{Gton/yr}} = 10\ \text{yr}.$$

6.6.1. Atmospheric Removal Processes

The most effective natural removal processes are the attachment of pollutants to atmospheric aerosols and removal by **dry deposition** and **wet deposition processes** (Figure 6.9). As conceptualized in the figure on the following page, wet deposition is the result of precipitation events (e.g., rain and snow) that remove particles and gases from the atmosphere. Dry deposition is the transfer of particles and gases to the Earth's surface in the absence of notable precipitation. A third process, **cloud deposition**, occurs when pollutant-laden clouds or fog droplets impact vegetation and other surfaces. Both advection and diffusion concepts are involved in deposition events and transport processes. Advection refers to transport with the mean fluid flow (horizontal movement in the atmosphere). In contrast, diffusion transports pollutants through the action of random motions, resulting in smoother concentration profiles. For dry deposition, we must also consider the effects of mountainous regions and areas that contain sharp contrasts in surface structure. Regardless, repeated cycles of transport and deposition can carry pollutants long distances through the atmosphere.

6.6.2. Example Deposition Processes

The physical processes of transport, atmospheric winds, and the formation of clouds and precipitation discussed above strongly influence the patterns and rates of deposition, while chemical reactions typically govern the forms of the compounds deposited. As highlighted below, the detailed study of common deposition processes will allow improved understanding of the temporal and spatial evolution of atmospheric chemistry and strengthen the concept of using such processes as indicators to evaluate natural and anthropogenic influences.

Acid Deposition

There is a strong relationship between emissions and the deposition of acid-forming substances that acidify tropospheric air. As generalized in Figure 6.9, acid substances in the troposphere occur from two main formation processes: sulfuric acid through oxidation of SO_2 and of nitric acid through the oxidation of NO_x. These formations are controlled by both gas- and liquid-phase

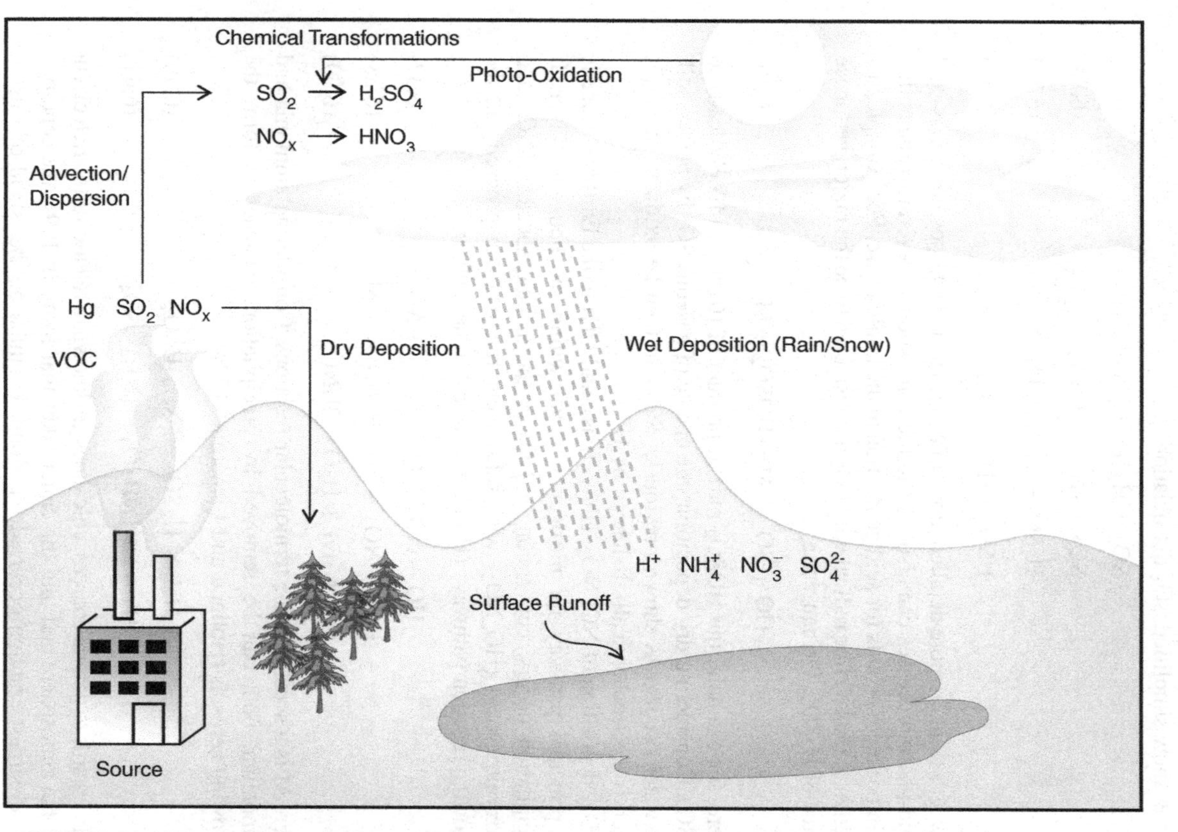

FIGURE 6.9 Schematic representation of natural removal processes including dry deposition and wet deposition processes.

chemical processes. If we dig a bit deeper, we can confirm that SO_2, SO_3, NO, and NO_2 are all involved in forming acid substances. Although SO_x and NO_x are acid anhydrides, acids are generated when dissolved in water. For example, we can show simplified SO_x dissociation:

$$SO_2 + H_2O \rightarrow H_2SO_3 \quad (6.32)$$

$$SO_3 + H_2O \rightarrow H_2SO_4 \quad (6.33)$$

$$H_2SO_4 \rightarrow H^+ + HSO_4^- \quad (6.34)$$

$$HSO_4^- \rightarrow H^+ + SO_4^{2-}. \quad (6.35)$$

Now let's think in more detailed terms of reactions in the troposphere. Gas- and liquid-phase processes can also lead to the generation of several oxidants that are important reactants for acid generation in the liquid phase. For example, the reaction of the HO free radical with SO_2 is considered to be the major gas-phase source of H_2SO_4. We can write

$$HO + SO_2 + M \rightarrow HOSO_2 + M, \quad (6.36)$$

where, for example, the $HOSO_2$ radical product of (6.36) ultimately leads to H_2SO_4. Considering dry deposition, the update of gaseous SO_2 by plants, soil, and water occurs. As shown previously, SO_2 can then be oxidized to SO_4^{2-}, which is electrochemically balanced by H^+.

Dry deposition of NO, NO_2, HNO_3 (gas), and NH_3 to the Earth's surface is the principal pathway for nitrogen removal within the boundary layer. If precipitation occurs, conversion of NO_x to the soluble species HNO_3 and peroxynitric acid (HO_2NO_2) with subsequent removal can proceed via the following reaction pathways (Holland and Lamarque, 1997):

$$HO + NO_2 + M \rightarrow HNO_3 + M \quad (6.37)$$

$$HO_2 + NO_2 + M \rightarrow HO_2NO_2 + M \quad (6.38)$$

$$N_2O_5 + H_2O \rightarrow 2HNO_3, \quad (6.39)$$

where M is a species in the termolecular reactions. Considering ammonia and ammonium, both can be removed by precipitation, or react to form the following aerosols (Holland and Lamarque, 1997):

$$NH_3 + H_2SO_4 \rightarrow (NH_4)HSO_4 \quad (6.40)$$

$$NH_3 + (NH_4)HSO_4 \rightarrow (NH_4)_2SO_4. \quad (6.41)$$

In regards to natural sources, volcanoes are known to release sulfur-rich ejecta into the atmosphere and into the crater lakes that form. In 1991, the century's second largest eruption occurred on Mount Pinatubo on the island of Luzon, Philippines. Accurate predictions by local volcanologists, combined with organized evacuations carried out by the authorities, saved many lives, but it still wrought havoc in the surrounding productive and populated areas by destroying thousands of buildings and rendering hundreds of square kilometers of prime agricultural lands barren overnight. The eruption is considered to be about 10 times larger than that of Mt. St. Helens in 1980. It was felt worldwide, ultimately releasing 17 million tons of SO_2 into the atmosphere—the largest volume ever recorded by modern instruments and injecting more dust and aerosols into the stratosphere since the Krakatoa eruption in 1883. Oxidation of the SO_2 from Pinatubo formed a global layer of sulfuric acid droplets, and, when combined with the dust ejecta, ultimately contributed to the lowering of global temperatures by about 0.5 °C, and ozone depletion temporarily increased substantially. After eruptions ended, a crater lake formed in the 1991 caldera. At first, the lake was small, hot, and highly acidic, with a minimum pH of 2 and a temperature of about 40°C. Abundant rainfall cooled and diluted the lake, lowering the temperature to 26°C and raising the pH to 5.5 by 2003 (Stimac, et al., 2004).

Mercury Deposition

Concerns over mercury (Hg) toxicity are not limited to isolated events. In fact, widespread atmospheric transport and wet/dry deposition of Hg has resulted in global contamination of sensitive ecosystems and their inhabitants. For example, monitoring studies concerning Lake Michigan and more pristine, inland lakes have revealed elevated levels of all three major forms of Hg (e.g., Landis and Keeler, 2002). These include Hg(0) (elemental Hg), RGM (reactive gaseous Hg, including, e.g., Hg(II)), and Hg(p) (particulate Hg). Nearly 95% of the total Hg in the atmosphere is in the form of H(0), which has a characteristic long atmospheric lifetime (~0.5–1 yr). Elemental mercury fluxes are the result of industrial activity, biological processes, and redox processes in terrestrial as well as aquatic environments. Although generally not susceptible to wet deposition processes, H(0) can be oxidized to RGM by O_3, H_2O_2, Cl_2, and OH species in the troposphere from oxidation by OH and halogen atoms. This involves a two-step mechanism,

$$Hg(0) + A \rightarrow HgA \tag{6.42}$$

$$HgA + B \rightarrow HgAB, \tag{6.43}$$

where A = e.g., Br, Cl, or OH and B = e.g., Br, Cl, or OH. For example, in the polar atmosphere, gaseous H(0) can be oxidized to a highly reactive form of mercury by bromine:

$$Hg(0) + Br \rightarrow HgBr \tag{6.44}$$

$$HgBr + OH \rightarrow HgBrOH, \tag{6.45}$$

which is released from sea-ice surfaces. Elemental mercury can also react with O_3:

$$Hg(0) + O_3 \rightarrow HgO + O_2. \tag{6.46}$$

In contrast, gaseous divalent Hg has been shown to be readily scavenged by precipitation. RGM demonstrates more local and regional effects and has a relatively short lifetime ($\sim$1 week or less). Hg(p) has also been shown to be influenced by wet deposition processes due to scavenging by cloud and precipitation occurrences. Hg(p) refers to Hg compounds that are associated with atmospheric particles, not pure particles of Hg. It also has local and regional effects with a moderate atmospheric lifetime ($\sim$1–2 weeks). All three forms of Hg can be dry-deposited with the rates dependent upon surface characteristics and meteorological conditions. Ultimately, the type of mercury emitted by sources dictates the impact this element has on the environment. Dedicated sources include waste incineration, coal-powered electrical generation, metallurgical processes, manufacturing, and other fuel combustion processes.

Deposition of Radionuclides

Recall our discussion in Chapter 4 on the alpha decay of Radium-226 to Radon-222 and its ability to be distributed through soils. Much of the Radon-222 concentration emanates from the soils to the atmosphere, with additional sources emanating from groundwater, and from a lesser extent, the oceans (Appleby and Piliposian, 2010). As was shown in Figure 4.10, Radon-222 decays via a number of short-lived isotopes to Lead-210. This readily occurs in the atmosphere, although Lead-210 has a relatively short lifetime. Individual Lead-210 atoms become attached to airborne particulate matter and are removed both by wet and dry deposition. An illustrated example of the global balance of Radon-222 and Pb-210 in the atmosphere is shown in Figure 6.10. To extend our discussion above, the Radon-222 inventory is balanced by the emission rate from soils (labeled as A_LF) and the rate of decay of Pb-210. Considering Pb-210, its inventory is balanced by the rate of production by Radon-222 decay and the rate of loss (A_EP) by fallout back to the Earth's surface. Ultimately, the distribution of Radon-222 entering the atmosphere from dedicated sources is influenced by advection, diffusion, and radioactive decay processes.

Deposition of Polycyclic Aromatic Hydrocarbons

Polycyclic aromatic hydrocarbons (PAHs) are produced from the incomplete combustion of fossil fuels and biomass burning. These compounds (see Figure 6.11 for common PAHs and their structural characteristics) are readily adsorbed to atmospheric particles and are eventually transported to the Earth's surface. As shown, these suspected carcinogens are organic compounds that typically contain two or more fused benzene and have been well-characterized in surface waters. Early-life exposure to PAHs adsorbed to fine-particulate matter has been linked to chronic respiratory conditions, including asthma. Photodegradation is an important sink for polycyclic aromatic. Studies have identified polar PAH photodegradation products, collectively known as polycyclic aromatic compounds (PACs), formed in the presence of various air pollutants such as SO_2, O_3, and NO_2. Their transport and deposition are

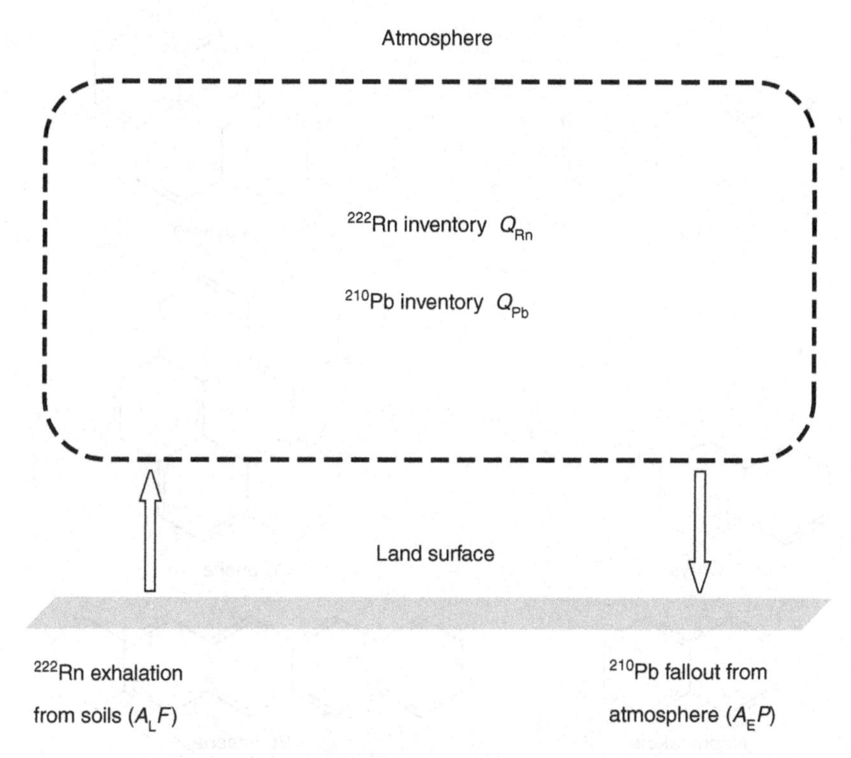

FIGURE 6.10 An illustrative example of the global balance of Radon-222 and Pb-210 in the atmosphere. *Modified from (Appleby and Piliposian, 2010) with permission from ILM Publications.*

influenced by advection, diffusion, and topographic barriers. PAH sources in ecosystems range from local point sources to diffuse regional and global sources.

Benzo[*a*]pyrene (BaP) and its photodegradation products have been extensively studied and are known to be carcinogenic. In fact, it is frequently used as a biomarker for combustion source air pollution in epidemiological studies and for risk assessment (Cohen, 2000). More specifically, they are used as biomarkers of exposure, in particular DNA adducts, as well as markers of early damage, including mutagenicity and cancer. As shown in Table 6.3, the principal natural sources of BaP in the United Kingdom, for example, are natural fires and open agricultural burning, while anthropogenic sources include the incomplete combustion of fossil fuels, coke oven emissions, aluminum smelters, and vehicle exhausts. Fortunately, the development of sensitive analytical techniques that specifically identify BaPs and its proposed photodegradation products in complex environmental and laboratory samples are allowing for more specific determination of the locations, lifetimes, and potential hazards created by having these compounds in environmental systems (Koeber, et al., 1999).

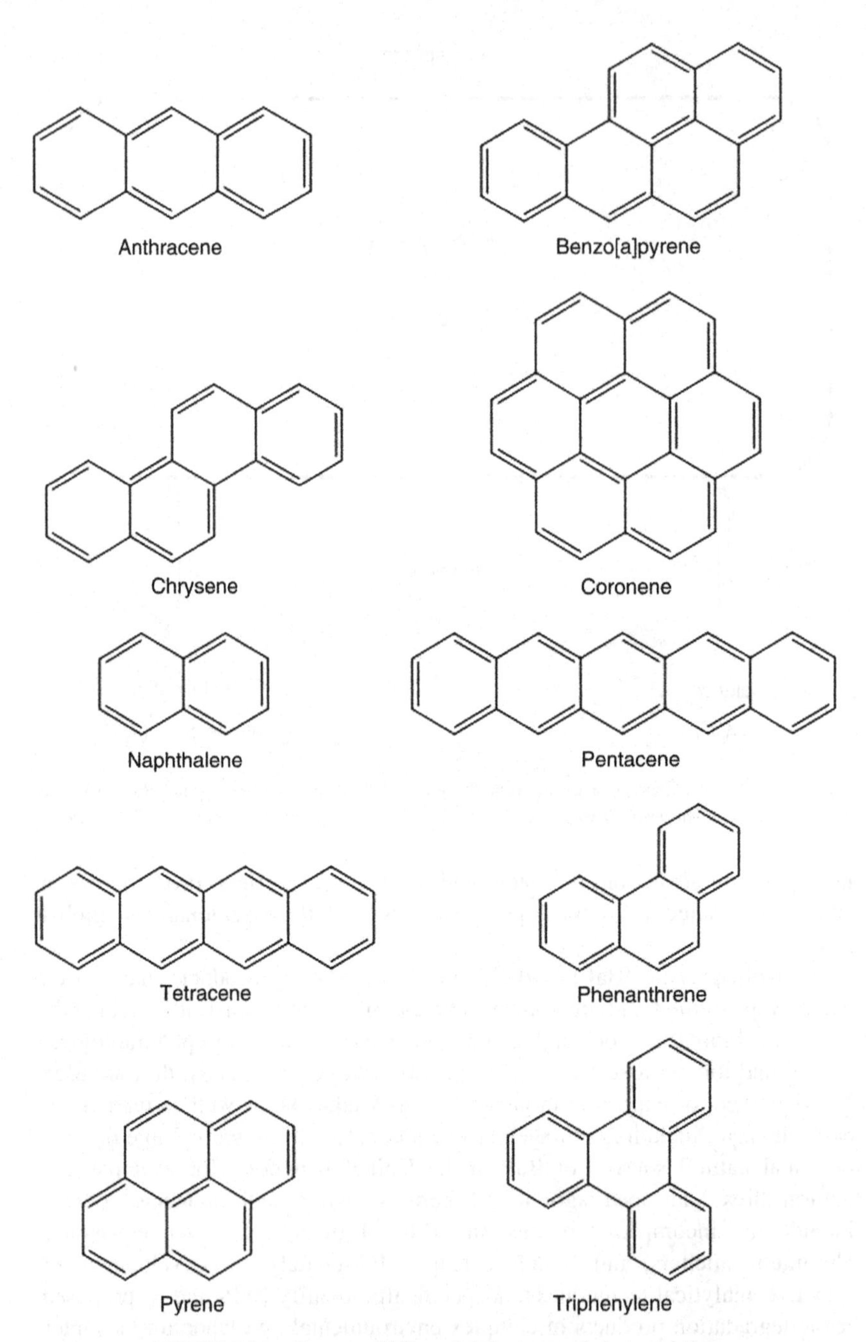

FIGURE 6.11 Common PAHs and their structural characteristics.

TABLE 6.3 Representative Benzo[*a*]pyrene Emission Sources in the United Kingdom 1990–2010

Sources	1990 (Tons)	1995 (Tons)	2010 (Estimated) (Tons)
Vehicles (gasoline and diesel)	9.9	7.7	4.0
Natural fires/agricultural burning	31	2.9	2.9
Aluminum production	1.9	1.4	0.03
Coke production	1.3	1.1	1.1
Industrial coal combustion	6.3	4.9	3.8
Creosote use	0.06	0.06	0.06
Domestic wood combustion	1.2	1.2	1.2

Data source: European Commission, Working Group on Polycyclic Aromatic Hydrocarbons, PAH Position Paper Annexes, July 27, 2001.

6.7. GLOBAL CLIMATE CHANGE: A PRIMER

As discussed in Chapter 1, life on Earth depends on energy emanating from the sun. Roughly half of the energy in the form of light reaching the Earth's atmosphere passes through the air and clouds to the surface. Roughly one-third of the solar energy that reaches the top of Earth's atmosphere is reflected directly back to space. The remaining two-thirds is absorbed by the Earth's surface, which is warmed to sustain life as we know it. In effect, a natural atmospheric "greenhouse" is produced. Without the natural **greenhouse effect**, the average temperature at the Earth's surface would be lower than the freezing point of water. Currently, many scientists agree that human activities, primarily the burning of fossil fuels and clearing of forests, are believed to greatly intensify the natural greenhouse effect, causing global temperatures to increase, generally referred to by many as global warming. However, as will be evident in our brief coverage, the causes and consequences of changing the natural atmospheric greenhouse are debatable and difficult to predict.

The greenhouse effect is the result of complex chemical and physical processes including direct radiation absorption and indirect radiative forcing when chemical transformations of a substance produce additional greenhouse gases, when a gas influences the **atmospheric lifetimes** of other gases, and/or when a gas affects atmospheric processes that alter the radiative balance of the Earth (U.S. Environmental Protection Agency, 2003). As was introduced in Example Problem 1.2, increased water vapor (the most dominant greenhouse gas) could lead to positive temperature **feedback**. This may in turn lead to erratic atmospheric circulation and thus result in altered precipitation patterns.

In contrast, long-lived gases, which do not respond physically or chemically to changes in temperature, are described as "forcing" climate change. Direct greenhouse gases include CO_2, CH_4, O_3, and N_2O, which do occur naturally in the atmosphere. Several classes of halogenated substances that contain fluorine, chlorine, or bromine are also considered greenhouse gases, but for the most part, are largely produced by human activities. These include chlorofluorocarbons (CFCs), hydrochlorofluorocarbons (HCFCs), hydrofluorocarbons (HFCs), perfluorocarbons (PFCs), and sulfur hexafluoride (SF_6). Gases that likely have an indirect global warming effect include CO, NO_x, and non-CH_4 volatile compounds (NMVOCs). Aerosols such as sea salt, industrial dust, those produced by SO_2, or elemental carbon emissions have also been shown to affect the absorptive characteristics of the atmosphere.

The Intergovernmental Panel on Climate Change (IPCC) developed a **Global Warming Potential (GWP)** concept to compare the ability of each greenhouse gas to trap heat in the atmosphere relative to other gases (IPCC, 2006). The GWP of a greenhouse gas is defined as the ratio of the time-integrated radiative forcing from the instantaneous release of 1 kilogram (kg) of a trace substance relative to that of 1 kg of a reference gas (over a 100-year period) (Shine, et al., 2005). The reference gas chosen by the IPCC was CO_2, and therefore GWP-weighted emissions are measured in teragrams of CO_2 equivalents (Tg CO_2 Eq.). The GWP ultimately depends on the time spent in the atmosphere by the gas (atmospheric lifetime), and on a gas's capacity to affect radiation, which describes the instantaneous effects on overall radiation of a rise in concentration of that gas. A unit that is often reported is MMTCDE, or million metric tons of carbon dioxide equivalents.

A list of representative greenhouse gases, their atmospheric lifetimes, and their GWPs is presented in Table 6.4. No GWP values are attributed to gases that are short-lived and spatially inhomogeneous in the atmosphere (e.g., water vapor, CO, and O_3). Interestingly, although CO_2 is one of the most prevalent

TABLE 6.4 GWP Values for Representative Greenhouse Gases

Gas	Atmospheric Lifetime	GWP
CO_2	50–200	1
CH_4	12–15	21
N_2O	120	310
HFC-23	264	11,700
CF_4	50,000	6,500
SF_6	3,200	23,900

Data source: (Landis and Keeler, 2002).

greenhouse gases, it demonstrates the least powerful GWP (1). Notice that CH_4 has a GWP of 21. This means that the emission of one Tg of CH_4 is the same as the emission of 21 Tg of CO_2 over a 100-year period.

A February 2002 report by the United Nations (UN) scientific panel investigating climate change declared that the world is "unequivocally" warming, and the warming is likely the result of emissions of carbon dioxide (CO_2) and other greenhouse gases from human-related activities. Scientists and researchers who participated in the 2008 International Conference on Climate Change in New York City, however, suggest that scientists should distance themselves from politically-motivated claims of "scientific consensus" on the causes of global warming. They believe that the high level of certainty claimed by the UN report cannot be justified based on climate models and current scientific evidence. So who is right? The answer to this question may well be answered with the help of statistics, ostensibly by determining how reliable the empirical analyses are upon which the evidence of global warming is founded.

Ultimately, several components of the climate system, notably the oceans and living organisms, affect atmospheric concentrations of greenhouse gases. A principal example of this was highlighted in Chapter 1, whereby discussion was focused on plants taking CO_2 out of the atmosphere and converting it into carbohydrates via photosynthesis. And given the natural variability of the Earth's climate, it is difficult to determine the extent of change that humans cause. Based on current computer-based models, rising concentrations of greenhouse gases generally produce an increase in the average global surface temperature on Earth. Rising temperatures have been postulated to produce changes in weather, sea levels, and land use patterns. Consider global sea surface temperature (SST) data, for example. Global SST is a major component of the observational record to model and assess climate change (Shine, et al., 2005). As with most environmental data sets, SST data are based on annual or monthly measurement series at given spatial levels (e.g., regional, country, or global). This data can be highly variable, both spatial and temporally, often introducing distortions. Not all monitoring stations provide complete records, with measurements often varying within a few degrees of latitude change. Proper statistical analysis of trends in the global data is thus paramount. This involves the search for a statistically-significant positive trend in the data, which is likely interpreted as evidence of global warming. Few studies, however, assess the statistical evidence for trends in global temperatures present in the results. There is thus a need to employ robust statistical methods to re-examine global SST data.

6.8. END OF CHAPTER PROBLEMS

6.1 Define the term "atmospheric boundary layer" and describe its role in pollutant dispersion.

6.2 Describe the conditions that lead to an unstable boundary layer.

6.3 The pressure at the surface of the Earth = 1.01325×10^5 Pa. Convert this to units of millibar (mbar) and report its new value.
6.4 The torr (Torr) is a non-SI unit of pressure with a ratio of 760 to 1 standard atmosphere and is approximately equal to the fluid pressure exerted by a millimeter (mm) of mercury (Hg) (1 Torr ≅ 1 mm Hg). If 1 mm Hg = 134 Pa, how many Torrs are there in 1013.25 mbar?
6.5 In 1999, scientists utilized a Global Positioning System (GPS) satellite on top of the world's highest mountain, Mount Everest, to reveal a revised elevation of 29,035 feet (8,850 meters). Calculate the estimated pressure (in bars) at this height.
6.6 Refer back to the Chapman mechanism provided in section 6.2. Comment on both the reactivity and stability of M and O_3 species. How do they differ? Are they dependent upon one another?
6.7 Provide explanation on why the cycle of reactions in the Chapman mechanism would lead to a slight warming of the stratosphere.
6.8 As discussed, the concentration of O_3 increases with altitude and so the concentration of O_3 is higher in the stratosphere than in the troposphere. Explain how this relates positively to the balance of life on the Earth's surface.
6.9 Visit the UV Index forecast map link provided in Textbox 6.1. Enter your zip code or city name and state and log the UV Index forecast for your community for seven days. Graph the UV Index forecast versus time (days). How does the UV index vary day-to-day? Record any noticeable trends or significant observations.
6.10 Consider the O-O_2 bond in ozone (O_3). Given that the enthalpy change for this reaction = $\Delta H = 105\ \text{kJ mol}^{-1}$, calculate how much energy is required to break (dissociate) this bond in kcal mol^{-1}.
6.11 Extending problem 6.10, calculate the λ_{max} needed for photodissociation of the O-O_2 to take place. Comment on the type of electromagnetic radiation required for this process.
6.12 For Example Problem 6.2 housed in the chapter text, provide a net reaction of O_3 to O_2.
6.13 Global anthropogenic activities are estimated to release approximately 30 Gtons of CO_2 per year to the atmosphere. It is estimated that 40% of this is absorbed, but the rest remains in the atmosphere. Compare this to Example Problem 6.3, where the flux in (75 Gtons) = flux out (75 Gtons).
   a) Does the steady-state approximation still apply after the anthropogenic addition?
   b) Calculate the overall mass balance in the atmosphere (reservoir) taking into consideration the anthropogenic influence (flux in).
6.14 Speculate on why persistent organic pollutants (POPs, for example, PCBs) have been found in relatively pristine Arctic regions despite the fact that known sources are thousands of miles away.

6.15 Based on our discussion of simple box models and knowledge of reaction rates, discuss ways in which temperature may affect the transport (particularly travel distance) of organic pollutants.

6.16 Like we have shown for SO_x, NO_x can dissolve in water to form acids that release H^+ ions and contribute to acid deposition. Show the simplified dissociation of NO_2 in moist air (H_2O). Hint: Oxygen plays a role in this process.

6.17 Review Equation 6.36 and our discussion of the reaction of the HO free radical with SO_2 in the troposphere. Propose a reaction mechanism for the ultimate formation of H_2SO_4.

6.18 List the various chemical and physical processes that dictate the content and patterns of atmospheric depositions.

6.19 Briefly describe what role temperature inversion has on pollutant transport. Provide an example situation (including geographic location) where temperature inversion may be physically evident.

6.20 Describe the factors that influence the rates of both wet and dry deposition.

6.21 In the atmosphere, elemental mercury Hg(0) can form a reactive species with Cl as follows:

$$Hg(0) + Cl \rightarrow HgCl_2$$

Produce a chemical reaction that results when $HgCl_2$ reacts with a charged Cl^-.

6.22 Deposition of sulfur compounds has been shown to be higher in the eastern portion of the United States. Provide reasoning for this occurrence.

6.23 Referring back to problem 6.17, comment on whether the same trend would likely apply for nitrogen deposition.

6.24 Although the atmospheric lifetime of N_2O falls in the range of CO_2, provide reasoning as to why its Global Warming Potential (GWP) is higher than CO_2.

6.9. REFERENCES

Appleby, P. G., & Piliposian, G. (2010). The origins of 210Pb in the atmosphere and its deposition on and transport through catchment lake systems. In G. Hanrahan (Ed.), *Modeling of pollutants in complex environmental systems, Volume II*. St. Albans, UK: ILM Publications.

Barry, R. G. (1987). *Atmosphere, weather, and climate* (5th ed.). London: Methuen Publishing.

Benkovitz, C. M., Schwartz, S. E., Jensen, M. P., Miller, M. A., Easter, R. C., & Bates, T. S. (2004). Modeling atmospheric sulfur over the northern hemisphere during the aerosol characterization experiment 2 experimental period. *Journal of Geophysical Research, 109*, 2004.

Carter, W. P. L. (1990). A detailed mechanism for the gas-phase atmospheric reactions of organic compounds. *Atmospheric Environment, 24A*, 481–515.

Cohen, A. J. (2000). Outdoor air pollution and lung cancer. *Environmental Health Perspectives, 108*, 743–750.

Elperin, T., Kleeorin, N., & Rogachevskii, I. (2000). Mechanisms of formation of aerosol and gaseous inhomogeneities in the turbulent atmosphere. *Atmospheric Research, 53*, 117–129.

Finlayson-Pitts, B. J., & Pitts, J. N. (2000). *Chemistry of the upper and lower atmosphere*. San Diego: Academic Press.

Godish, T. (2003). *Air quality* (4th ed.). Boca Raton, FL: CRC Press.

Hartmut, H. (2003). Kinetics of aqueous phase reactions relevant for atmospheric chemistry. *Chemical Reviews, 103*, 4691–4716.

Holland, E. A., & Lamarque, J.-F. (1997). Modeling bio-atmospheric coupling of the nitrogen cycle through NOx emissions and NOy deposition. *Nutrient Cycling in Agroecosystems, 48*, 7–24.

IPCC (2006). *2006 IPCC guidelines for national greenhouse gas inventories*. The national greenhouse gas inventories programme, the intergovernmental panel on climate change. In H. S. Eggleston, L. Buendia, K. Miwa, T. Ngara, & K. Tanabe (Eds.), Hayama, Kanagawa, Japan.

Kasting, J. F. (1993). Earth's early atmosphere. *Science, 259*, 920–926.

Koeber, R., Bayona, J. M., & Niessner, R. (1999). Determination of benzo[*a*]pyrene diones in air particulate matter with liquid chromatography mass spectrometry. *Environmental Science & Technology, 33*, 1552–1558.

Landis, M. S., & Keeler, G. J. (2002). Atmospheric mercury deposition to Lake Michigan during the Lake Michigan mass balance study. *Environmental Science & Technology, 36*, 4518–4524.

National Oceanographic & Atmospheric Administration. <http://www.esrl.noaa.gov/gmd/ccgg/trends/> Accessed 29.05.2010.

Pfister, G. G., Emmons, L. K., Hess, P. G., Lamarque, J.-F., Orlando, J. J., Walters, S., Guenther, A., Palmer, P. I., & Lawrence, P. J. (2008). Contribution of isoprene to chemical budgets: A model tracer study with the NCAR CTM MOZART-4. *Journal of Geophysical Research, 113*, 1–21.

Prather, M. J. (1994). Lifetimes and eigenstates in atmospheric chemistry. *Geophysical Research Letters, 21*, 801–804.

Seinfeld, J. H., & Pandis, S. N. (2006). *Atmospheric chemistry and physics: From air pollution to climate change*. Hoboken, NJ: John Wiley & Sons.

Sharkey, T. D., Wiberley, A. E., & Donohue, A. R. (2008). Isoprene emission from plants: Why and how? *Annals of Botany, 101*, 5–18.

Shine, K. P., Fuglestvedt, J. S., Hailemariam, K., & Stuber, N. (2005). Alternatives to the global warming potential for comparing climate impacts of emissions of greenhouse gases. *Climate Change, 68*, 281–302.

Stimac, J. A., Goff, F., Counce, D., Larocque, A. C. L., Hilton, D. R., & Morgenstern, U. (2004). The crater lake and hydrothermal system of Mount Pinatubo, Philippines: Evolution in the decade after eruption. *Bulletin of Volcanology, 66*, 149–167.

Stull, R. B. (1988). *An introduction to boundary layer meteorology*. Dordrecht: Kluwer Academic Publishers.

Stutz, J., Alicke, B., Ackermann, R., Geyer, A., White, A., & Williams, E. (2004). Vertical profiles of NO_3, N_2O_5, O_3, and NO_x in the nocturnal boundary layer: 1. Observations during the Texas air quality study. *Journal of Geophysical Research, 109*, D12306.

U.S. Environmental Protection Agency. (2003). *Inventory of U.S. greenhouse gas emissions and sinks: 1990–2001*. EPA 430–R–03–004. Washington, D.C: U.S. Environmental Protection Agency. April 15, 2003.

Valsaraj, K. T., & Thibodeauz, L. J. (2010). On the physiochemical aspects of the global fate of long-range atmospheric transport of persistent organic pollutants. *Journal of Physical Chemistry Letters, 1*, 1694–1700.

Wiedinmyer, C., Tie, X., & Guenther, A. (2006). Future changes in biogenic isoprene emissions: How might they affect regional and global atmospheric chemistry? *Earth Interactions, 10*, 1–19.

Zawar-Reza, P., Sturman, A., & Khan, B. (2010). Airshed modeling in complex terrain. In G. Hanrahan (Ed.), *Modeling of pollutants in complex environmental systems* (pp. 438). St. Albans, UK: ILM Publications.

Chapter 7

Abstracts and Keywords

Abstract

This chapter presents a more detailed investigation of atmospheric pollutants including the identification of sources and precursor compounds. We begin with a primer on indoor air pollution and its correlation to ambient levels. Next, both primary and secondary contributions and formation of ambient particulate matter are presented. Following this, the photochemical mechanism of ozone formation is presented, including the roles of volatile organic and nitrogen compounds. Next, the formation of peroxyacetyl nitrate (PAN), a compound involved in summer smog formation, is detailed. Subsequent detail on the formation conditions for photochemical smog and its related health effects is covered. Following this, information on example heterogeneous atmospheric reactions is presented. Finally, a short section will involve discussion on predictive atmospheric models.

Keywords

Air pollutants, Indoor air pollution, Particulate matter, Secondary organic aerosols (SOA), Photochemical smog, Ozone formation, Heterogeneous reactions

Chapter 7

Air Pollutants and Associated Chemical and Photochemical Processes

Environmental experts warn that anywhere from 80–90% of 30 world heritage sites around China are being menaced by air pollution, with the biggest threats coming from SO_2 emissions and photochemical smog production (Carmichael, et al., 2002). When temperature inversions limit the mixing layer above Beijing, the iconic structures of the Forbidden City appear as ghostly silhouettes (Figure 7.1). Limited visibility may be the most immediate indicator of China's poor air quality, but by far the greatest damage done by polluted air is to human health, to plants, to the Imperial architecture, and to economic activity.

FIGURE 7.1 Photochemical smog covering the Forbidden City (Imperial Palace), Beijing, China.

Researchers are taking both standard and non-standard approaches to assessing the health and economic impacts of such pollution episodes. They are attempting to identify all sources and atmospheric concentration of offending primary and secondary pollutants in the region through comprehensive monitoring and modeling activities. Environmental modeling efforts have been particularly useful and are a valid way to estimate and predict local population exposures, economic costs, and also provide an avenue for policy development to aid in controlling pollution episodes.

7.1. INTRODUCTION

In Chapter 6, a brief description of air pollutants and their reaction mechanisms were covered in relation to physical and chemical atmospheric processes. In this chapter, a more detailed investigation of both primary pollutants and secondary pollutants will be provided. A **primary pollutant** is one that is emitted into the atmosphere directly from the source of the pollutant while still retaining much of its original chemical form. For example, SO_2, one of a group of highly reactive gases known as oxides of sulfur, is directly emitted from fossil fuel combustion processes (e.g., power plants and industrial facilities) and other processes (e.g., metal ore extraction). Related, nitrogen dioxide falls under a group referred to in previous chapters as oxides of nitrogen (NO_X). For example, NO_2 forms directly from emissions from cars, trucks, buses, power plants, and off-road equipment. Table 7.1 shows a complete list of primary pollutants, their sources (both natural and anthropogenic), and potential human health effects. **Secondary pollutants** are formed by the lower atmospheric reactions of precursors or primary pollutants where a defined chemical change results. For example, O_3 is formed from photochemical reactions of NO_x and VOCs. Other examples include acid droplet formation from SO_2 and NO_2 and organic aerosol formation from VOCs in gas-to-particle reactions. Thus, the proper identification of precursor compounds and the elucidation of underlying reaction mechanisms are essential in any secondary pollutant investigation and/ or mitigation procedure.

7.1.1. Indoor Air Pollution: A Primer

Although the primary focus of this chapter is on ambient air pollution, a brief examination of **indoor air pollution** is warranted given the increasing relevance of such hazards on human health, and their possible correlation to ambient levels. This is especially true for rural locations in developing countries where wood and coal are used for heating, and in large metropolitan areas around the globe where residents are exposed to ambient particles modified by building structure and air ventilation (Yli-Tuomi, et al., 2008). Using real-time particle monitors, scientists are able to simultaneously monitor both outdoor and indoor

TABLE 7.1 Primary Air Pollutants, Common Sources, and Potential Health Effects

Pollutant	Natural Sources	Anthropogenic Sources	Human Health Effects
Carbon monoxide (CO)	Volcanic eruptions, forest fires, photochemical reactions in the troposphere, decaying organic matter.	Motor vehicle exhaust, electrical generating facilities, and other incomplete combustion processes.	Blocks the ability of hemoglobin to transport oxygen to the cells of the body (acute). At moderate concentrations, angina, impaired vision, and reduced brain function may likely result.
Nitrogen oxides (NO_x)	Lightning storms and biological decay processes.	Primarily from fossil fuel combustion in electrical utilities, high-temperature operations, and operation of motor vehicle exhaust.	Eye, skin, and respiratory tract irritation. Pulmonary edema, bronchitis, bronchiolitis, and emphysema in moderate to high levels.
Sulfur oxides (SO_x)	Biological decay processes, volcanic eruptions.	Burning of fuel (e.g., coal and oil) containing sulfur and during metal smelting and other industrial processes.	Respiratory illness, alterations in the lungs' defenses, and aggravation of existing heart or lung disease in susceptible populations.
Particulate matter (PM)	Pollen, bacterial, and fungal spores, dust, soot ash.	Electricity generation, industrial processes, fossil fuel combustion, waste disposal.	Decreased lung function, irritation of the airways, aggregated asthma, and heart problems. Small particles less than 10 μm in diameter pose the greatest risk.
Volatile organic compounds (VOCs)	Trees and other plants emit hydrocarbons (terpines).	Solvent use, vehicular exhaust, industrial processes.	Eye, nose, and throat irritation, headaches, loss of coordination, damage to liver, kidney, and central nervous system. Some are suspected or known to cause cancer in humans.

environments to quantity $PM_{2.5}$ concentrations. For example, studies have demonstrated that ammonium nitrate aerosol undergoes **phase transitions** during transport into and within residences, shifting the equilibrium of the nitrate formation reaction (Hering and Avol, 1996; Lunden, et al., 2003),

$$HNO_{3(g)} + NH_{3(g)} \leftrightarrow NH_4NO_{3(aq)},$$

to the left toward the gas phase. These gases are known to rapidly diffuse toward and absorb to surfaces inside the residence. This further drives the equilibrium toward the gas phase. After taking gas-phase measurements of ammonia (NH_3) and nitric acid (HNO_3) inside residences, investigators found a sizeable amount of remaining ammonia with low to undetectable levels of nitric acid (Lunden, et al., 2003). Shortly, we will discuss phase transitions of water molecules in the atmosphere, including discussion on humidity, clouds, and water vapor. The latter, in particular, affects the phase of other atmospheric particles (e.g., aerosols).

Example Problem 7.1

Describe possible factors that affect the fate of NH_4NO_3 as described in the indoor air example above.

Answer: In general, phase transitions in a given system are the result of some external condition such as temperature, pressure, and other factors. NH_4NO_3 is a chemically-reactive species, and its indoor concentration will depend on conditions such as temperature, relative humidity, and gas concentrations inside the residence.

As evident from the discussion above, indoor air pollutants arise from indoor and outdoor sources. Indoor sources specifically include those from combustion (e.g., tobacco products, oil, wood, and coal), building materials (e.g., asbestos-containing insulation and pressed wood products), household cleaning products, radon infiltration, and those biologic (e.g., mold, dust) in their makeup. For example, both the U.S. EPA and the UK's Environment Agency identify secondhand cigarette smoke as a major source of indoor air pollution. The chemical composition and number of toxic and carcinogenic agents in cigarette smoke may surprise you. In fact, a study in 1996 identified 4,800 compounds in cigarette smoke (Green and Rodgman, 1996). A representative list according to functional group is provided in Table 7.2. The carcinogenic potential of these compounds has been assessed, with the International Agency for Research on Cancer identifying 11 proven human carcinogens and another eight that are likely to be carcinogenic in humans. For example, chronic inhalation of formaldehyde from burnt cigarettes can lead to tissue damage and cancer of the upper respiratory tract. There are also many opportunities for exposure to carbon monoxide (CO) in indoor environments given the multitude of combustion processes that occur on a regular basis.

TABLE 7.2 A Representative List of Chemical Compounds in Tobacco Smoke

Functional Groups	Number of Compounds in Tobacco Smoke
Alcohols	157
Aldehydes	106
Aliphatics (saturated)	113
Aliphatics (unsaturated)	178
Amides and imides	227
Amines	150
Anhydrides	10
Carboxylic acids	69
Esters	456
Inorganics and metals	111
Ketones	461
Lactones	135
Monocyclic aromatics	138
Nitriles	101
N-nitrosamines	18
Pesticides	25
Phenols	188
Polycyclic aromatics	317
Pyrazines	55
Pyrroles and indoles	88
Sulfur compounds	37

Data source: (Hoffmann, et al., 2001).

Inhalation of CO (even at low concentrations) can result in deleterious health effects. Acute effects of CO are due to its striking affinity for hemoglobin and resulting impairment of oxygen transport. At moderate concentrations, angina, impaired vision, and reduced brain function may result. At higher concentrations, CO exposure can be fatal. More specifically, exposure to an atmosphere of only 0.1% has been shown to convert over half of the hemoglobin in blood to its useless CO derivative, which has been reported to cause death within one hour (Carr, 1965).

7.2. AMBIENT PARTICULATE MATTER: PRIMARY AND SECONDARY CONTRIBUTIONS AND FORMATION

Particulate matter (PM), originating from anthropogenic (e.g., smoke, soot, and fly-ash) or biogenic sources (e.g., pollen and spores), has a pronounced effect on the chemistry of the atmospheric, air quality, and human health. Particulate matter in the atmosphere is commonly referred to as aerosol, although this term accurately applies to a stable suspension of fine solid or liquid particles dispersed in a gas. It is a complex mixture of extremely small particles and liquid droplets, including components such as acids, organic substance, metals, and soil or dust particles. Two particle types are encountered: primary PM and secondary PM. As expected, primary PM is emitted directly to the atmosphere in solid or liquid form, for example, emissions emanating from fuel combustion and industrial processes. Secondary PM is formed in the atmosphere through condensation/deposition of gaseous particles, including precursor pollutants such as NO_X, SO_2, and NH_3. Additionally, **secondary organic aerosols (SOA)** are formed when oxidized products of VOCs condense. They comprise a substantial portion of the organic mass fraction of atmospheric aerosols, thus strongly influencing air quality and climate change (Griffin, et al., 1999). Scientists have suggested that a potentially important source of SOAs is the oxidation products of biogenic volatile organic compounds (BVOCs). For example, recall from Chapter 6 that isoprene is the principal biogenic hydrocarbon contributing to O_3 formation. Gas phase oxidation of BVOCs has been shown to lead to a number of semi-volatile compounds capable of portioning into the particle phase (Carlton, et al., 2009). For instance, Figure 7.2 displays the chemical structures of compounds

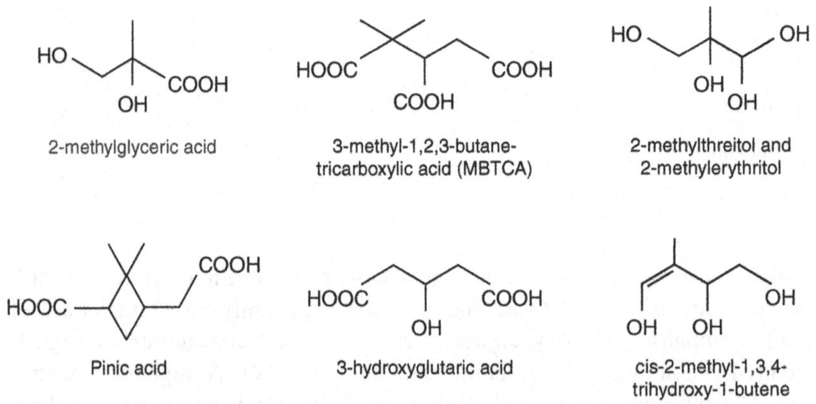

2-methylglyceric acid

3-methyl-1,2,3-butane-tricarboxylic acid (MBTCA)

2-methylthreitol and 2-methylerythritol

Pinic acid

3-hydroxyglutaric acid

cis-2-methyl-1,3,4-trihydroxy-1-butene

FIGURE 7.2 Chemical structures of common isoprene oxidation products.

identified as isoprene, monoterpenes, and sesquiterpene oxidation products found in high Arctic aerosols (Fu, et al., 2009). The aerosol concentration in Arctic regions has a strong seasonal variation commonly characterized by a winter maximum and summer minimum.

In terms of size distribution, PM is operationally defined as 1) $PM_{2.5}$ (fine particles; aerodynamic particle diameter <2.5 μm), and 2) PM_{10} (inhalable course particles; aerodynamic particle diameter >2.5 μm to <10.0 μm). Fine particles are of particular concern given that they can easily be deposited deeply in the human lungs, thus increasing the risk for cardiorespiratory disease (Pope, et al., 2002). The U.S. EPA has set National Ambient Air Quality Standards (NAQS) at levels of 15.0 $\mu g/m^3$ and 150 $\mu g/m^3$ for $PM_{2.5}$ and PM_{10}, respectively. Ultimately, aerosol size distributions are determined by source, chemical, and physical processes, and removing (e.g., deposition) conditions. More on that shortly. Understanding aerosol formation and transformation dynamics of particles in the troposphere is a much more complex task. To better comprehend such processes we will concentrate on combustion sources. For example, diesel combustion processes from large vehicles along major roadways result in the formation of carbonaceous agglomerates (soot) and metallic ash. Soot particles are believed to form with a homogenous process, by either physical or chemical condensation. After soot formation, particle coagulation and surface growth occur, in due course leading to larger size particles. Surface growth itself is a heterogeneous process, one involving: 1) deposition of gas-phase hydrocarbons on the surfaces, 2) adding carbon atoms on the surfaces, and 3) desorption reactions from particles. Primary organic aerosols, PAHs, and ionic species are also important components of the fine-particle fractions emitted from diesel combustion, with the latter commonly including sulfate, nitrate, and ammonium. Recall from Chapter 6 that several PAHs and associated degradation products have high carcinogenic and mutagenic potential. Exhaust emissions of PAHs are typically distributed between particulate and semi-volatile phases.

Immediately following combustion and emission of PM from the diesel source, chemical and physical processes dominate the dynamics of PM behavior in the troposphere and influence their characteristics. Continuing our brief description above, these processes include particle formation involving **nucleation** of gaseous precursors and/or their condensation onto pre-existing particles, gas-particle partitioning of primary semi-volatile PM with **atmospheric dilution**, and further secondary particle formation by means of photochemical reactions (Ning and Sioutas, 2010). In general terms, nucleation refers to the process of forming particles into small crystalline structures. The dilution process occurs as the hot vapors and particles from the exhaust are quickly cooled, with subsequent formation of new particles by nucleation of resulting vapors and/or their condensation onto pre-existing particles. Note that increasing the dilution ratio increases the concentration of ultrafine particles ($PM_{2.5}$). For example, nucleation of H_2SO_4 vapor during dilution is an important source of ultrafine particles. The overall concentration of H_2SO_4 is based on an estimate of the SO_3 level in the exhaust.

For a more specific example, let's consider the heavy diesel combustion processes that occur in the Los Angeles basin on a daily basis. Los Angeles has a high ambient concentration of sodium and chloride due to the saturated marine air that surrounds the region. The H_2SO_4 vapor that results from diesel combustion subsequently reacts with sea salt aerosols leading to the formation of secondary ammonium chloride particles, as follows:

$$H_2SO_4 + 2NaCl \rightarrow Na_2SO_4 + 2HCl \quad (7.1)$$

$$HCl + NH_3 \rightarrow NH_4Cl. \quad (7.2)$$

Moreover, photochemically-reactive organic aerosols including unsaturated hydrocarbons, alcohols, aldehydes, and acids present will form secondary pollutants during the intense summer periods that exist.

Research Application IV: Particulate Matter and the 2008 Beijing Olympics

Many of us remember the magnificent display of pageantry and athleticism at the 2008 Beijing Olympics. However, Beijing is particularly susceptible to high particulate matter (PM) levels due to frequent dust storm events and rapid urbanization and motorization. Moreover, Beijing is surrounded by mountain terrain, and as we learned previously, this situation can lead to trapping of air pollutants in associated valleys and depression zones in the region. As a result, the Chinese government worked to improve air quality for the Olympics.

In an attempt to study the success of air quality improvements, Wang et al., (2009) collected size-fractionated particulate matter samples ($PM_{2.5}$ and PM_{10}) for a two-week period prior to the Olympics, during the two-week period of the Olympics, and for a four-week period following the Olympics. More specifically, the objectives of this study were three-fold:

1. Determine Beijing's atmospheric PM pollution during Olympic and non-Olympic periods, during both source control and non-source control periods;
2. Assess the influence of meteorology on PM concentrations;
3. Determine the source regions that influenced Beijing's PM concentrations and the effectiveness of source control strategies.

It was found that mean $PM_{2.5}$ and PM_{10} were lower by 31% and 35% during the Olympic period compared to the non-Olympic period. Interestingly, though, the PM concentrations were not statistically different between the source control and non-source control periods. Therefore, variation in PM concentrations was likely due to meteorological parameters (air masses from the south and precipitation) (Figure 7.3). For example, such parameters accounted for 40% of the total variation in PM_{10} concentration, compared to only 16% for source control. Strikingly, it was also found that $PM_{2.5}$ and PM_{10} concentrations during the Olympic period exceeded the WHO 24-hour guideline 100% and 81% of the time, respectively. Thus, although efforts were made to reduce PM concentrations, Beijing still suffered more particulates than were present at previous Olympic host cities (e.g., Atlanta, Sydney, and Athens).

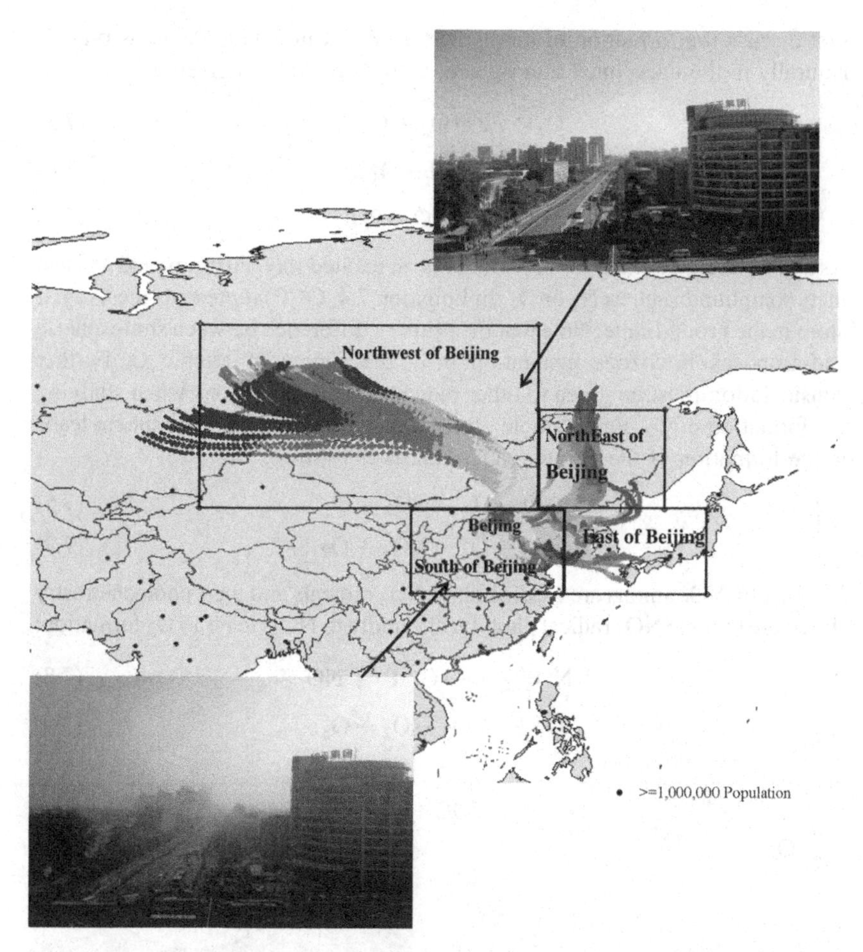

FIGURE 7.3 A map of the source regions that influenced Beijing's particulate matter (PM) concentrations during the 2008 Beijing Olympic Games. Also shown are representative 4-day air mass back trajectories. Modeled back trajectory analysis is a commonly used tool for studying how short-term variability in air pollution depends on transport into a given location. *Reprinted with permission from Wang, et al., (2009).*

7.3. PHOTOCHEMICAL MECHANISM OF OZONE (O_3) FORMATION: THE ROLE OF VOCs AND NO_X

As briefly discussed in Chapter 6, knowledge of the tropospheric chemistry of NO_X and VOCs is needed to fully understand the photochemical mechanism of ozone formation. Also recall that the hydroxyl (OH) radical is the key reactive species in the troposphere, reacting with most organic compounds apart from the chlorofluorocarbons (CFCs) and specified halons (Atkinson, 2000). Let's first discuss the formation of the hydroxyl radical in detail. Ozone is present naturally in the unpolluted atmosphere, with its photolysis given by

$$O_3 + hv \rightarrow O_2 + O(^1D) \quad (7.3)$$

$$O(^1D) + M \rightarrow O(^3P) \quad (7.4)$$

$$O(^1D) + H_2O \rightarrow 2OH. \quad (7.5)$$

As shown in Equations 7.3–7.5, $O(^1D)$ is an excited oxygen atom, and M is an inert compound such as N_2 or O_2. In Equation 7.4, $O(^3P)$ represents the oxygen atom in the ground state. Note that the primary difference between stratospheric and tropospheric ozone generation is in the source of atomic O. Further consideration must be given to other radical-initiated reactions when studying the formation of O_3. For example, the presence of NO in the troposphere leads to the formation of the nitrate (NO_3) radical as follows:

$$NO + O_3 \rightarrow NO_2 + O_2 \quad (7.6)$$

$$NO_2 + O_3 \rightarrow NO_3 + O_2. \quad (7.7)$$

Ultimately, NO_3 anions are oxidized into NO_3 radicals and their photochemistry (be aware that the NO_3 radical photolyzes rapidly to NO_2) leads to O_3 formation:

$$NO_2 + hv \rightarrow O(^3P) + NO \quad (7.8)$$

$$O(^3P) + O_2 \rightarrow O_2 \rightarrow O_3. \quad (7.9)$$

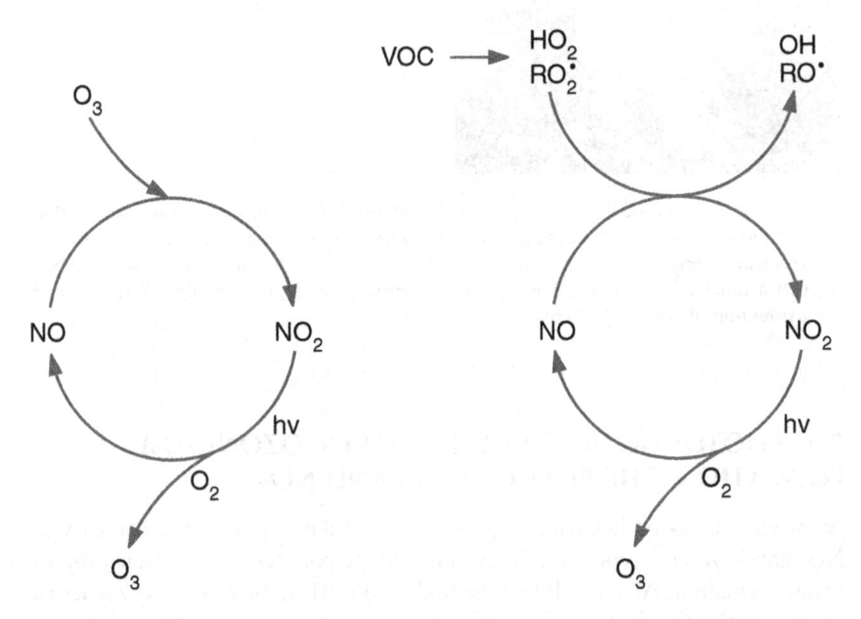

FIGURE 7.4 Reactions involved in NO to NO_2 conversion and O_3 formation in a) NO-NO_2-O_3 systems (absence of VOCs), and b) NO-NO_2-O_3 systems in the presence of VOCs. *Modified from Atkinson (2000) with permission from Elsevier.*

Let's now consider two scenarios. First, if only the reactions in Equations 7.8 and 7.9 were considered, a steady-state O_3 concentration will occur, given that O_3 will be destroyed by NO according to the following reaction (Figure 7.4a):

$$O_3 + NO \rightarrow NO_2 + O_2. \quad (7.10)$$

However, high ambient VOC concentrations can lead to additional O_3 formation by transforming NO to NO_2 by peroxyl radicals such as $HO_2^{\cdot}$ and $RO_2^{\cdot}$ as follows (Figure 7.4b):

$$NO + HO_2^{\cdot} \rightarrow NO_2 + OH^{\cdot} \quad (7.11)$$

$$NO + RO_2^{\cdot} \rightarrow NO_2 + RO^{\cdot}. \quad (7.12)$$

The NO_2 produced is then utilized to form O_3. The total amount of O_3 generated is therefore dependent on the number of NO to NO_2 conversions likely to occur during the degradation process.

TEXTBOX 7.1

O_3 as an air pollutant should not be confused with the "ozone layer" in the stratosphere. As discussed in Chapter 1, ozone occurs naturally in the stratosphere—up to 50 times higher than in most polluted cities. This is beneficial to life on Earth given that stratospheric ozone blocks harmful UV radiation.

As discussed in Chapter 6, VOCs are largely removed from the troposphere by wet and dry deposition processes. However, complete understanding of the degradation/transformation reactions of VOCs in the troposphere is best represented by the scheme presented in Figure 7.5. Moreover, the ability to make unequivocal measurements of the concentrations of the O_3 precursors, the reactive NOx compounds and VOCs, is of vital concern. As shown in Figure 7.5, intermediate radicals including alkyl or substituted alkyl radicals ($R^{\cdot}$), alkyl peroxy or substituted alkyl peroxy radicals ($RO_2^{\cdot}$), and alkoxy or substituted alkoky radicals ($RO^{\cdot}$) are intimately involved. In due course, reactions of alkoky radicals determine the products formed and the amount of NO converted to NO_2 by $RO_2^{\cdot}$ and $HO_2^{\cdot}$ radicals, and hence the extent of O_3 formation during high ambient VOC concentrations (Ryerson, et al., 2001).

If we consider the major products of NO_x oxidation as a whole, HNO_3, peroxyacetyl nitrate (PAN), and other organic nitrates are prevalent. In contrast to the discussion above, low ambient VOC concentrations increase the relative importance of the $OH + NO_2$ reaction, thereby favoring HNO_3 formation (Jenkin, et al., 1997). Also notice in Figure 7.5 that the degradation of VOCs by the reactions described above leads to a variety of oxygenated products, including carbonyl compounds, organic nitrates, carboxylic acids, alcohols, and, in some cases, multi-functional compounds containing two or more of these functionalities (Jenkin, et al., 2002). Carbonyl compounds (e.g.,

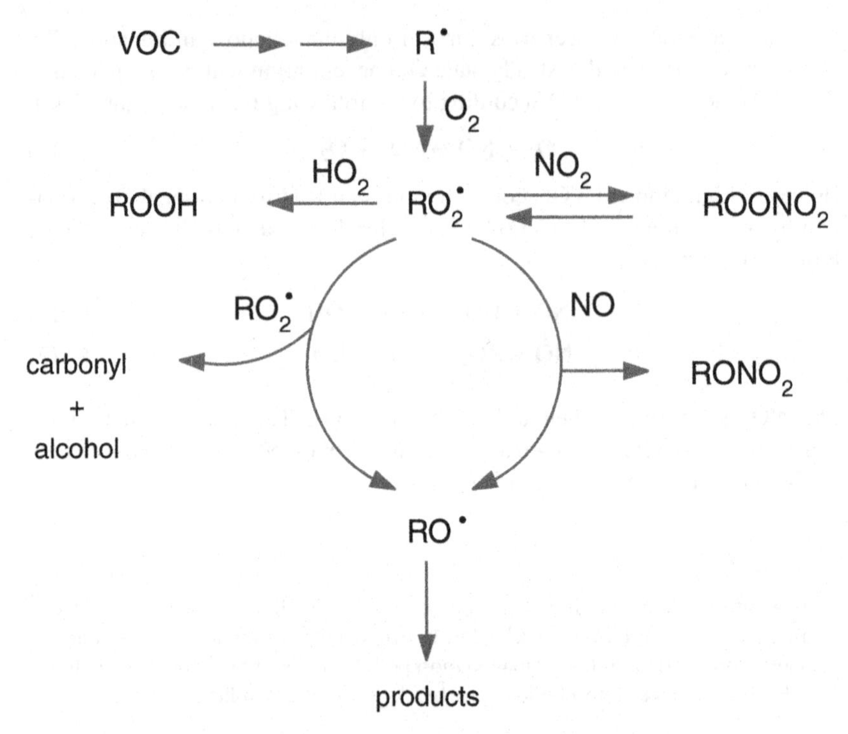

FIGURE 7.5 General tropospheric degradation/transformation reactions of VOCs. *Modified from Atkinson (2000) with permission from Elsevier.*

aldehydes and ketones), for example, are formed from the catalyzed reactions of NO with $RO_2^{\cdot}$. These compounds are highly reactive and known to decisively participate in O_3 formation.

The simplest example of VOC degradation is that of methane (CH_4). The chemistry of the clean, unpolluted troposphere is dominated by the chemistry of CH_4 and its degradation products, formaldehyde (HCHO) and carbon monoxide (CO). In the presence of NO, the following reaction scheme takes place (Jenkin, et al., 1997):

$$OH + CH_4 \rightarrow H_2O + CH_3^{\cdot} \quad (7.13)$$

$$CH_3^{\cdot} + O_2 \overset{M}{\rightarrow} CH_3O_2^{\cdot} \quad (7.14)$$

$$CH_3O_2^{\cdot} + NO \rightarrow CH_3O^{\cdot} + NO_2 \quad (7.15)$$

$$CH_3O^{\cdot} + O_2 \rightarrow HCHO + HO_2 \quad (7.16)$$

$$HO_2 + NO \rightarrow OH + NO_2. \quad (7.17)$$

As shown in the above scheme, reaction with methane is an example loss process for the OH radical concentration in the clean atmosphere. Therefore, increases in tropospheric CH_4 concentrations will reduce the OH radical concentration. What about the total quantity of O_3 potentially formed as a byproduct of the complete OH-initiated and NO_x-catalyzed oxidation processes? Considering the catalyzed oxidation of CO_2 and H_2O using methane, ethane, and ethene (Jenkin, et al., 2002),

$$CH_4 + 8O_2 \rightarrow CO_2 + 2H_2O + 4O_3 \tag{7.18}$$

$$C_2H_2 + 14O_2 \rightarrow 2CO_2 + 3H_2O + 7O_3 \tag{7.19}$$

$$C_2H_4 + 12O_2 \rightarrow 2CO_2 + 2H_2O + 6O_3, \tag{7.20}$$

the production of 4, 7, and 6 molecules of O_3 are formed, respectively. The number of O_3 molecules produced is equivalent to the number of reactive bonds in the parent molecule, i.e., the number of C-H and C-C bonds, which are eventually broken during the complete oxidation to CO_2 and H_2O (Jenkin, et al., 2002).

Example Problem 7.2

Based on knowledge of the reaction mechanisms provided in this section, what species and reaction comparisons dictate the degree of net photochemical formation of O_3 versus net photochemical loss of O_3 in the troposphere?

Answer: The NO concentration dictates the degree of net photochemical formation of O_3 versus net photochemical loss of O_3.

$$HO_2 + NO \rightarrow OH + NO_2$$

versus those for

$$HO_2 + HO_2 \rightarrow H_2O_2 + O_2$$

and

$$HO_2 + O_3 \rightarrow OH + 2O_2.$$

Moreover, it is also determined by the rate of the reaction of the $RO_2^{\cdot}$ radical with NO:

$$RO_2^{\cdot} + NO \rightarrow RO^{\cdot} + NO_2.$$

Compared to those for the reactions of the $RO_2^{\cdot}$ radical with the HO_2 radical:

$$RO_2^{\cdot} + HO_2 \rightarrow ROOH + O_2.$$

TEXTBOX 7.2

Free radicals play an important role in the catalytic self-cleaning of the atmosphere. For example, in the daytime, photolytically-formed OH radicals govern the oxidation capability of the atmosphere. During the night, nitrate radicals (NO_3) are the main atmospheric oxidants.

7.4. PEROXYACETYL NITRATE (PAN) FORMATION AND NO_X TRANSPORT

Vitally important information concerning the transformation of NO_x during photochemical episodes can be gained from quantitative atmospheric measurements of NO, NO_2, **peroxyacetyl nitrate (PAN)**, and nitrate. PAN (Figure 7.6), in particular, is formed in the atmosphere in the oxidative degradation of numerous organic compounds (e.g., *trans*-butene, propene, *n*-butane, and toluene). PAN is involved in summer **smog** formation and plays a significant role in transporting NO_x in the troposphere, particularly allowing for NO_x photochemistry and/or deposition of nitrogen to remote environments (Dassau, et al., 2004). The direct precursor of the peroxyacetyl radical is acetyl, CH_3CO, which adds molecular oxygen under atmospheric conditions (Seefeld, et al., 1997):

$$HO^{\cdot} + CH_3CHO \rightarrow H_2O + CH_3CO^{\cdot} \tag{7.21}$$

$$CH_3CO^{\cdot} + O_2 + M \rightarrow CH_3C(O)OO^{\cdot} + M. \tag{7.22}$$

In the polluted atmosphere, there are two competing reactions for the acetylperoxy radical,

$$CH_3C(O)OO^{\cdot} + NO_2 + M \rightleftharpoons CH_3C(O)OONO_2 + M \tag{7.23}$$

and

$$CH_3C(O)OO^{\cdot} + NO \rightarrow CH_3C(O)O^{\cdot} + NO_2, \tag{7.24}$$

where the latter reaction is followed by the fast reaction,

$$CH_3C(O)O^{\cdot} \rightarrow CH_3^{\cdot} + CO_2. \tag{7.25}$$

As shown, NO_2 to PAN conversion is intensified under hydrocarbon-rich conditions with the formation compounds representing a temporary sink of NO_2. These compounds have been shown to be relatively stable in cold air masses and

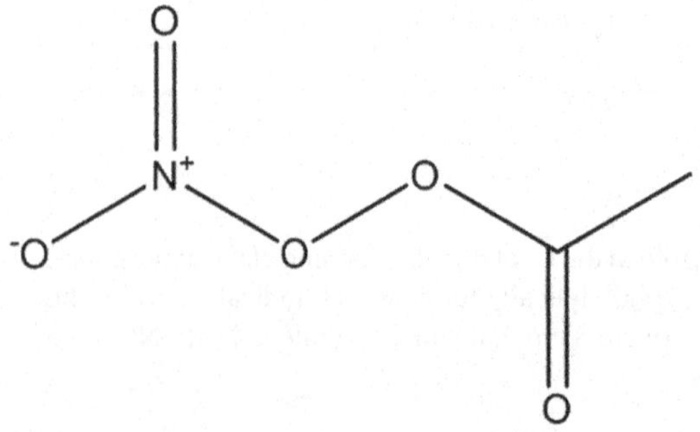

FIGURE 7.6 Chemical structure of peroxyacetyl nitrate (PAN). It is considered a NO_x reservoir compound that has the ability to transport NO_x to remote environments.

at high NO_2:NO ratios, and may thus serve as a transport vehicle of NO_x to warmer climates (Tanner, et al., 1988). Studies have shown that diurnal PAN profiles were consistent with those expected for a secondary pollutant of photochemical origin, with late afternoon maxima on smog episode days. For example, PAN maxima in the range of 32–51 ppb have been previously recorded in the Los Angeles, California, area during this timeframe (Glavas and Schurath, 1985). Moreover, persistence of 10–20 ppb of PAN overnight has been shown to accelerate smog formation on the following day(s) of a multi-day episode.

TEXTBOX 7.3

Photochemical smog is a general term used to describe a mixture of air pollutants that result from the effects of UV light and oxidation of hydrocarbons and other volatile organic compounds often associated with (but not limited to) automotive exhaust. As will be evident in this chapter, photochemical smog formation proceeds through a sequence of reactions, all involving a free radical mechanism. Tropospheric ozone has been and continues to be a significant air pollution problem and is the primary constituent of smog.

7.5. PHOTOCHEMICAL SMOG: FORMATION CONDITIONS AND HEALTH EFFECTS

As outlined above, photochemical smog is a condition that is developed when primary pollutants, i.e., NO_x and VOCs, interact under sunlight to produce a mixture of hazardous secondary pollutants. A variety of conditions are necessary for the formation of photochemical smog:

1. Emission rates of the sources of NO_x and VOCs;
2. The time of day;
3. Meteorological factors;
4. Topography.

Let's use Los Angeles as an example city when discussing these conditions in greater detail. The climate of the Los Angeles basin is semi-arid with extended periods of warm and abundant sunlight. It is surrounded by mountains to the north and the east, which serve as barriers to air movement, thus reducing dispersion. This topography can play a significant role in creating **temperature inversions** given that cold air will typically flow from mountain peaks to the lower valleys present in the basin. Photochemical smog is impacted by inversion layers because it is in essence, trapped, when the warm air masses move over the valleys. In addition, Los Angeles is home to millions of light- and heavy-duty motor vehicles, which contribute significantly to the emissions of NO_x and VOCs, especially during early morning traffic hours. As the day progresses, and the sunlight becomes more intense, nitrogen-containing compounds are broken down, and their byproducts form increasing amounts of PAN and O_3. Note that cold weather smog formation events have also been demonstrated. For example, in London and other northern European cities, smoke from coal and other industrial sources has mixed with fog generated from the North Sea to produce conditions of reduced visibility and sulfurous emissions for more than a century. In fact, London was the site of dense smog caused by heavy fog and coal combustion during the winter of 1952, which killed just under 12,000 people.

Ozone, the main oxidant in photochemical smog, has been shown to have deleterious effects on the environment, on human health, and even on buildings of aesthetic or cultural merit. For example, ozone's phytotoxic characteristics have been shown to reduce the growth and yield of agricultural crops and forests. During normal gas exchange, ozone enters a plant's leaves through stomata. As a strong oxidant, ozone (or secondary products resulting from oxidation by ozone) causes numerous types of symptoms including chlorosis and necrosis (Novak, et al., 2003). In regards to human health, O_3 and PAN have been shown to cause eye irritation, wheezing, coughing, and more advanced respiratory problems (e.g., asthma). In cities around the world, smog-derived acids are eroding precious artifacts, including the Parthenon temple in Athens, Greece, and the Taj Mahal in Agra, India. In an effort to combat these problems, various regulations have been created by governmental bodies. For example, the Clean Air Act requires the U.S. EPA to set National Ambient Air Quality Standards (NAAQS) for tropospheric (ground-level) ozone (Table 7.3).

TABLE 7.3 National Ambient Air Quality Standards for Ground-Level Ozone

Primary Standards		Secondary Standards	
Level	Averaging Time	Level	Averaging Time
0.075 ppm (2008)	8-hour[1]	0.075 ppm (2008)	8-hour[1]
0.080 ppm (1997)	8-hour[2]	0.080 ppm (1997)	8-hour[2]
0.12 ppm	1-hour[3]	0.12 ppm	1-hour[3]

[1] *To attain this standard, the 3-year average of the fourth-highest daily maximum 8-hour average ozone concentrations measured at each monitor within an area over each year must not exceed 0.075 ppm (effective May 27, 2008).*
[2] *To attain this standard, the 3-year average of the fourth-highest daily maximum 8-hour average ozone concentrations measured at each monitor within an area over each year must not exceed 0.08 ppm.*
[3] *The standard is attained when the expected number of days per calendar year with maximum hourly average concentrations above 0.12 ppm is ≤ 1.*
Data source: U.S. EPA NAAQS (http://www.epa.gov/air/ozonepollution/standards.html)

Primary standards are set to protect public health, including "sensitive" populations such as asthmatics and the elderly. Secondary standards are set to protect public welfare, including harm to animals, crops, vegetation, and buildings.

As mentioned above, the Los Angeles basin is particularly susceptible to high levels of photochemical smog-related compounds. For example, consider the data presented in Table 7.4. Listed are measured maximum national 8-hour ozone averages over a 10-week period during the months of June–August 2010. Tropospheric ozone is routinely measured by UV photometric ozone analyzers, which measure the concentration of ozone in ambient air at the 254-nm emission line of a mercury lamp. In Table 7.4, values in bold represent averages that exceed the national ambient air quality standard. Note that California's air quality standards tend to be more stringent than the national standards, i.e., the air is deemed to be unhealthy at lower concentrations. As a result, the number of days above the state standard is a more sensitive indicator of the number of days of unhealthy air. In January 2010, the U.S. EPA proposed to strengthen the national ambient air quality standards for ground-level ozone. The proposed revisions are based on scientific evidence about ozone and its effects on people and sensitive trees and plants. Based on updated findings about ozone and its effects, they proposed that the level of the 8-hour primary standard be lowered within the range of 0.060–0.079 ppm to provide protection to children and "sensitive" populations (e.g., those with decreased lung function and respiratory problems).

7.6. ALDEHYDES IN THE ATMOSPHERE

There has been considerable interest in ethanol and other biomass-derived alcohol fuels in both developing and the more developed countries as part of efforts to curb CO levels in urban areas (Grosjean, 1983). This has led to increased ethanol-fueled vehicular use and a whole host of consequences including increased aldehyde emissions into the atmosphere. As mentioned in section 7.2, formaldehyde (HCHO) is a degradation product of CH_4 in the clean troposphere and is released from the combustion and photooxidation of biogenic and anthropogenic hydrocarbons. It is the most abundant carbonyl compound in the atmosphere and plays a key role as a free radical source. Carbonyl compounds are molecules containing the carbonyl group, C=O. In addition, acetaldehyde (CH_3CHO) has been shown to be abundant in the atmosphere and also released from the combustion and photooxidation of hydrocarbons.

Both formaldehyde and acetaldehyde levels are important since they influence photochemical smog processes. The photolysis of acetaldehyde proceeds as follows:

$$CH_3CHO \xrightarrow{h\nu} CH_3 + CHO + 2O_2 \rightarrow CH_3O_2 + HO_2 + CO. \qquad (7.26)$$

Recall above that the direct precursor of the peroxyacetyl radical is acetyl, CH_3CO, which adds molecular oxygen under atmospheric conditions,

TABLE 7.4 Maximum National 8-Hour Average Ozone Levels[1] in the Los Angeles Basin (July–August, 2010)

Day	June 27	July 4	July 11	July 18	July 25	August 1	August 8	August 15	August 22	August 29
Sunday	0.074	**0.076**	0.074	**0.077**	0.064	0.071	0.071	**0.080**	0.064	0.058
Monday	**0.080**	**0.080**	0.060	**0.079**	0.068	**0.080**	**0.081**	**0.084**	0.064	0.066
Tuesday	**0.084**	0.075	0.057	0.071	0.061	**0.087**	**0.079**	**0.080**	0.073	0.074
Wednesday	**0.076**	0.075	0.057	0.074	0.067	**0.084**	0.070	0.055	**0.081**	**0.079**
Thursday	**0.085**	0.071	0.070	0.070	**0.077**	**0.084**	**0.078**	0.072	**0.076**	**0.083**
Friday	**0.084**	0.071	0.071	0.065	**0.085**	**0.080**	0.068	0.070	0.069	0.068
Saturday	**0.078**	0.071	0.066	0.062	**0.077**	**0.078**	0.068	0.063	0.070	0.065

[1] *Values in ppm*

Data source: California Protection Agency Air Resources Board (http://www.arb.ca.gov/homepage.htm)

ultimately forming PAN as shown in section 7.4. Formaldehyde and acetaldehyde are hydrogen bond acceptors making them considerably soluble in water. If we now consider the gas-phase photochemistry of formaldehyde,

$$HCHO + OH + O_2 \rightarrow HO_2 + CO + H_2O \quad (7.27)$$

$$HCHO + h\nu \rightarrow HCO + H \quad (7.28)$$

$$HCHO + O_2 \rightarrow HO_2 + CO \quad (7.29)$$

$$H + O_2 \rightarrow HO_2, \quad (7.30)$$

we can see that it can lead via reaction with OH or via photolysis to the net formation of one or two HO_2 radicals. Studies have also shown that photolysis of HCHO in the aqueous phase is a dominant source of hydrogen peroxide in wet deposition processes (e.g., Kieber, et al., 1999). In fact, atmospheric deposition is a significant source of HCHO to aquatic systems, since concentrations in rainwater have been shown to be up to three orders of magnitude higher than in surface waters (Kieber, et al., 1999). Levels reported exhibit distinct seasonal oscillation, with higher concentrations during the summer. This pattern is similar to other rainwater parameters including pH, nitrate, sulfate, and ammonium. The intercorrelation of HCHO with these parameters suggests a possible anthropogenic input, most likely from the oxidation of hydrocarbons generated from the burning of fossil fuels.

7.7. HETEROGENEOUS ATMOSPHERIC REACTIONS

It is obvious from discussions above that gas-phase oxides of, for example, nitrogen and volatile organic compounds, play a central role in air pollution, whether through their reactions in air to form O_3, or in the formation of particles and a horde of other secondary pollutants. The chemistry of these compounds is well known and represented by numerous airshed models being used today. However, **heterogeneous reactions** on surfaces found in the boundary layer have not been fully studied, yet they affect a multitude of atmospheric processes. For example, particulates can modify the chemical balance of the atmosphere by providing a reactive substrate on and in which reactions can occur. Example processes will be detailed below.

TEXTBOX 7.4

An **airshed** is a geographic area (domain) whereby pollutant emissions are contained and subsequently deposited to receptor locations (e.g., watersheds).

On a basic level, we will refer to heterogeneous chemistry as the interactions between gases and aerosols in the atmosphere. If a reaction is slow in the gas phase, and has a high thermodynamic tendency to react, it is a good candidate for a laboratory study to seek a heterogeneous catalyst (Fairbrother, et al., 1997). Laboratory simulations under conditions typically found in the atmosphere are quite common and necessary endeavors to help elucidate complex reaction mechanisms. Let's first consider the hydrolysis of NO_2:

$$2NO_2 + H_2O \rightarrow HONO + HNO_3. \quad (7.31)$$

A number of studies (e.g., Finlayson-Pitts and Pitts 2000) have shown that HONO is a major OH radical source when compared to other well-known sources of OH such as the photolysis of O_3 and HCHO, as discussed earlier. The HONO produced is subsequently released to the gas phase. The HNO_3 remaining on the surface can react with gaseous NO to form NO_2 and HONO:

$$NO_2 + HNO_3 \rightarrow HONO + NO_2. \quad (7.32)$$

Ultimately, irradiation of surface-adsorbed nitrate can result in **renoxification** of HNO_3 back into the gas phase in the form of NO_x product species (Chen, et al., 2011). Next, consider the reaction of SO_2 and HNO_3 to form HONO:

$$SO_2 + HNO_3 \rightleftharpoons H_2SO_4 + HONO. \quad (7.33)$$

This reaction has a high thermodynamic potential. Further, in H_2SO_4-rich aerosols, formaldehyde has been shown to have a high thermodynamic potential to reduce nitric acid:

$$HNO_3 + 2H_2CO \rightleftharpoons 2HONO + CO_2 + H_2O. \quad (7.34)$$

The uptake of O_3 on metal oxides associated with **mineral dust** has also been increasingly studied, with the reaction postulated to proceed by the following scheme (Chen, et al., 2011):

$$O_3 + S \rightarrow O_2 + O - S \quad (7.35)$$

$$O_3 + O - S \rightarrow O_2 + O_2 - S \quad (7.36)$$

$$O_2 - S \rightarrow O_2 + S \quad (7.37)$$

$$O - S + O - S \rightarrow O_2 + 2S, \quad (7.38)$$

where S represents an active site or surface-bound species. For instance, iron oxide (Fe_2O_3) shows catalytic decompositions toward O_3, while aluminum oxide (Al_2O_3) is deactivated by O_3 exposure (Chen, et al., 2011). Solar radiation plays a major role in these processes by changing the product distribution, the reaction extent, the reaction rate, and the reaction mechanism of atmospheric gases on the surface of mineral dust (Chen, et al., 2011).

TEXTBOX 7.5

Mineral dust aerosols are soil particles that have been mobilized by strong wind currents and entrained into the atmosphere. Common clays and minerals in atmospheric dust include calcite, chlorite, gypsum, kaolinite, mica, opal, and quartz.

7.7.1. The Role of Transition Metals and Other Atmospheric Species in Catalyzed and Uncatalyzed Oxidation of Sulfur(IV) Oxides

Several studies have highlighted the importance of transition metal concentrations in various atmospheric aqueous phases (e.g., cloud, rain, fog, and wet aerosols) (Brandt and van Eldik, 1995). For example, transition metal ions have been shown to play an important role in the heterogeneously catalyzed autoxidation of sulfur(IV) oxides. The oxidation process of sulfur(IV) oxides plays a crucial role in atmospheric chemistry (e.g., acid deposition) as well as in industrial processes (e.g., desulfurization of plume gases) (Brandt and van Eldik, 1995). More on such processes shortly. First, let's review example chemical forms of transition metal ions and associated oxidation states. Most transition metal ions contain partially occupied *d* subshells, which are responsible for several characteristics of transition metals including their ability to exhibit more than one stable oxidation state. For example, in aqueous solution, iron is largely found under two oxidation states, +II and +III, with iron(II) typically more soluble than iron(III). In the atmospheric liquid phase, iron under its +II oxidation state is mainly present at Fe^{2+}. Iron(III) in its simplest form occurs as a monomer and can precipitate and form oxides. However, the partitioning between iron(III) species depends on pH, initial iron(III) concentration, and ionic strength (Deguillaume, et al., 2005). If a fixed initial iron(III) concentration is present, partitioning between different chemical forms is given by the following equilibria (Brandt and van Eldik, 1995):

$$Fe^{3+} + H_2O \rightleftharpoons Fe(OH)^{2+} + H^+ \quad (7.39)$$

$$Fe(OH)^{2+} + H_2O \rightleftharpoons Fe(OH)_2^+ + H^+. \quad (7.40)$$

Note that at typical atmospheric pH (between pH 3–5) the most abundant iron species depends on the ionic strength (Deguillaume, et al., 2005). For example, $Fe(OH)_2^+$ is the predominant species for ionic strength near zero, whereas the $Fe(OH)^{2+}$ species becomes prevalent as the ionic strength increases. Owing to their reactivity, photolysis of iron(III) complexes is known to be a potential source of HO_x radicals in the aqueous phase (Deguillaume, et al., 2005). Consider the following photochemical reaction:

$$Fe^{3+} + h\nu + H_2O \rightarrow Fe^{2+} + OH + H^+. \quad (7.41)$$

Here, the OH radical is produced by the photolysis of the ferric ion. Copper and manganese have also been associated with atmospheric aqueous phase chemistry. Given the acidic nature of atmospheric aqueous phases, and the presence of potential ligands, manganese(III) predominates. Copper is mainly present in the +II oxidation state but is reduced to copper(I) in the atmospheric liquid phase. However, both copper and manganese can be reoxidized by the presence of increasing amounts of dissolved oxygen.

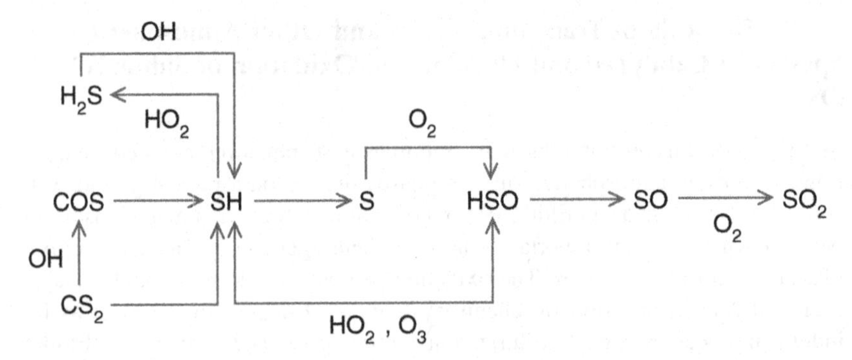

FIGURE 7.7 Schematic of the transformation of sulfur species in the troposphere. COS = oxidation state −2; CS_2 = oxidation state −2; H_2S = oxidation state −2; SH = oxidation state −1; S = oxidation state 0; HSO = oxidation state +1; SO = oxidation state +2; SO_2 = oxidation state +4.

To expand our discussion, Figure 7.7 depicts the transformation of the most important sulfur species in the atmosphere. As shown, there is an obvious tendency of transformation toward the higher oxidation state. And as previously mentioned, it is the conversion of sulfur(IV) to sulfur(VI) that is directly contributing to the acidity of rain. This conversion in the atmospheric liquid phase occurs via (Deguillaume, et al., 2005):

1. Direct oxidation of sulfite by hydrogen peroxide, ozone, organic peroxides, or pernitric acid;
2. An indirect oxidation pathway initiated by OH, Cl_2^-, and NO_3 radicals and propagated by SO_3^-, SO_4^-, and SO_5^- radicals;
3. Oxidation by dissolved oxygen, catalyzed by transition metals (in particular, iron).

Considering the latter conversion process in more detail, the following chain mechanism can be written (Deguillaume, et al., 2005):

$$\mathrm{Fe(III)} + \mathrm{HSO_3^-} \rightarrow \mathrm{Fe(III)} + \mathrm{SO_3^-} + \mathrm{H^+} \tag{7.42}$$

$$\mathrm{SO_3^-} + \mathrm{O_2} \rightarrow \mathrm{SO_5^-} \tag{7.43}$$

$$\mathrm{SO_5^-} + \mathrm{Fe(II)} + \mathrm{H^+} \rightarrow \mathrm{HSO_5^-} + \mathrm{Fe(III)} \tag{7.44}$$

$$\mathrm{HSO_5^-} + \mathrm{Fe(II)} \rightarrow \mathrm{SO_4^-} + \mathrm{OH^-} + \mathrm{Fe(III)} \tag{7.45}$$

$$\mathrm{SO_4^-} + \mathrm{Fe(II)} \rightarrow \mathrm{SO_4^{2-}} + \mathrm{Fe(III)}. \tag{7.46}$$

In addition, iron in solution under its +III oxidation state forms an $[Fe(SO_4)]^+$ complex with sulfate ions (Deguillaume, et al., 2005),

$$[\mathrm{Fe(SO_4)}]^+ + h\nu \rightarrow \mathrm{Fe^{2+}} + \mathrm{SO_4^-}, \tag{7.47}$$

which is photolyzed forming iron(II) and sulfate radical.

As listed above, mechanisms for the uncatalyzed oxidation of sulfur(IV) by O_2, O_3, H_2O_2, radicals, or NO_x have also been proposed in the scientific literature. For example, consider the reaction of sulfur(IV) oxides with OH· radicals,

$$HSO_3^- + OH^{\cdot} \rightarrow SO_3^{\cdot -} + H_2O, \tag{7.48}$$

which results in the formation of the sulfite radical, $SO_3^{\cdot -}$. In addition, we find

$$SO_3^{2-} + OH^{\cdot} \rightarrow SO_3^{\cdot -} + OH^-. \tag{7.49}$$

Note that proposed reaction mechanisms for the homogeneous transition metal catalyzed autoxidation of sulfur(IV) oxides have also been based on radical mechanisms (Deguillaume, et al., 2005):

$$SO_3^{2-} + M^{n+} \rightleftharpoons SO_3^{\cdot -} + M^{(n-1)+} \tag{7.50}$$

$$SO_3^{\cdot -} + O_2 \rightleftharpoons SO_5^{\cdot -} \tag{7.51}$$

$$SO_5^{\cdot -} + HSO_3^- \rightleftharpoons HSO_5^{\cdot -} + SO_3^{\cdot -} \tag{7.52}$$

$$SO_5^{\cdot -} + SO_3^{2-} \rightleftharpoons SO_5^{2-} + SO_3^{\cdot -} \tag{7.53}$$

$$SO_5^{\cdot -} + HSO_3^- \rightleftharpoons HSO_4^- + SO_4^{\cdot -} \tag{7.54}$$

$$SO_5^{\cdot -} + SO_3^{2-} \rightleftharpoons SO_4^{2-} + SO_4^{\cdot -} \tag{7.55}$$

$$SO_3^{2-} + HSO_5^- \rightleftharpoons HSO_4^- + SO_4^{2-} \tag{7.56}$$

$$SO_3^{2-} + SO_5^{2-} \rightleftharpoons 2SO_4^{2-} \tag{7.57}$$

$$SO_4^{\cdot -} + HSO_3^- \rightleftharpoons HSO_4^- + SO_3^{\cdot -} \tag{7.58}$$

$$SO_4^{\cdot -} + SO_3^{2-} \rightleftharpoons SO_4^{2-} + SO_3^{\cdot -}. \tag{7.59}$$

7.8. EXAMPLE PREDICTIVE ATMOSPHERIC MODELS

As stated in the beginning of this chapter, recent developments in atmospheric chemistry have shed light on the importance of secondary pollutants and their complex formation processes. For example, photochemical air pollution episodes are associated with non-linear reactions between chemistry, meteorology, and ozone-precursor relationships, thus presenting challenges for air quality management. Revisiting Chapter 2 modeling concepts, it should be apparent that associated tools might enhance our understanding of these complex processes and contribute greatly to difficult decision-making. For instance, multiple linear regression (MLR) and principal component regression (PCR) models are routinely used to predict and forecast O_3 and PM concentrations. In regards to O_3, predictor variables that have some linear correlation with O_3 can be generated and used for prediction. Variables such as the ratio of NO_2 and NO, temperature, and wind velocity can be used, for example, in the prediction of maxima hourly values for O_3.

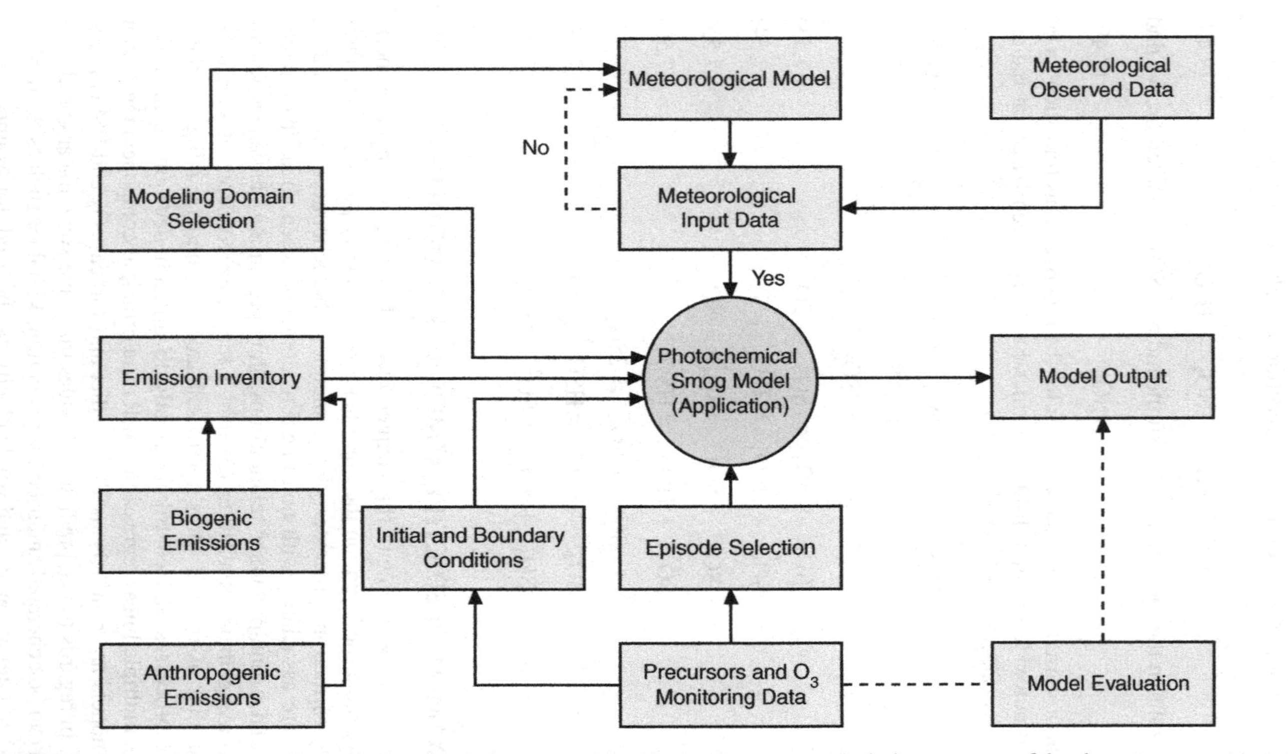

FIGURE 7.8 An example framework for a photochemical smog model with significant meteorological components. Other important components include precursor and O_3 monitoring data, boundary conditions, and emission inventories.

For non-linear processes, **time series analysis** is a commonly used modeling technique. As also discussed in Chapter 2, environmental datasets often contain non-linearities and temporal, spatial, and seasonal trends, as well as non-Gaussian distributions. Given the complexity, we are limited in our discussion of time series analysis and its various modeling approaches. But in general, time series analysis attempts to address the above concerns through the following operations:

1. The identification of the phenomenon represented by sequences of observations;
2. The characterization of patterns in a given dataset;
3. Forecasting or prediction of future values of a given time series variable or set of variables.

Typically, the question asked is: Are there any noticeable trends or fluctuations within the dataset? For example, trend analysis can detect any underlying directions (an upward or downward tendency) of O_3 concentrations measured over, say, a one-week period. Seasonality is also readily revealed. For instance, many environmental parameters and pollutants (e.g., rainfall, PM concentrations, and O_3 levels) are seasonal and exhibit a high degree of seasonality in their formation and distribution. Finally, when forecasting, predicted pollutants (e.g., O_3 concentration) is modeled as a function of the previous O_3 concentration or as a function of the position in time.

Photochemical smog models extend knowledge about the relationships between chemical mechanisms and meteorological conditions. As discussed, O_3 and its precursors undergo transport, diffusion, and wet and dry deposition that are mostly dependent upon meteorology. Thus, the development of models that have significant meteorological components (as that depicted in Figure 7.8) are warranted in studying such concepts. For example, selected models should be able to stimulate the heat island effects that are present, for example, in large cities, because temperature is an important factor in photochemical reactions. In tropical areas, high moisture content and cloudy conditions may significantly affect photochemical smog formation (Zhang and Oanh, 2002). Other considerations when selecting models include the characteristics of the terrain or domain, the temporal and spatial scale of the air pollution episode, the quality of input data, and the availability of computational facilities.

7.9. END OF CHAPTER PROBLEMS

7.1 Provide reasoning why the control of secondary pollutants is generally more problematic than that of primary pollutants?

7.2 Which of the following is not a primary pollutant?

a. Particulate matter
b. Sulfuric acid droplets

c. Sulfur dioxide
d. Carbon monoxide

7.3 List five potential examples of biogenic air pollutants.
7.4 List the major sources of indoor air pollutants in residential properties.
7.5 It is often convenient to refer to the family of oxidized nitrogen species as NO_y. Based on our discussions in Chapters 6 and 7, list the species that would fall under this larger categorical definition. In other words, list the total reactive gas-phase nitrogen oxides (NO_y) that can serve as indicators of NO_x emissions.
7.6 Based on discussion in the main text, rapid interconversion of NO and NO_2 during the daytime results in important byproduct formation. Discuss two byproducts/conversion processes that result.
7.7 Based on discussion in the main text, explain whether it is possible to make inferences about the production rate of O_3 based on ambient NO_x and VOC.
7.8 Discuss key factors that would contribute to the efficiency of NO_2-PAN conversion in the atmosphere.
7.9 In section 7.6 we learned that photolysis of HCHO in the aqueous phase is a dominant source of hydrogen peroxide (H_2O_2) in wet deposition processes. Carrying on from Equation 7.30, provide a simple reaction that would produce gas-phase H_2O_2.
7.10 As also discussed in the main text, a variety of gas-phase reactions will result in the formation of HONO if they undergo heterogeneous catalysis under atmospheric conditions. Considering the presence of HNO_3-rich aerosols, propose an additional three reactions that will likely result in the formation of HONO.
7.11 According to laboratory experiments, the gas-phase reaction of ammonia and nitric acid will likely result in the formation of HONO (on HNO_3-rich sulfuric acid aerosols). Write the proposed balanced reaction that occurs.
7.12 In our study of transition metal ions and atmospheric chemistry, we learned that the photolysis of iron(III) complexes is known to be a potential source of OH radicals. Write a complete reaction showing the photolysis of $Fe(OH)^{2+}$.
7.13 Studies have shown that the initial step in the copper(II)-catalyzed free radical oxidation of sulfur(IV) is the reduction of copper(I) by sulfur(IV). Propose a complete reaction showing this step.
7.14 Studies have also proposed a first-step manganese-catalyzed sulfur(IV) oxidation mechanism where manganese(II) forms a hydrogen sulfite complex [$Mn(HSO_3)^+$]. Propose a complete reaction showing this step.
7.15 Again consider the first-step manganese-catalyzed sulfur(IV) oxidation reaction above in problem 7.14. Propose further mechanisms for the complete autoxidation of hydrogen sulfite. Hint: In your answer, ponder the formation of relevant radicals and the role dissolved molecular oxygen may play.

7.16 Nitrous acid, HNO_2, formed in the gas phase, dissociates into the nitrite ion. Given its formation below, propose a reaction mechanism for the reaction of the nitrite ion with sulfur(IV).

$$NO + OH^{\cdot} \rightarrow HNO_2$$

$$HNO_2 \rightleftharpoons H^+ + NO_2^-$$

7.17 List and describe the important factors influencing the various SO_2 pathways discussed in the text.

7.18 As discussed in section 7.7, irradiation of surface-adsorbed nitrate can result in renoxification of HNO_3 back into the gas phase in the form of NO_x product species. Briefly discuss the significance of this in terms of NO_x chemistry.

7.10. REFERENCES

Atkinson, R. (2000). Atmospheric chemistry of VOCs and NOx. *Atmospheric Environment, 34*, 2063–2102.

Brandt, C., & van Eldik, R. (1995). Transition metal-catalyzed oxidation of sulfur(IV) oxides: Atmospheric-relevant processes and mechanisms. *Chemical Reviews, 95*, 119–190.

Carlton, A. G., Wiedinmyer, C., & Kroll, J. H. (2009). A review of secondary organic aerosol (SOA) formation from isoprene. *Atmospheric Chemistry and Physics, 9*, 4987–5005.

Carmichael, G. R., Streets, D. G., Calori, G., Usamann, M., Jacobson, M. Z., Hansen, J., & Ueda, H. (2002). Changing trends in sulfur emissions in Asia: Implications for acid deposition, air pollution, and climate. *Environmental Science & Technology, 36*, 4707–4713.

Carr, D. E. (1965). *The Breath of life*. New York: W. W. Norton.

Chen, H., Navea, J. G., Young, M. A., & Grassian, V. H. (2011). Heterogeneous photochemistry of trace atmospheric gases with components of mineral dust aerosol. *Journal of Physical Chemistry A, 115*, 490–499.

Dassau, T. M., Shepson, P. B., Bottenheim, J. W., & Ford, K. M. (2004). Peroxyacetyl nitrate photochemistry and interactions with the Arctic surface. *Journal of Geophysical Research, Atmospheres, 109*, 334–365.

Deguillaume, L., Leriche, M., Desboeufs, K., Mailhot, G., George, C., & Chaumerliac, N. (2005). Transition metals in atmospheric liquid phases: Sources, reactivity, and sensitive parameters. *Chemical Reviews, 105*, 3388–3431.

Fairbrother, H., Sullivan, D. J. D., & Johnston, H. S. (1997). Global thermodynamic atmospheric modeling: Search for new heterogeneous reactions. *Journal of Physical Chemistry A, 101*, 7350–7358.

Finlayson-Pitts, B. J., & Pitts, J. N. (2000). *Chemistry of the upper and lower atmosphere: Theory, experiments and applications*. Burlington, MA: Academic Press.

Fu, P., Kawamura, K., Chen, J., & Barrie, L. A. (2009). Isoprene, monoterpene, and sesquiterpene oxidation products in the high Arctic aerosols during late winter to early summer. *Environmental Science & Technology, 43*, 4022–4028.

Glavas, S., & Schurath, U. (1985). Peroxyacetyl nitrate forming potential of five prototype hydrocarbons. *Environmental Science & Technology, 19*, 950–955.

Green, C. R., & Rodgman, A. (1996). The tobacco chemists' research conference: A half-century of advances in analytical methodology of tobacco and its products. *Recent Advances in Tobacco Science, 22*, 131–304.

Griffin, R. J., Cocker, D. R., Seinfeld, J. H., & Dabdub, D. (1999). Estimate of global atmospheric organic aerosol from oxidation of biogenic hydrocarbons. *Geophysical Research Letters, 26*, 2721–2724.

Grosjean, D. (1983). Distribution of atmospheric nitrogenous pollutants at a Los Angeles area smog receptor site. *Environmental Science & Technology, 17*, 13–19.

Hering, S. V., & Avol, E. (1996). Indoor-outdoor concentration ratios for fine particle mass and inorganic ions in twelve southern California homes. In *Proceedings of the second colloquium on particulate air pollution and human health*, Park City, UT, May 1–3, 1996, pp. 3–52.

Hoffmann, D., Hoffmann, I., & El-Bayoumy, K. (2001). The less harmful cigarette: A controversial issue. *Chemical Research in Toxicology, 14*, 768–785.

Jenkin, M. E., Saunders, S. M., Derwent, R. G., & Pilling, M. J. (2002). Development of a reduced speciated VOC degradation mechanism for use in ozone models. *Atmospheric Environment, 36*, 4725–4734.

Jenkin, M. E., Saunders, S. M., & Piling, M. J. (1997). The tropospheric degradation of volatile organic compounds: A protocol for mechanism development. *Atmospheric Environment, 31*, 81–104.

Kieber, R. J., Rhines, M. F., Willey, J. D., & Avery, G. B. (1999). Rainwater formaldehyde: Concentration, deposition and photochemical formation. *Atmospheric Environment, 33*, 3659–3667.

Lunden, M. A., Thatcher, T. L., Hering, S. V., & Brown, N. J. (2003). Use of time- and chemically-resolved particulate data to characterize the infiltration of outdoor $PM_{2.5}$ into a residence in the San Joaquin Valley. *Environmental Science & Technology, 37*, 4724–4732.

Ning, Z., & Sioutas, C. (2010). Atmospheric processes influencing aerosols generated by combustion and the inference of their impact on public exposure: A review. *Aerosol and Air Quality Research, 10*, 43–58.

Novak, K., Skelly, J. M., Schaub, M., Krauchi, N., Hug, C., Landolt, W., & Bleuler, P. (2003). Ozone air pollution and foliar injury development on native plants of Switzerland. *Environmental Pollution, 125*, 41–52.

Pope, C. A., Burnett, R. T., Thun, M. J., Calle, E. E., Krewski, D., Ito, K., & Thurston, G. D. (2002). Lung cancer, cardiopulmonary mortality, and long-term exposure to fine particulate air pollution. *Journal of the American Medical Association, 287*, 1132–1141.

Ryerson, T. B., Trainer, M., Holloway, J. S., Parrish, D. D., Huey, L. G., Sueper, D. T., Frost, G. J., Donnelly, S. G., Schauffler, S., Atlas, E. L., Kuster, W. C., Goldan, P. D., Hüber, G., Meagher, J. F., & Fehsenfeld, F. C. (2001). Observations of ozone formation in power plant plumes and implications for ozone control strategies. *Science, 292*, 719–723.

Seefeld, S., Kinnison, D. J., & Kerr, J. A. (1997). Relative rate study of the reactions of acetylperoxy radicals with NO and NO_2: Peroxyacetyl nitrate formation under laboratory conditions related to the troposphere. *Journal of Physical Chemistry A, 101*, 55–59.

Tanner, R. L., Miguel, A. H., de Andrade, J. B., Gaffney, J. S., & Strelt, G. E. (1988). Atmospheric chemistry of aldehydes: Enhanced peroxyacetyl nitrate formation from ethanol-fueled vehicular emissions. *Environmental Science & Technology, 22*, 1026–1034.

Wang, W., Primbs, T., Tao, S., & Simonich, S. L. M. (2009). Atmospheric particulate matter pollution during the 2008 Beijing Olympics. *Environmental Science & Technology, 43*, 5314–5320.

Yli-Tuomi, T., Lanki, T., Brunekreef, B., & Pekkanen, J. (2008). Determination of the sources of indoor $PM_{2.5}$ in Amsterdam and Helsinki. *Environmental Science & Technology, 42*, 4440–4446.

Zhang, B.-N., & Oanh, N. T. K. (2002). Photochemical smog pollution in the Bangkok metropolitan region of Thailand in relation to O_3 precursor concentrations and meteorological conditions. *Atmospheric Environment, 36*, 4211–4222.

Chapter 8

Abstracts and Keywords

Abstract

This chapter presents a condensed view of soil chemistry. It begins by focusing on soil formation, composition, and structure. Next, soil organic matter is described in detail, including humic and fulvic acid substances. Related, the sorption of metals and hydrophobic organic compounds in soils is presented. Topics will include ion exchange and complexation processes. Lastly, information on the causes of soil acidity and reactions between soil and wastes and pesticides is covered.

Keywords

Soil chemistry, Soil horizons, Organic matter, Humic substances, Fulvic acid, Metal sorption, Pesticides

Chapter 8

Soil Chemistry

What is a soil scientist? What do they do? How do they contribute to the field of Environmental Chemistry? Soil scientists (or pedologists) study the upper portion of the Earth's crust in terms of its physical and chemical properties, distribution, genesis and morphology, and plant and animal material. Like any applied science discipline, a strong background in the physical and biological sciences and mathematics is advantageous. Advanced knowledge of soil chemistry is vital in studying the accumulation and transport of contaminants, managing soils for crop

FIGURE 8.1

production, watershed rehabilitation, hazard remediation, and erosion control management. Graduates may go on to be employed by universities, research institutions, mining industries, state or federal conservation services, consulting firms, and regulatory and land management agencies.

8.1. INTRODUCTION TO SOIL CHEMISTRY

Soil chemistry is a young but growing branch of soil science with varying definitions and degrees of interpretation. In general terms it is the study of the chemical characteristics of soils. However, physical concepts such as the study of particle density and size, soil depth and pore spaces, bulk density, and soil moisture and temperature complement pure chemical concepts. In this chapter, attention will be given to such processes as weathering, adsorption, precipitation, complexation, and ion exchange, both in the soil solution and the solid-liquid interface. In addition, we will explore soil biochemistry, the causes of soil acidity, soil buffering, and reactions between soil and wastes, pesticides, and metals. It is important to remember that this chapter is only an overview to an ever-growing field of information with a wide variety of applications in agriculture and the environment.

8.2. SOIL FORMATION, COMPOSITION, AND STRUCTURE

What is soil? Let's begin with an accurate definition of soil developed by Joffe in 1936, but one that is still relevant today (Joffe, 1936):

The soil is a natural body, differentiated into horizons of mineral and organic constituents, usually unconsolidated, of variable depth, which differs from the parent material below in morphology, physical properties and constitution, chemical properties and composition, and biological characteristics.

Any journey into the study of soil chemistry must begin with the kinetics of soil formation, i.e., knowledge of the variety of reactions that are likely to occur during the transformation of **soil parent material** to a **soil profile** (Barshad, 1965). Soil parent material is the material that soil develops from, and may be solid rock that has decomposed, or alluvial material that has been deposited by natural forces (e.g., wind, water, or ice). As expected, the chemical composition of the parent material plays an essential role in influencing soil properties, especially during the early stages of soil formation. The soil profile is the vertical display of **soil horizons** arising from continued weathering. Soils horizons are composed of a wide range of materials from organic rich surface layers to underlying layers high in rock matter. Complete descriptions of the master soil horizons are found in Table 8.1. In-depth coverage of soil organic matter is provided in subsequent sections. As can be surmised from the descriptions above and information contained in Table 8.1, climate and topography have significant impacts on soil formation. For example, the parent material necessitates relatively undisturbed soil horizon processes for proper vertical development. Heavy precipitation shuffling across the surface strips parent material away, hindering soil formation. Topography helps determine water runoff; its orientation affects the microclimate, which in turn affects vegetation growth and cover.

Structured soil in solid form consists of **aggregates** as depicted in Figure 8.2. Soil aggregates are groups of soil particles that bind to each other by

TABLE 8.1 Master Soil Horizons and Associated Properties

Soil Horizon*	Characteristics
O	The top portion of the soil profile dominated by high organic matter content (or humus) consisting of partially decomposed litter (e.g., leaves, twigs, moss, and lichens) that has accumulated on the surface.
A	A surface horizon that is characterized by a mix of mineral (inorganic products of weathering) and organic matter. Eluviation (the removal of inorganic and organic matter by leaching) occurs in this horizon.
E	A horizon dominated by the removal of silicate clay, iron, and aluminum, ultimately leaving a layer of sand and silt particles. No major rock structure present.
B	A horizon dominated by illuviation, a process where fine particle material is accumulating in a downward fashion through, for example, percolation of soil drainage water. Often associated with increased concentrations of carbonate precipitates and residual concentrations of iron and aluminum oxides.
C	A horizon composed of unconsolidated parent material, either created at source or transported into its present location. C horizons are characteristically composed of sediments, saprolite, and bedrock.
R	A horizon characteristically composed of hard bedrock (e.g., granite, basalt, limestone, and sandstone).

** Note: Base symbols to which other characters are added to complete the designation.*

clay particles and organic matter. The space between the aggregates provides **pore space** for retention and exchange of air and water. This allows soil to drain appropriately, while retaining enough moisture to promote and uphold healthy plant growth. As to be expected, sandy soils (sand particles range from 0.05 mm for fine sand to 2.0 mm for course sand) have limited structure but are typically free-draining. Those soils with higher clay content (clay particles are typically 0.001–0.002 mm in size) demonstrate increased soil structural characteristics but do suffer from decreased drainage ability (Williams, et al, 1983). Ultimately, the amount of soil pores and the pore size relate to the **drainage capacity** of the soil. Also, as expected, when soil structure varies in different soil types and the horizons of these soil types, water flow and contaminant transport in soils will be impacted. For example, the extent of groundwater contamination caused by pesticides largely controlled by water flow is strongly dependent upon the composition of soil structure components resulting from soil formation and management.

TABLE 8.2 Chemical Compositions and Physical Properties of Selected Clays in Soil

Type	Chemical Formula	Particle Size (mm)	Surface Area (m^2/g)	Cation Exchange Capacity (meq/100 g)
Kaolinite	$Al_2(OH)_4Si_2O_5$	0.10–5.0	10–25	5–15
Illite	$K_{0\text{-}2}Al_4(Si_{8\text{-}6}Al_{0\text{-}2})O_{20}(OH)_4$	0.10–2.5	100–200	15–40
Montmorillonite	$Al_2(OH)_2Si_4O_{10}$	0.10–1.0	650–800	75–100

Soils that are predominately composed of clay are called fine textured soils, while those dominated by larger particles are termed coarse textured soils. Clays are typically characterized as layered silicates, metal oxides and hydroxides, amorphous and allophones, and crystalline chain silicates. The most important property of clay fractions is that of cation adsorption and exchange, given that this property largely determines the storage capacity of a soil for plant nutrients (Reynolds, et al., 2002). Table 8.2 provides relevant information on selected clay types, including the **cation-exchange capacity (CEC)**: a measure of the quantity of sites on soil surfaces that can retain cations by electrostatic forces. These cations are in turn easily exchangeable with other cations in the soil solution.

TEXTBOX 8.1

ME or meq represents 1 milligram of exchangeable H^+. For example, in soil with a CEC of 1, every 100 grams of soil contain an amount of negative sites equal to the amount of positive ions in 1/1000th of a gram of H^+ (or its equivalent). CEC estimates are made by determining the extractable cations (K^+, Ca^{2+}, Mg^{2+}, and Na^+) and estimating H^+ from soil and buffer pH measurements.

Negative surface charge on clay surfaces makes them extremely reactive. They are developed in two ways: 1) isomorphic substitution (permanent charge) and 2) pH dependent charge through deprotonation of surface functional groups. For example, substitution of a cation of lower valence (e.g., Al^{3+}) for one of higher valence (e.g., Si^{4+}) in the tetrahedral layer results in permanent charge. In regards to pH dependent charge, as the pH of the soil environment increases, for example, weak acid functional groups (e.g., carboxylic acids) donate a proton and generate negative charge as follows:

$$COOH + OH^- \rightleftharpoons COO^- + H_2O \tag{8.1}$$

Exchange capacity is routinely measured in milligram equivalents, abbreviated ME or meq. One meq of negative charge on a clay particle is neutralized by one meq of associated cation. Note that cmolc/kg (centimoles of charge per kilogram of dry soil) may also be reported. Universally, the more clay and organic matter in the soil, the higher the CEC. Organic matter has an estimated CEC of about 150 meq/100 g.

Example Problem 8.1

Using concepts discussed in Textbox 8.1, a clay material with a CEC of 1 meq/100 g is capable of adsorbing 1 mg of H^+ (or its equivalent) for every 100 g of clay. What would the equivalent amount of Mg be in mg?

Answer: Mg has a +2 charge (2 charges) and an atomic weight of 24.31. Therefore, Mg with an atomic weight of 24.31 has an equivalent weight of 24.31/2 = 12.1. Hence, 12.1 mg is the weight of 1 meq of Mg.

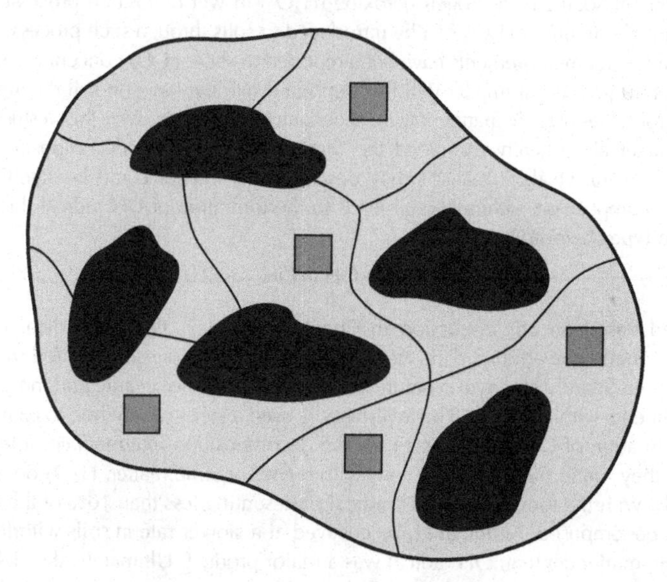

FIGURE 8.2 Schematic representation of a typical soil aggregate held together by organic matter and clay. The space between the aggregates provides pore space for retention and exchange of air and water.

8.3. SOIL ORGANIC MATTER AND BIOCHEMICAL ASPECTS

Let's reflect on the importance of **soil organic matter (SOM)** in more detail. Any discussion on SOM must include biochemical concepts, as they play vital roles in decomposition of complex organic compounds. Recall from Chapter 1 that the carbon cycle illustrates how carbon is distributed through the atmosphere, biosphere, hydrosphere, and pedosphere. As also mentioned, dead organic matter (OM) of the soil is colonized by microorganisms, which obtain energy for growth from the oxidative decomposition of organic molecules (Federle, et al., 1986). It is during the decomposition process whereby inorganic elements are converted from organic compounds, a process called **mineralization** (Nierop, et al., 2006). For example, organic nitrogen and phosphorus are mineralized to ammonia and orthophosphate, respectively, thereby providing nutrients for biotic uptake and growth. In addition, carbon is converted to carbon dioxide. The residue carbon substrate used by the microorganisms is incorporated into their cell substance (biomass), which is called **immobilization** (Alexander, 1977). The incorporated minerals are immobilized and freed after the organisms die off.

Research Application X—Hydrogen Peroxide Decomposition in Soils: A Role for Organic Matter?

Recall from Chapter 7 that the photolysis of HCHO in the aqueous phase is a dominant source of hydrogen peroxide (H_2O_2) in wet deposition processes. A substantial amount of H_2O_2 can be introduced to soils through such processes. In fact, rainwater measurements have been reported to show H_2O_2 concentrations as high as 40 μM. What impact will this deposited oxidant have on soil chemistry? What role does organic matter play in its decomposition? To answer this, a study by Petigara et al. critically examined the mechanisms of H_2O_2 decomposition in soils (Petigara, et al., 2002). As they described, its impact could be significant if its decomposition within the soil leads to the formation of OH radical through Fenton-type chemistry:

$$H_2O_2 + M^{n+} \rightarrow OH + OH^- + M^{n+1}$$

And based on our discussion in Chapters 3 and 7, the OH radical is an exceptionally powerful oxidant. As a result, oxidative processes initiated by OH could significantly modify the nature and speciation of the organic and inorganic constituents within the soil. The investigators used a sensitive method to examine the formation of OH radical as a possible decomposition intermediate. Interestingly, they found that in surface soils with higher organic matter, H_2O_2 decayed rapidly, with the formation of OH radical representing less than 10% of the total H_2O_2 decomposed. Although H_2O_2 decayed at a slower rate in soils with lower organic matter content, OH radical was a major product. Ultimately, the yield of OH radical varied widely and depended on numerous soil parameters. More detail on the mechanisms behind such processes can be found in the body of the original published study referenced above.

8.3.1. An Introduction to Humic Substances

The process of humification results in the formation (primarily through microbial degradation) of **humus** from decaying organic matter of plant, animal, and microbial origin. The biological composition, the main fraction of natural humic matter, is the humic substances, which largely contain **humic acids** and **fulvic acids** (Martin, et al., 1998). Humic acids (see proposed model structure in

Figure 8.3) are heterogeneous, high molecular weight organic materials that demonstrate a variety of components including quinone, phenol, catechol, and sugar moieties (Stevenson, 1994). The functional groups that contribute most to surface charge and reactivity of humic acids are believed to be phenolic and carboxylic groups. In fact, by looking at the model structure proposed by multiple research groups, the matrix appears to be substituted by alkyl chains or aromatic moieties (both acting as monosubstituents) that correspond to esterifed aliphatic monocarboxylic acids, alkanol, or aromatic acids. In all cases, the consideration of fatty acid and alkanol moieties indicates a possible bacterial input (Grasset & Ambleá, 1998). As expected, the characteristics of humic acids change with soil depth, which in turn affect their sportive behavior in soil.

Example Problem 8.2

What is the significance of the presence of carboxylate and phenolate groups in humic acids?

Answer: For example, when a carboxyl group is deprotonated, its conjugate base, a carboxylate anion, is produced. In addition, the phenol molecule has a tendency (in a slightly acidic environment) to lose the H^+ ion from the hydroxyl group, resulting in the highly water-soluble phenolate anion. Thus, complexation with metal ions (e.g., Ca^{2+}, Fe^{2+}, Fe^{3+}, and Mg^{2+}) is possible. (See section 8.4 for greater detail on the interaction of metals with humic acids.)

Fulvic acids (see proposed model structure in Figure 8.4) can be characterized as those components of SOM that can form water-soluble and water-insoluble complexes with metal ions and hydrous oxides, interact with clay materials, and combine with alakanes, fatty acids, and dialkyl phthalates (Ogner and Schnitzer, 1970; 1971). Like humic acids, fulvic acids contain acidic functional groups, primarily carboxylic acid and phenolic hydroxyl groups, which give them the ability to react with an assortment of species including free radicals, metals, and biological enzyme systems. As depicted in the model structure, the molecular weight of fulvic acids is generally thought to be less than those of humic acids. Regardless, laboratory studies have confirmed that a complex mixture of humic and fulvic acids can exist in the same soil deposit.

8.4. SORPTION OF METALS AND HYDROPHOBIC ORGANIC COMPOUNDS IN SOILS

Sorption of heavy metals in soils has important implications with respect to plant nutrition and soil contamination as a result of numerous industrial activities, vehicular emissions, and landfilling of fly ash from incineration processes. The term sorption was described earlier in this book; here we describe it in terms of the binding of metals in soils by chemical processes. One of the most important chemical mechanisms of metal sorption is adsorption, which will be covered in detail below.

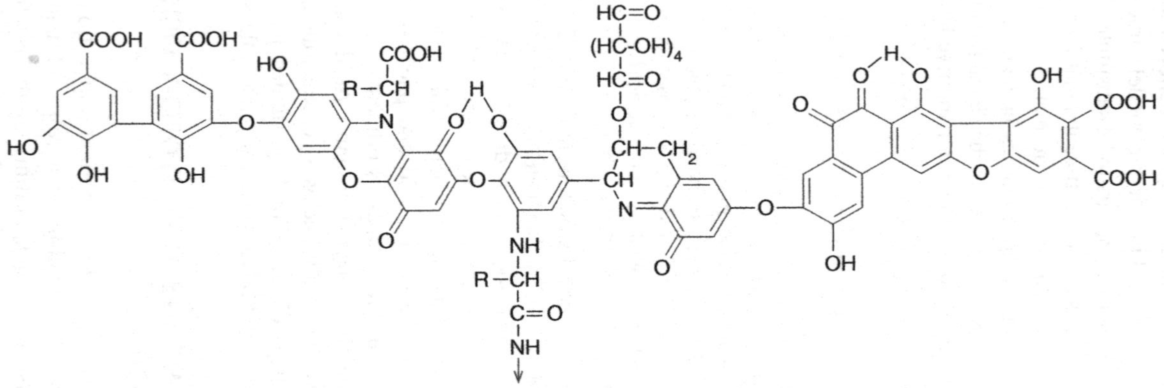

FIGURE 8.3 Model structure of humic acids. In regards to structure, aliphatic dicarboxylic acids are believed to act as bridges between alkyl chains in the matrix, while fatty acids, alcohols, and aromatic acids correspond mainly to monosubstituents of the matrix.

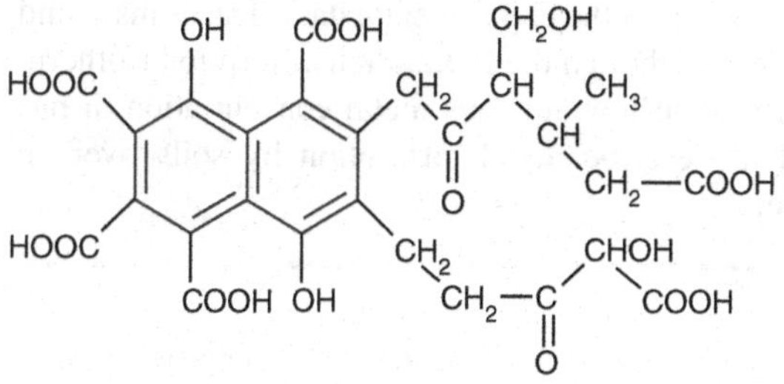

FIGURE 8.4 Model structure of fulvic acid. As with humic acids, fulvic acid contains many reactive functional groups, including carboxyls, hydroxyls, carbonyls, phenols, quinones, and semiquinones.

8.4.1. Metal Adsorption in Soils

The migration of metals in the soil environment is directly related to their partitioning between soil and soil solution. The presence of trace metals in the solid phase is partially the result of adsorption to components of the soil, a process that is dictated by pH, metal concentration, and other properties of the soil and solution (Lee, et al., 1996). As previously defined in Chapter 4, the equilibrium distribution between the soil phase and water phase is expressed as

$$K_d = \frac{C_s}{C_l}, \tag{8.2}$$

where C_s = the concentration of metal in the solid phase (mg metal/kg sorbing material) and C_l = the concentration (mg metal/L of solution) of the metal in the solution phase. K_d is the partition coefficient. The ultimate goal in any sorption study is to determine the sorption capacity of a given metal or group of metals. Equilibrium sorption is described by a **sorption isotherm**, which is a plot of the relationship between the amount of metal sorbed and the equilibrium concentration of the metal. In other words, it is the activity of the free metal in solution. A simple linear relationship between sorbed and solution phases is expressed as

$$S = K_d C. \tag{8.3}$$

Here, K_d alone describes the sorption process and is independent of metal solution concentration. This relationship is depicted in Figure 8.5. However, in most applications the isotherm is not linear over the range of interest of metal concentration (Antoniadis, et al., 2007). This curvilinear sorption behavior is depicted in Figure 8.6. This relationship is based on a linearized **Langmuir equation**, given by

$$\frac{C}{S} = \frac{C}{S_{\max}} + \frac{1}{S_{\max} K_a}, \tag{8.4}$$

where C is the equilibrium concentration, S is the amount of metal ion sorbed (mg/g), $S_{\max}$ is the maximum sorption capacity of the soil, and K_a is the sorption equilibrium constant. In addition, we can describe metal adsorption by the **Freundlich equation**:

$$\log S = (N)\log C + \log K, \tag{8.5}$$

where N and K are constants fitted from the experimental data. The terms C and S were defined previously above. Although the Freundlich adsorption isotherm is valid only for metal adsorption at low aqueous metal concentration, it has often been incorrectly used to describe metal adsorption by soils over an extended concentration range.

Example Problem 8.3

Referring to Figure 8.6, if we were to plot C/S as a function of C, describe what the slope and y-intercept would be characterized as.

Answer: The slope would be the reciprocal of the sorption capacity, S_{max}, and the intercept.

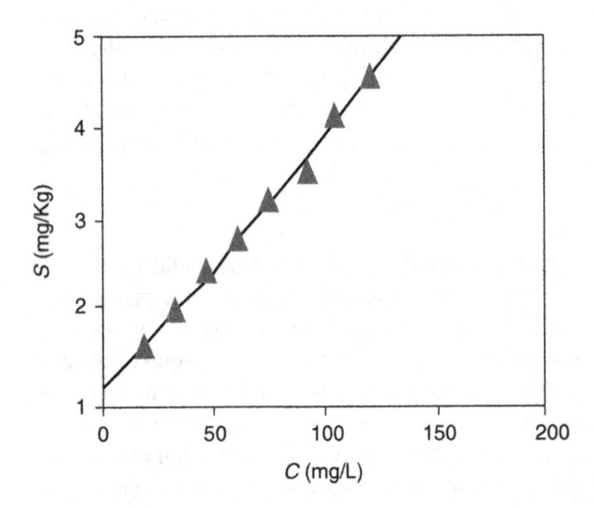

FIGURE 8.5 Simple linear sorption isotherm depicting the relationship between the amount of metal sorbed and the equilibrium concentration of the metal.

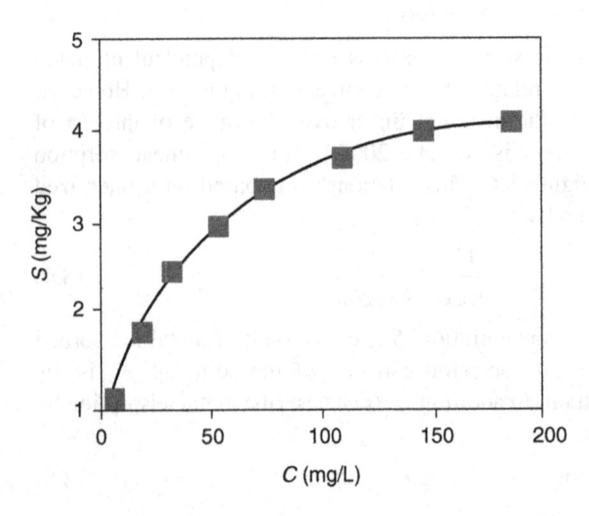

FIGURE 8.6 An example curvilinear sorption isotherm for metal adsorption based on a linearized Langmuir equation.

The processes of **ion exchange** and **surface complexation** are considered to be the two primary adsorption mechanisms for metals. And as introduced earlier, the association of heavy metals with soils depends upon many soil parameters, most significantly on organic matter and clay content, which contribute greatly to these two mechanisms. Recall that the net charge of soil particles is nearly always negative given the nature of clay materials and organic matter. Therefore, ion exchange is expected to be an important adsorption process for cations, but will not likely affect the charge of the solid surface. In this process, a dissolved ion is attracted electrostatically by a charged particle surface. By and large, ions with higher valency will exchange for those demonstrating lower valency: for example, $Al^{3+} > Ca^{2+} > Mg^{2+} > K^{+} = NH_4{}^{+} > Na^{+}$. For ions of the same charge, the cation with the smallest hydrated radius is strongly absorbed since it tends to move close to the site of charge (Sverjensky, 2006). For example, the hydrated radius for $K^{+} = 0.53$ nm, while that for $Na^{+} = 0.79$ nm. Therefore, based on the fact above, K^{+} will exchange for Na^{+} on available exchange sites.

As we have learned, the surfaces of soils contain key functional groups, for example, $-OH$, $-COOH$, $-SH$, and $-NH_2$, that can interact with metal ions (forming an **inner-sphere complex** with the ligand that is present). Note that anions may also form surface complexes; this most likely involves the inner-sphere complexation of an anionic ligand to, for example, Fe or Al atoms found on oxide surfaces. In short, surface complexation involves the direct coordination of the adsorbing ion with the surface, so that the ion becomes part of the surface, for example, the exchange of K^{+} for Mg^{2+}, which forms a surface complex in the presence of a carboxylate group:

$$-COO^{-} - - - - - - - - - - - - K^{+} + Mg^{2+} = -COOMg^{+} + K^{+}, \quad (8.6)$$

where the sign "=" is a surface group site available for binding. As discussed, clay minerals, various oxide surfaces, and solid-phase humic substances possess both larger surface areas and available reactive sites. Recall that many humic acids have two or more carboxylate and/or phenolate groups, enabling them to form chelate complexes with ions. The formation of chelate complexes is an important aspect of the biological role of humic acids in regulating bioavailability of metal ions in soils (Piccolo, 2002). Finally, sorption to soil organic matter is also a dominant sorption process for hydrophobic organic compounds (HOC), thus playing a role in controlling the fate of such pollutants in terrestrial environments. For example, studies have shown the sorption of such compounds as naphthalene, phenanthrene, carbazole, and anthracene on clay and humic acid substances (Xing, 2001; Feng, et al., 2006).

TEXTBOX 8.2

An **inner-sphere complex** is one that is formed when covalent bonds between constituent ions are formed.

8.4.2. The Role of Weak Organic Acids

Although their soil concentrations are generally low, non-humified, weak organic acids (e.g., acetate, citric, tartaric, oxalic, and formic acids; see Figure 8.7 for example structures) from the decay of plants, animals, and microbial mass have also been shown to have chelating characteristics (Wasay, et al., 2001). These relatively simple carboxylic acids have been found to be present in most soil systems with concentrations typically expressed in the range of 1–20 nM (Strobel, 2001). Concentration issues aside, what roles do organic acids play in soil? Do they show strong affinity mineral surfaces? First, they perform many roles in shaping soil properties and the surrounding environment, including (vanHees, et al., 2003):

1. Mobilization of nutrients (e.g., Fe, N, and P) by plants and microorganisms;
2. Detoxification of metals by plant roots;
3. Promoting the dissolution of primary minerals via weathering;
4. Forming an important source of labile carbon for soil microorganisms.

Considering the second question, a variety of organic acids show a strong affinity for certain mineral surfaces, and these have been used extensively as model anions for sorption studies (vanHees, et al., 2003). They are particularly important due to their metal chelating/complexing properties for mobilization of metals. For

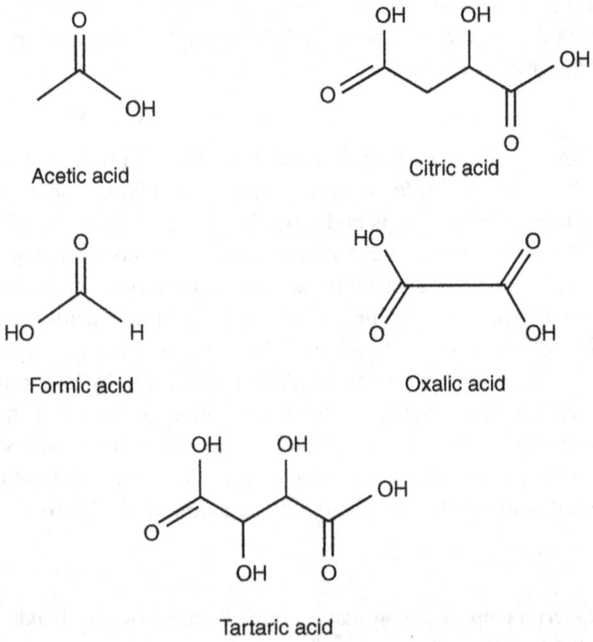

Acetic acid

Citric acid

Formic acid

Oxalic acid

Tartaric acid

FIGURE 8.7 Example non-humified organic acids found in soil.

example, organic acids have been shown to influence the rate of cadmium release from soils and increase the solubility of cadmium in bulk soil through the formation of soluble cadmium-organic acid complexes (Krishnamurti, et al., 1997). Sources of organic acids are now well recognized; they comprise anthropogenic and biogenic emissions and chemical transformations of precursors.

8.5. PESTICIDE ADSORPTION IN SOILS

As discussed in Chapter 4, pesticides are also divided into numerous classes based on chemical nature, the most widely reported being organochlorine and organophosphorus compounds. Common examples of each (and their structural characteristics) are depicted in Figures 8.8 (organochlorine) and 8.9 (organophosphorus). As expected, the extent of adsorption depends on soil properties and those of the compound, which include size, shape, configuration, molecular structure, solubility, polarity, polarizability, and the acid-base nature of the pesticide molecule (Pignatello and Xing, 1996).

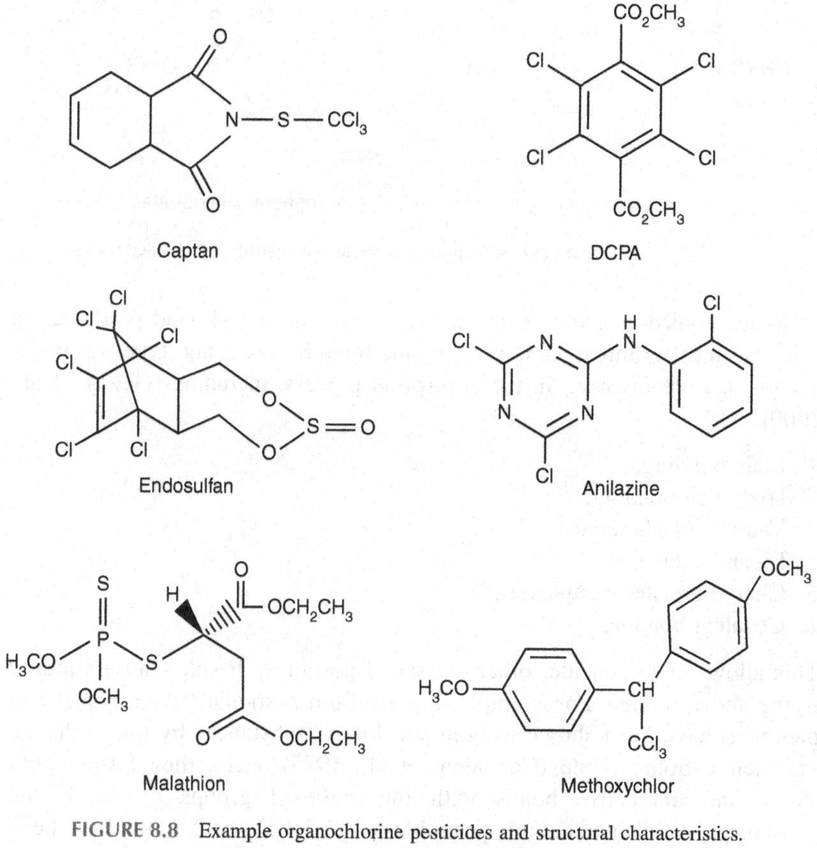

Captan

DCPA

Endosulfan

Anilazine

Malathion

Methoxychlor

FIGURE 8.8 Example organochlorine pesticides and structural characteristics.

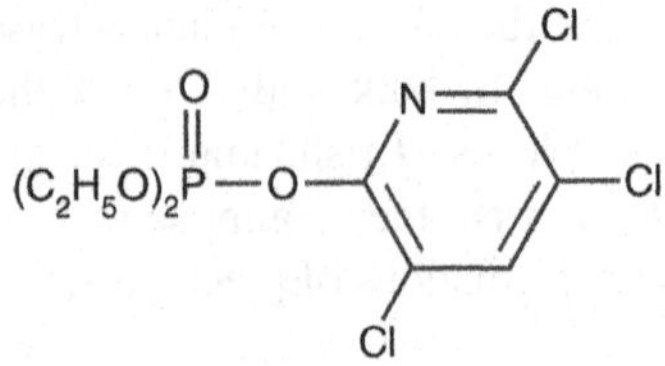

Chlorpyrifos

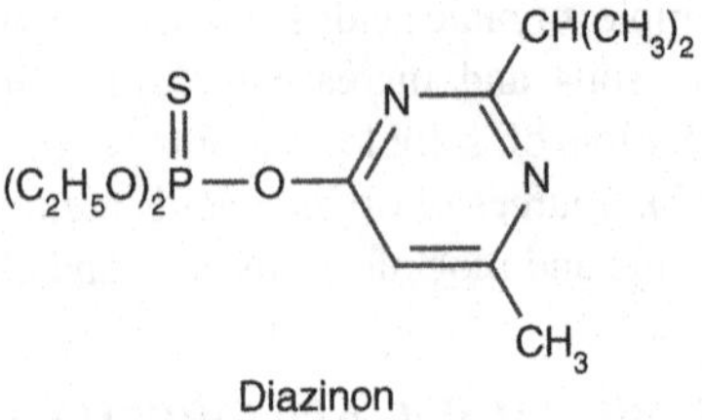

Diazinon

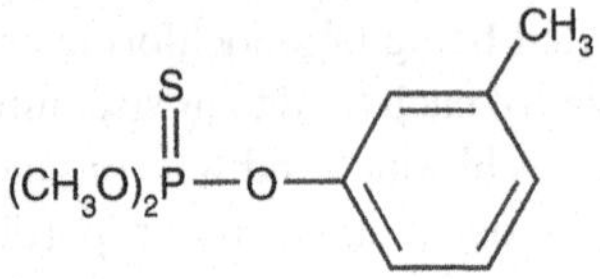

OP-2

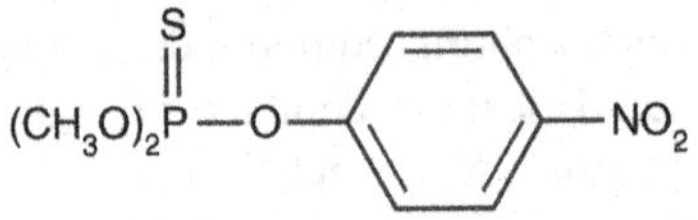

Methyl Parathion

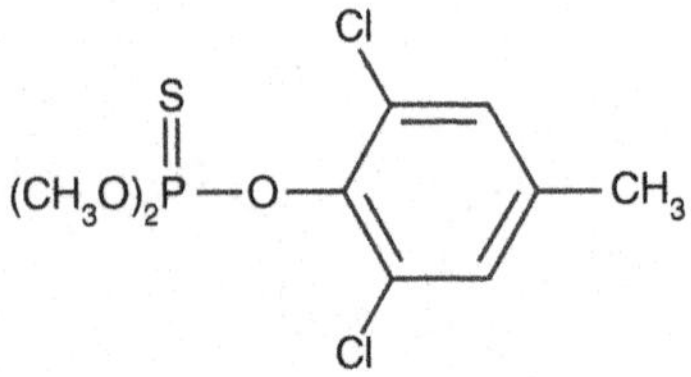

Tolclofos

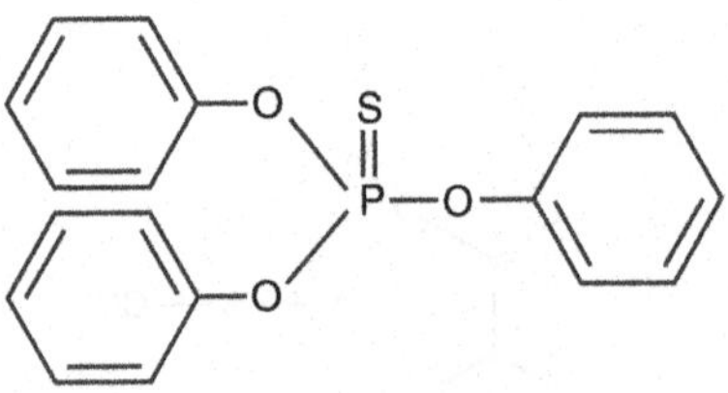

Triphenyl Phosphate

FIGURE 8.9 Example organophosphorus pesticides and structural characteristics.

When considering the environmental significance of bound pesticides in soils, we must examine some very unique binding forces and types of mechanisms that are involved in the adsorption process, including (Gevao, et al., 2000):

1. Ionic bonding;
2. Hydrogen bonding;
3. Van der Waals forces;
4. Ligand exchange;
5. Charge-transfer complexes;
6. Covalent bonding.

This allows us to consider other classes of pesticides besides those featured in the above figures. For example, bipyridilium pesticides (e.g., daquat and paraquat) have been shown to bind soil humic substances by ion exchange via their cationic group (Carringer, et al., 1975). Here, they form highly stable and unreactive bonds with the carboxyl groups of the humic substances. Acidic and anionic pesticides (e.g., 2,4-D and 2,4,5-T) have been shown to interact with soil organic matter by hydrogen bonding at pH values below their pK_a in non-ionized forms through −COOH and −COOR groups (Gan, et al., 1995). Finally, it should not be surprising that pesticides may undergo binding by several different mechanisms simultaneously. What might be some of the environmental factors that are capable of influencing the fate and binding of pesticides in soils? Will they be similar to metals and other xenobiotics of interest? As demonstrated by many researchers in the field, the environmental significance of bound pesticide residues depends on a variety of factors including (Gevao, et al., 2000; Carringer, et al., 1975; Gan, et al., 1995):

1. Concentration;
2. Soil pH;
3. Rate and mode of pesticide application to soil;
4. Soil type and depth;
5. Microbial degradation;
6. Photochemical reactions;
7. Soil amendments with organic and inorganic fertilizers.

8.6. ACID AND BASE CHARACTERISTICS

The issue of pH was a much discussed topic in previous sections with obvious consequences, for example, in the adsorption of inorganic and organic substances. Hence, soil acidity is one of the most commonly measured properties. The occurrence of relevant acids and bases is a result of numerous biotic and abiotic processes including the decomposition of plant and animal tissues by microbes, wet and dry deposition of non-humic organic acids, runoff from industrial processes, acidified rain, and the weathering of pyrite in soil from mining sites. Carbonate minerals can also act as buffers, for example:

$$2\,CaCO_3 + 3H^+ \rightleftharpoons 2Ca^{2+} + HCO_3^- + H_2CO_3. \qquad (8.7)$$

Here, carbonate and bicarbonate ions are released into the water phase upon the addition of H^+. What would happen if bulk acids were released into soil due to an industrial spill or acid mine drainage? What effects might acid deposition have on soil properties? Let's consider the latter in greater detail. As expected, soil plays a vital role in the biogeochemical processes that establish the watershed reaction to acid deposition. Within the soil, the depletion of base cations, the biological transformation of nitrogen, and the retention or release of sulfate comprise the major processes governing the export of acidic anions that ultimately leads to stream acidification (Sullivan, et al., 2004). Considering the former (base cations), introduction of H^+ ions from acid deposition (or from decomposition processes within the soil) can promote the mobilization and leaching of base cations. For instance, lowered soil pH results in facilitating the dissolution of aluminum (particularly Al^{3+}) from mineral sources. The dissolution of aluminum ion produces a moderate acidic environment; hence the properties of acid soils are controlled by aluminum chemistry. Raising the pH of soils is typically accomplished by some form of neutralization. For example, liming agents such as $CaCO_3$ or powdered limestone are added to soils (mainly in agriculture) to produce a neutralization reaction. In the case of $CaCO_3$ we can write:

$$CaCO_3 + 2HCl \rightarrow CaCl_2 + H_2O + CO_2. \tag{8.8}$$

As shown, this is a highly efficient process with one molecule of $CaCO_3$ treating two molecules of HCl. Liming is also believed to help improve soil structure, but the mechanisms behind this process are unclear.

8.7. END OF CHAPTER QUESTIONS

8.1 How does soil particle size influence, for example, water infiltration?

8.2 Describe two scenarios in soil that would cause a decrease in soil cation-exchange capacity (CEC).

8.3 Alternatively, describe two scenarios that would cause an increase in soil CEC. Use agricultural practices as a guide.

8.4 Assuming that a clay material with a CEC of 1 meq/100 g is capable of adsorbing 1 mg of H^+ (or its equivalent) for every 100 g of clay, calculate the equivalent amount of Ca in mg.

8.5 Describe two effects that cannot be simulated when using the simplistic nature of the K_d model for metal adsorption.

8.6 Refer back to Equation 8.5. Simplify the relationship to show how it is possible that the single term K_d can be used to describe metal adsorption data. Comment on its appropriateness in true experimental studies.

8.7 The pH is the single most important parameter that governs the adsorption of metal ions in soils, and therefore electrostatic attraction varies depending on pH value. Based on attraction differences, describe the pH conditions necessary for the adsorption of both anion and cations. In your discussion, use oxide surface and solid-phase humic substance characteristics to aid your explanation.

8.8 In some arid regions, high concentrations of sodium may result in the saturation of the soil's cation exchange capacity (CEC) with sodium. As a result, a high soil pH likely results. Provide an example hydrolysis reaction that may account for this (hint: OH^- will be produced).

8.9 List other metals that may be common to evaporative soil salts in arid conditions.

8.10 As mentioned in the text, the dissolution of aluminum ion produces a moderate acidic environment. Produce a hydrolysis reaction to correlate with this statement.

8.11 What property related to cation exchange capacity (CEC) will occur as a result of the high immobilization affinity of Al^{3+}?

8.12 Nitrogen fertilizers containing ammonium are known to increase soil acidity. Provide chemical justification for this statement.

8.8 REFERENCES

Alexander, M. (1977). *Soil microbiology.* New York: John Wiley & Sons.

Antoniadis, V., McKinley, J. D., & Zhuairi, W. Y. W. (2007). Single-element and competitive metal mobility measured with column infiltration and batch tests. *Journal of Environmental Quality, 36*, 53–60.

Barshad, I. (1965). Chemistry of soil development. In F. E. Bear (Ed.), *Chemistry of the Soil* (2nd ed.). New York: Reinhold Publishing Corporation.

Carringer, R. D., Weber, J. B., & Monaco, T. J. (1975). Adsorption-desorption of selected pesticides by organic matter and montmorillonite. *Journal of Agricultural and Food Chemistry, 23*, 569–572.

Federle, T. W., Dobbins, D. C., & Thornton-Manning, J. R. (1986). Microbial biomass, activity and community structure in subsurface soils. *Groundwater, 24*, 365–374.

Feng, X., Simpson, A. J., & Simpson, M. J. (2006). Investigating the role of mineral-bound humic acid inphenanthrene sorption. *Environmental Science & Technology, 40*, 3260–3266.

Gan, J., Koskinen, W. C., Becker, R. L., & Buhler, D. D. (1995). Effect of concentration on persistence of alochlor in soil. *Journal of Environmental Quality, 24*, 1162–1169.

Gevao, B., Semple, K. T., & Jones, K. C. (2000). Bound pesticide residues in soils: A review. *Environmental Pollution, 108*, 3–14.

Grasset, L., & Ambleá, S. (1998). Structure of humin and humic acid from an acid soil as revealed by phase transfer catalyzed hydrolysis. *Organic Geochemistry, 29*, 881–891.

Joffe, J. S. (1936). *Pedology.* New Brunswick, NJ: Rutgers University Press.

Krishnamurti, G. S. R., Huang, P. M., & Van Rees, K. C. J. (1997). Kinetics of cadmium release from soils as influenced byorganic: Implications in cadmium availability. *Journal of Environmental Quality, 26*, 271–277.

Lee, S.-Z., Allen, H. E., Huang, C. P., Sparks, D. L., Sanders, P. F., & Peijnenburg, W. J. G. M. (1996). Predicting soil-water partition coefficients for cadmium. *Environmental Science & Technology, 30*, 3418–3424.

Martin, D., Srivastava, P. C., Ghosh, D., & Zech, W. (1998). Characteristics of humic substances in cultivated and natural forest soils of Sikkim. *Geoderma, 84*, 345–362.

Nierop, K. G. J., Preston, C. M., & Verstraten, J. M. (2006). Linking the B ring hydroxylation pattern of condensed tannins to C, N and P mineralization: A case study using four tannins. *Soil Biology and Biochemistry, 38*, 2794–2802.

Ogner, G., & Schnitzer, M. (1970). Humic substances: Fulvic acid-dialkyl phthalate complexes and their role in pollution. *Science, 170*, 317–318.

Ogner, G., & Schnitzer, M. (1971). Chemistry of fulvic acid, a soil humic fraction, and its relation to lignin. *Canadian Journal of Chemistry, 49*, 1053–1063.

Petigara, B. R., Blough, N. V., & Mignerey, A. C. (2002). Mechanisms of hydrogen peroxide decomposition in soils. *Environmental Science & Technology, 36*, 639–645.

Piccolo, A. (2002). The supramolecular structure of humic substances. A novel understanding of humus chemistry and implications in soil science. *Advances in Agronomy, 75*, 57–134.

Pignatello, J. J., & Xing, B. (1996). Mechanisms of slow sorption of organic chemicals to natural particles. *Environmental Science & Technology, 30*, 1–11.

Reynolds, W. D., Bowman, B. T., Drury, C. F., Tan, C. S., & Lu, X. (2002). Indicators of good soil physical quality: Density and storage parameters. *Geoderma, 110*, 131–146.

Stevenson, F. J. (1994). *Humus chemistry: Genesis, composition, reactions*. New York: John Wiley & Sons.

Strobel, B. W. (2001). Influence of vegetation on low molecular weight carboxylic acids in soil solution: A review. *Geoderma, 99*, 169–198.

Sullivan, T. J., Fernandez, I. J., Herlihy, A. T., Driscoll, C. T., McDonnell, T. C., Nowicki, N. A., Snyder, K. U., & Sutherland, J. W. (2004). Acid-base characteristics of soils in the Adirondack Mountains, New York. *Soil Science of America Journal, 70*, 141–152.

Sverjensky, D. A. (2006). Prediction of the speciation of alkaline earths adsorbed on mineral surfaces in salt solutions. *Geochimicaet Cosmochimica Acta, 70*, 2427–2453.

vanHees, P. A. W., Vinogradoff, S. I., Edwards, A. C., Godbold, D. L., & Jones, D. L. (2003). Low molecular weight organic acid adsorption in forest soils: Effects on soil solution concentrations and biodegradation rates. *Soil Biology & Biochemistry, 35*, 1015–1026.

Wasay, S. A., Barrington, S., & Tokunaga, S. (2001). Organic acids for the in situ remediation of soils polluted by heavy metals: Soil flushing in columns. *Water, Air and Soil Pollution, 127*, 301–314.

Williams, J., Prebble, R. E., Williams, W. T., & Hignett, C. T. (1983). The influence of texture, structure and clay mineralogy on the soil moisture characteristic. *Australian Journal of Soil Research, 21*, 15–32.

Xing, B. (2001). Sorption of naphthalene and phenanthrene by soil humic acids. *Environmental Pollution, 111*, 303–309.

Chapter 9

Abstracts and Keywords

Abstract

This chapter presents an overview of environmental toxicology and hazardous waste characterization. It begins with a definition of xenobiotics and examples with chemical characterization. Next, an evaluation of xenobiotic exposure pathways and mechanisms of action within the human body are presented. Topics in this section include toxicity testing and examination of dose-response relationships. Next, biotransformation processes related to xenobiotic metabolism is presented. Subsequently, hazardous waste identification, characterization, and minimization concepts are covered. Lastly, information on risk assessment activities in relation to hazardous waste is offered.

Keywords

Toxicity, Xenobiotics, Dose-response relationships, Exposure, Hazardous waste, Risk assessment, Biotransformation

Chapter 9

Environmental Toxicology and Hazardous Waste Characterization

Polystyrene (trade name Styrofoam) is a hard and transparent synthetic resin produced by the polymerization of styrene. Styrene (the monomer) is a two-carbon chain with a benzene ring attached to one of the carbons (Figure 9.1). Originally developed to mold and insulate homes and buildings, polystyrene has since been widely employed in the food service industry as rigid trays, disposable eating utensils, bowls, and foamed cups. When in contact with warm food and drink, alcohol, oil, and acidic foods, styrene and benzene are leached from the containers, thus posing a health threat to humans. The Foundation for Advancements in Science and Education estimates that one-quarter to one-third of all the matter in landfills is Styrofoam. The substance takes centuries, if not millennia, to biodegrade. Fortunately, numerous U.S., Canadian, European, and Asian cities have banned polystyrene food packaging.

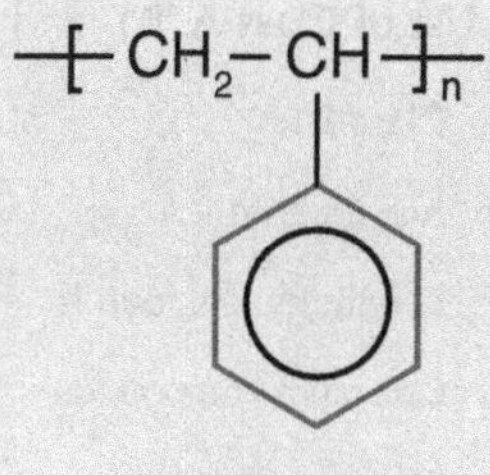

$$-[CH_2-CH]_n-$$

FIGURE 9.1

9.1. INTRODUCTION TO ENVIRONMENTAL TOXICOLOGY

Environmental toxicology is an interdisciplinary field of study that examines the underlying biological and chemical mechanisms by which **xenobiotics** interact with the biosphere, including humans. The three types of xenobiotic substances that produce injurious effects on human health that are of highest concern are classified and defined as follows:

1. **Carcinogens**: agents that may produce uncontrolled cell growth (cancer), either by themselves or in combination with other substances;
2. **Mutagens**: substances that induce genetic changes (e.g., mutations) in the DNA framework;
3. **Teratogens**: substances that can cause physical defects in a developing embryo.

The U.S. EPA further classifies carcinogens as Class A (a substance with adequate human data indicating that the substance is known to cause cancer in humans), Class B1 (a substance with a moderate amount of human data, with sufficient animal data to indicate its potential to cause cancer), and Class B2 (a substance that has limited to no human data but sufficient animal data to indicate its potential to cause cancer. Examples of each are provided in Table 9.1. For example, formaldehyde (Class B1) is produced directly from natural gas or methane. In addition, it can be produced from methanol by one of two processes: 1) catalytic vapor-phase oxidation:

$$CH_3OH + 1/2\ O_2 \rightarrow CH_2O + H_2O,$$

or 2) a combination oxidation-dehydrogenation process:

$$CH_3OH \rightarrow CH_2O + H_2$$

Regardless of the production method, the U.S. EPA has classified formaldehyde as a probable human carcinogen under conditions of unusually high or

TABLE 9.1 Representative List of Class A, B1, and B2 Carcinogens

Class A	Class B1	Class B2
4-Aminobiphenyl	Acrylonitrile	1,3-Butadiene
Asbestos	Beryllium compounds	2,4-Toluene Diisocyanate
Arsenic compounds	Cadmium compounds	Acetaldehyde
Benzene	Diethyl sulfate	Dioxin
Benzidine	Ethylene oxide	Dichloromethane
Bis(chloromethyl)ether	Formaldehyde	Lead compounds
Chloromethyl methyl ether	Styrene oxide	N-Nitrosodimethylamine
Chromium VI particulates	Tetrachloroethylene	Polychlorinated biphenyls
Coke oven emissions	Trichloroethylene	Vinyl bromide

prolonged exposure. Even at low concentrations (approximately 0.1 ppm) and short-term exposure, formaldehyde can cause irritation of the respiratory tract. The Class B2 chemical 1,3-butadiene is an intermediate organic compound used commercially to produce various types of rubber, resin, and plastics. Most of the industrial exposure to this compound occurs in the processing of petrochemicals, including associated wastewater as the result of production. A more diffuse environmental source of 1,3-butadiene is the incomplete combustion of fuels from mobile sources (e.g., automobile exhaust). Although the specific mechanisms of 1,3-butadiene-induced carcinogenesis are unknown, the U.S. EPA is certain that the carcinogenic effects are mediated by genotoxic metabolites of 1,3-butadiene.

Example Problem 9.1

Tetrachloroethylene (Class B1 carcinogen) is produced from ethylene via 1,2-dichloroethane. In fact, when 1,2-dichloroethane is heated to 400°C with chlorine gas, tetrachloroethylene can result. Complete and balance this reaction:

$$ClCH_2CH_2Cl + Cl_2 \rightarrow \_\_\_ + \_\_\_.$$

Answer:

$$ClCH_2CH_2Cl + 3\ Cl_2 \rightarrow Cl_2C = CCl_2 + 4\ HCl$$

Environmental toxicology encompasses a variety of scientific disciplines, including input from such fields as analytical chemistry, environmental modeling, statistics, molecular genetics, microbiology, organic chemistry, biochemistry, soil science, limnology, risk assessment, and evolutionary biology. It uses both basic and applied knowledge to better evaluate toxicological effects of xenobiotics at each step, from the introduction to the biosphere to the final series of effects on ecosystems and human populations. Ultimately, organisms may be adversely affected by a large number of xenobiotics, where both short- and long-term accumulation will inexorably lead to harm.

TEXTBOX 9.1

A **xenobiotic** is a substance that is foreign to a given environment, typically represented by a toxic chemical. It also relates to substances that are present in much higher concentrations than are usually demonstrated or natural compounds taken up by other organisms.

In this chapter, a broad overview of environmental toxicology and related concepts in relation to ecosystems and human health will be covered. Unlike many of the current textbooks available, we will treat environmental toxicology and environmental chemistry as complimentary subjects, not disparate disciplines. Studies on the chemical mechanisms of toxicity, routes of exposure, dose-response relationships, and toxicity testing methods will complement material directed toward using toxicity concepts in the characterization, assessment, and management of hazardous waste.

9.2. EVALUATING EXPOSURE PATHWAYS AND MECHANISMS OF ACTION

Before we delve into more complicated mechanistic concepts, it is important to consider the **absorption** of xenobiotics on a basic level: the process whereby xenobiotics gain entry to the body. The most important factors influencing absorption are:

1. Route of exposure;
2. Concentration of the substance;
3. Chemical and physical properties of the substance.

More specific detail on the factors influencing toxicity will be discussed shortly. There are three main exposure routes: inhalation (entry via the respiratory tract); ingestion (entry via the gastrointestinal tract); and dermal contact (entry via the skin). We must also consider injection (entry for drug delivery) as

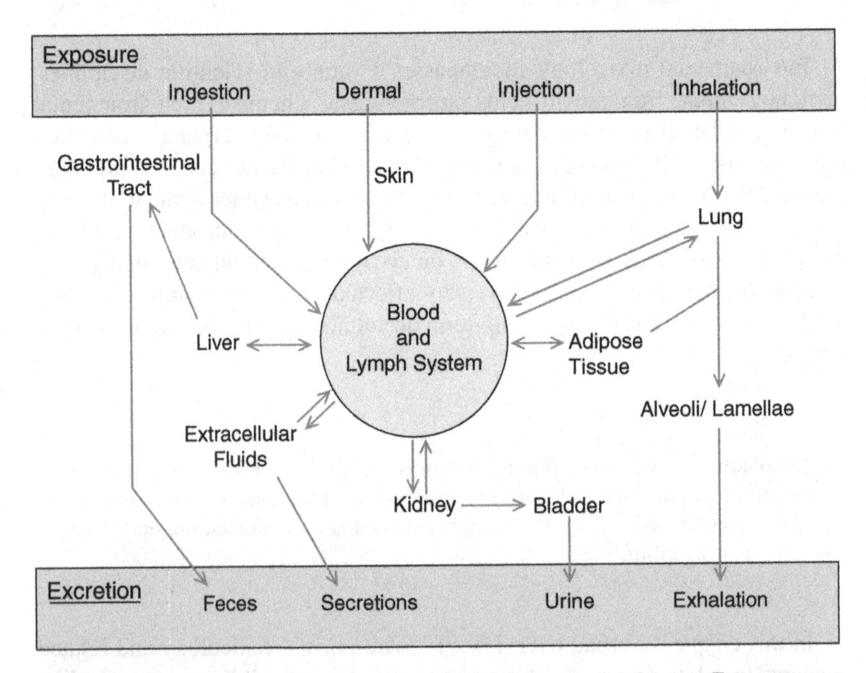

FIGURE 9.2 A general overview of the three main exposure routes into humans. Exposure routes and translocation processes are designated by arrows.

a possible exposure route. A generalized overview of these processes is shown in Figure 9.2. Known pathways (exposure routes) and translocation processes are designated by arrows. As will be evident in our discussion of hazardous waste characterization and risk assessment, establishing a direct pathway provides a defined link between the hazard and suspected health effects on a given individual or populations. Do note that substantial variability occurs in the doses received by individuals exposed to a toxic substance through inhalation or ingestion. This is a result of individual differences in anatomical and physiological factors controlling absorption and consequent internal organ redistribution (Cuddihy, et al., 1979).

9.2.1. Toxic Effects and Factors Influencing Toxicity

Toxicity can result from a number of adverse cellular, biochemical, or macromolecular changes including DNA modification/damage, cell replacement, impairment of an enzymatic system, and disruption of protein synthesis (Pelkonen, et al., 1997). Xenobiotics may also act indirectly to induce toxicity by the alteration of physiological mechanisms, modification of biochemical functions, and interfering with general nutrition. The toxicity of a given substance depends on numerous identified and highly studied factors including:

1. Form and intrinsic chemical activity;
2. Dose-response relationship;
3. Exposure route;
4. Ability to be absorbed by the body;
5. Metabolism;
6. Distribution within the body and target organs affected;
7. Excretion;
8. Presence of other chemicals;
9. Age, sex, and ethnicity.

Toxicologists categorize toxic effects of substances according to the site(s) of the toxic response or effect. Single site effects occur at specific **target organs** within the body, which are largely affected by dosage and route of exposure. For example, due to its high metabolic potential, and the capacity to rid xenobiotics from the blood, the liver represents an organ with high susceptibility to the toxic effects of chemicals. And because of its central role in xenobiotic metabolism, the liver is particularly susceptible to associated drugs (e.g., certain antibiotics and non-steroidal anti-inflammatory drugs). Many organic solvents (e.g., toluene and benzene) also contribute to liver toxicity due to metabolic processes that occur as a result. We will examine the metabolism of toluene shortly. Alternatively, toxic effects can occur at multiple sites within the body. This is referred to as **systemic toxicity** and includes the following types: acute toxicity, subchronic toxicity, chronic toxicity, developmental toxicity, and genetic toxicity.

TEXTBOX 9.2

For a substance to have **systemic toxic effects**, it must be absorbed by the body and distributed by the circulation to sites in the body where it exerts toxic effects.

LD_{50} is expressed in milligram of chemical per kilogram of body weight (mg/kg). The lower the LD_{50}, the lower the lethal dose and thus more toxic the substance.

9.2.2. Toxicity Testing

The process of **acute toxicity testing** estimates the human hazard potential of a substance by determining its systemic toxicity in a test system (e.g., animal) following a prescribed oral, dermal, and/or inhalation exposure(s), depending upon the projected routes of human exposure to a particular substance. The toxicity experimental design consists of positioning organisms in growth chambers, treating the units with the dose-controlled xenobiotic of interest, and monitoring the number of organisms over a defined time period. The traditional method in analyzing these data is to apply statistical tests based on analysis of variance followed by pairwise tests to detect treatment concentrations that differ from the **control group**. As detailed in Section 2.4, hypothesis testing relies on the ability to distinguish statistically significant differences between two different conditions. In toxicity testing, one compares a control group and other test groups with subsequent testing of the null hypothesis (no difference between the groups). This null hypothesis is rejected if the difference between the control treatment and any other test treatment is statistically significant. The use of risk ratios in estimating risk due to exposure is covered in section 9.4.1.

The **dose-response relationship** listed above is the foundation for all toxicity measurements, although it is used differently to assess acute effects and chronic effects. It describes the relationship between exposure and the observed response (health effect). In other words, it estimates how different levels of exposure to a xenobiotic alter the likelihood and acuteness of health effects. A **dose-response curve** (Figure 9.3) is a simple graph relating the magnitude of a **stressor** (e.g., concentration of a pollutant or amount of a drug) to the response of the **receptor** (e.g., animal under study). In such a graph, the response is usually mortality. Under optimum circumstances, the entire shape of the dose-response curve is determined, from which it is possible to define the dose that results in a percentage (%) response. Typically, it is the logarithm of the dose that is plotted on the *x*-axis, and in such case, the curve is characteristically sigmoidal. The caption under Figure 9.3 explains the various points on the curve in more detail.

Assessments are typically based on the **median lethal dose (LD_{50})** value: an estimate of the dose of a test substance that kills 50% of the test animals (e.g., rats and mice) via the oral and/or dermal route. A related concept, the **lethal concentration (LC_{50})**, is typically used to assess dosing via inhalation.

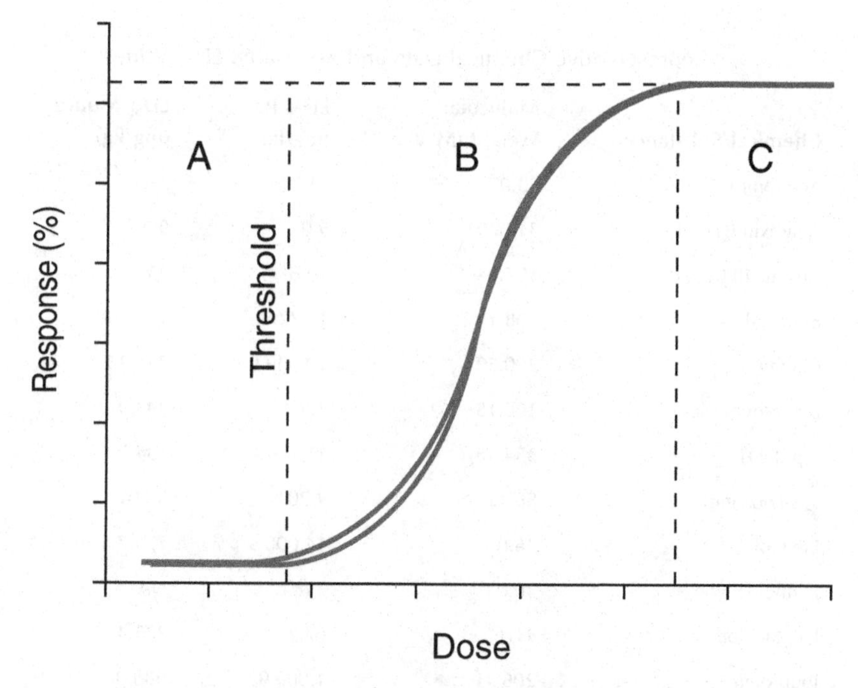

FIGURE 9.3 A simple dose-response curve relating the magnitude of a stressor to the response of the receptor. The first area of the curve (A) is the no-effect range, typically referred to as the threshold dose. Area B of the curve demonstrates a range of increasing effect with increasing dose. The stronger the effect, the steeper this curve will be. Finally, the last area of the curve (C) houses the maximum effect range.

For such experiments, the concentration of the substance in air that kills 50% of the test animals in a given time is of interest. A representative list of chemical substances and associated LD_{50} values in both rats and mice is given in Table 9.2. Students must understand that the LD_{50} does not address the non-lethal toxic effects that may result. For example, a substance may demonstrate a low LD_{50} value (e.g., botulinum toxin = 10 pg/kg) but could potentially bring about deleterious effects at very low exposure levels. Botulinum toxin is one of the most toxic substances on Earth. In fact, just one gram of crystalline toxin is considered enough to kill nearly one million people. Finally, given the obvious concerns of animal testing, *in vitro* methods (e.g., cytotoxicity assays) and quantitative structure-activity relationship (QSAR) models for the prediction of acute systemic toxicity are also performed. QSAR models aim to study the relationships between molecular intrinsic chemical features, physico-chemical property predictions, and biological properties (Fourches, et al., 2010). They generate mathematical models that connect experimental measures with a set of chemical descriptors determined from the molecular structure for a set of compounds. For example, results obtained from valid QSAR studies may

TABLE 9.2 Representative Chemical Data and Associated LD_{50} Values

Chemical/Substance	Molecular Weight (MW)	LD_{50} Rat (mg/kg)	LD_{50} Mouse (mg/kg)
Acrylonitrile	53.07	81.7	27.1
Aflatoxin B1	312.29	5.0	9.1
Arsenic III trioxide	197.84	19.8	45.5
Benzylalcohol	108.15	1,232.9	1,579.0
Captan	300.59	10,009.6	7,003.7
p-Cresol	108.15	206.6	343.9
p,p′ DDT	354.48	113.4	134.7
1,4-Dioxane	88.12	4,203.3	5,701.4
Ethanol	46.08	14,008.3	7,787.5
Formaldehyde	30.03	798.8	NA
Heptachlor	41.1	67.2	373.3
Ibuprofen	206.31	1,008.9	980.0
Kelthane	370.48	574.2	418.6
Mercury II chloride	271.49	1.0	10.0
Nicotine	162.26	50.3	24.3
Nitrobenzene	123.12	640.2	NA
Parathion	291.28	2.0	6.1
Potassium cyanide	65.12	9.8	8.5
Propylene glycol	76.11	20,016.9	23,974.7
Styrene	104.16	4,999.7	315.6
1,2,4-Trichlorobenzene	181.44	756.6	765.7
Toluene	92.15	5,003.7	NA
Triethylene melamine	204.27	1.0	14.9
Zinc II chloride	136.27	350.2	350.2

Data source: Registry of Cytotoxicity Data, Federal Institute for Consumer Health Protection and Veterinary Medicine (http://www.bgvv.de)

indicate the presence or absence of a certain dangerous property or characteristic. Results of QSAR studies may be substituted for animal testing when conditions are appropriate.

Refer back to Figure 9.3. Let's assume that the *y*-axis designation represents animal mortality responses (%) taken from experimental trials with a representative xenobiotic. Mortality data are used to determine the LD_{50} (through either advanced statistical methods or extrapolation on the dose-response curve), which is the first step in discerning the xenobiotic relative toxicity. A very simple calculation can be performed to assess the % mortality, that being

$$\% \text{ Mortality} = \frac{(X_i - X_s) \times 100}{X_i}, \tag{9.1}$$

where X_i is the number of the initial experimental test animals and X_s is the number of survivors. In Chapter 2, we learned the importance of obtaining statistically-valid results. The precision of such a procedure is thus dependent on the number of test animals used. With small test populations, valid results require the use of moderately large doses so that the effect will occur regularly enough to be readily demonstrated (Zbinden and Flury-Roversi, 1981). But even with large numbers of animals there are substantial variations of the test results, because as we can surmise from previous discussion, the numerical value of the LD_{50} is influenced by a number of factors, including animal species and strain, age, sex, diet, temperature, and other experimental conditions. Through standardization of the test animals and the experimental conditions the unpredictability of LD_{50}, determinations can be diminished but never completely eliminated. In addition, statistical outliers are not always biological outliers. The reasons for declaring a data point to be an outlier should be statistically persuasive and, if possible, pre-specified in the experimental protocol. This is not always possible in animal studies. Related, a "significant" statistical test ($p < 0.05$) does not always specify biological significance.

The question then becomes: How can animal LD_{50} data be applied to humans? This is an obvious question with a not so definite answer. However, we can make general assumptions and correlations. In general, if the immediate toxicity is similar in all of the different animals tested, the degree of immediate toxicity will likely be similar for humans. When the LD_{50} values are different for various animal species, health professionals make approximations when estimating the probable lethal dose for humans. Calculations are adapted and used when translating animal LD_{50} values to probable lethal dose values for humans. Safety factors are customarily included in such calculations to allow for the unpredictability between individual reactions to a chemical. When comparing between individuals, weight has an obvious influence given that LD_{50} values are reported in mg of substance/kg of body weight.

Example Problem 9.2

The LD_{50} of a representative chemical substance is 100 mg/kg. What would be the lethal amount of this substance for a child weighing 10 kg and an adult weighing 70 kg?

Answer: The lethal amount for the child would equal 100 mg/kg × 10 kg = 1,000 mg. For the adult, the lethal amount would equal 100 mg/kg × 70 kg = 7,000 mg.

In regards to radiation, the **absorbed dose**, which is a measure of the amount of energy imparted to matter (e.g., human tissue) by the radioactive substance, is routinely used to describe the effects of radiation. The units of absorbed dose are the **rad** or **gray**. The gray, the SI unit of absorbed dose, is equivalent to one joule of energy absorbed by a kg of representative tissue (J/kg). Note that one gray is equal to 100 rad. Recall from Chapter 4 that all types of radiation emitted by radionuclides may harm living cells by ionizing atoms or molecules within the cells, with exposures typically occurring through inhalation and/or ingestion. Although the quantity of energy absorbed per unit mass is essential to determining the unfavorable impact of radiation on the absorbing tissue (and ultimately on the health of the person being exposed), the impact also depends upon the type of radiation absorbed. And as we also learned in Chapter 4, different types of radiation are more or less efficient in producing damage in tissue, given the same amount of energy absorbed per unit mass. Thus, doses produced from internal or external exposure are unique for each radionuclide. More detailed discussion of radiation and its health effects can be found in dedicated sources (e.g., Alexander 1988).

9.2.3. Xenobiotic Metabolism: Biotransformation

When we think of the word "metabolism" in reference to xenobiotics, we are referring to the transformation of a substance by enzymatic reactions. A preponderance of changes deactivates the substance, assuaging its biological activity and conceivably eliminating it from the human body. Unfortunately, the transformation process may produce reactive intermediates or products that modify cellular constituents, ultimately producing harm. Entire books can be devoted to xenobiotic metabolism and the enzymatic reactions that occur as it proceeds. A condensed summary of the major processes with example metabolic pathways of common xenobiotics follows.

Biotransformation can be classified as Phase I or Phase II, which contains a number of enzymatic reactions. In simplest terms, it is a process whereby a substance is transformed from one chemical to another by a given chemical reaction that occurs in the body. For brevity purposes, we will only concentrate on a few pathways and showcase the metabolism of toluene and the pesticide methoxychlor as example applications. During Phase I metabolism, the addition of a single atom of oxygen as hydroxyl, ketone, or epoxides to the

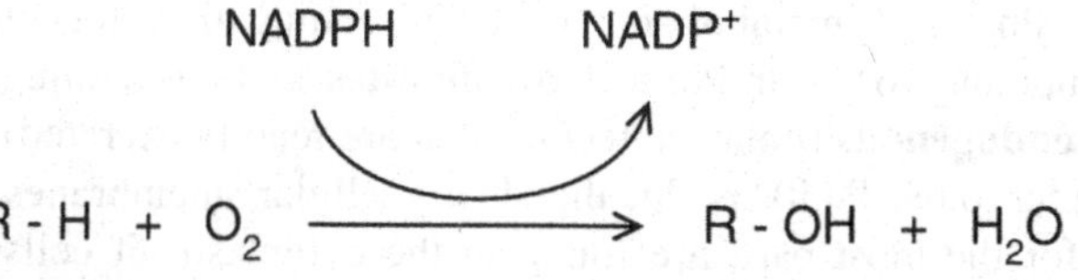

$$R\text{-}H + O_2 \xrightarrow{NADPH \rightarrow NADP^+} R\text{-}OH + H_2O$$

FIGURE 9.4 General reaction of monooxygenases, whereby one atom of molecular oxygen (O_2) is added to substrate (R), and one atom of oxygen in converted to water. Note that R is a substrate such as a xenobiotic.

lipophilic substance renders it more hydrophobic. The primary oxidative enzymes (**monooxygenases**; Figure 9.4) are a family of enzymes referred to as cytochrome P450 monooxygenases (P450s), which are membrane-bound and heme-iron containing enzymes that use $NADPH + H^+$ as a co-substrate by forming a complex with a cytochrome P450 reductase (Guengerich, 2006). The reductases function by transferring electrons from NADPH to cytochrome P450 during the oxidation of xenobiotics (Rand, 1995). Over 30 human 450 isoenzyme systems have been identified to date. As an example, CYP 1A1 (Cytochrome P450 1A1) plays an important role in detoxification of PAHs. Example reactions catalyzed by the cytochrome P450 monooxygenase system include hydroxylation, epoxidation, N, S, and O-dealkylation, deamination, sulfoxidation, and desulfuration. Note that oxidation can also occur without cytochrome P450. For example, alcohol dehydrogenase, an enzyme that uses cofactor NAD as a coenzyme, converts alcohols to aldehydes or ketones:

$$RCH_2OH + NAD^+ \rightarrow RCHO + NADH + H^+.$$

Note that alcohol dehydrogenase can use either NAD or NADP as a coenzyme, but the reaction proceeds at a much slower rate with NADP. Furthermore, aldehyde dehydrogenase helps to alleviate some of the toxic effects of aldehyde initiation in the body. More specifically, this enzyme catalyzes the formation of acids from aliphatic and aromatic aldehydes:

$$RCHO + NAD^+ \rightarrow RCOOH + NADH + H^+.$$

The acids are then available as substrates for conjugating enzymes. The process of O-dealkylation is demonstrated as follows:

$$R\text{-}O\text{-}CH_3 \rightarrow R\text{-}OH + HCHO,$$

a common reaction in the metabolism of drugs, insecticides, and other xenobiotics.

TEXTBOX 9.3

The term **lipophilic** refers to the tendency of molecular entities to dissolve in fat-like solvents. Lipophilic molecules are generally electrically neutral and characteristically non-polar.

During Phase II metabolism (covalent conjugation for elimination), parent compounds or their Phase I **metabolites** undergo conjugation reactions with **endogenous** (polar molecules that are readily excreted) substrates. While cytochrome P450 is localized in cellular membranes, Phase II enzymes, for the most part, are found in the cytoplasm of cells. Activated metabolites are conjugated with charged species such as glutathione (GSH), sulfate, glycine, or glucuronic acid. Such reactions are catalyzed by a large group of broad-specificity transferases, which in combination can metabolize hydrophobic compounds that contain nucleophilic or electrophilic groups. Nucleophiles react with positively-charged (or partially-positive) atoms. Characteristically, they will have either lone pairs or pi bonds that can be used to form new bonds to electrophiles. Electrophilic atoms will characteristically 1) have positive charge, 2) have partial positive charge, 3) be polarizable, or 4) have an empty orbital.

One of the most important families in Phase II biotransformation is the glutathione S-transferases (GSTs), which are able to catalyze the detoxification of a number of electrophilic xenobiotics and/or their metabolites generated during oxidative metabolism by microsomal cytochrome P450-dependent monooxygenases (Jakoby, 1978). Glutathione adducts are further metabolized in the kidneys to derivatives referred to as mercapturic acids of the parent xenobiotic. Mercapturic derivatives can then be found in the urine upon excretion. Other Phase II reactions include sulfate conjugation, glycoside conjugation, methylation, and acetylation. Considering the latter process, a less hydrophilic metabolite (e.g., an amine) is converted into an amide. More specifically, acetylated derivatives of foreign exogenous amines are acetylated by *N*-acetyl transferase, the acetyl donor being coenzyme A (CoA). For example, acetylation of benzidine (an aryldiamine associated with the development of bladder cancer in occupationally exposed individuals) proceeds as follows:

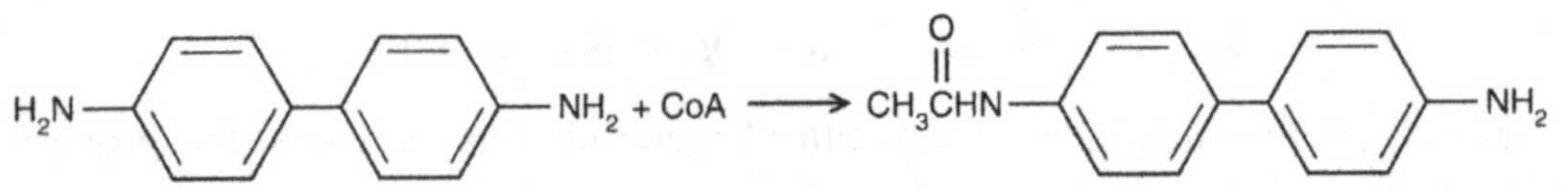

$$H_2N-C_6H_4-C_6H_4-NH_2 + CoA \longrightarrow CH_3\overset{O}{\overset{\|}{C}}HN-C_6H_4-C_6H_4-NH_2$$

FIGURE 9.5

As a result, the physiological consequence is to "deactivate" benzidine or its Phase I metabolites. As with acetylation, the process of methylation also results in a less hydrophilic metabolite upon conversion.

Metabolism of Toluene and Associated Health Effects

Like many lipophilic solvents, toluene is absorbed from the respiratory tract, the gastrointestinal tract, and through the skin. While a sizeable portion of toluene (20–30%) is exhaled unchanged via the lungs, a greater proportion is

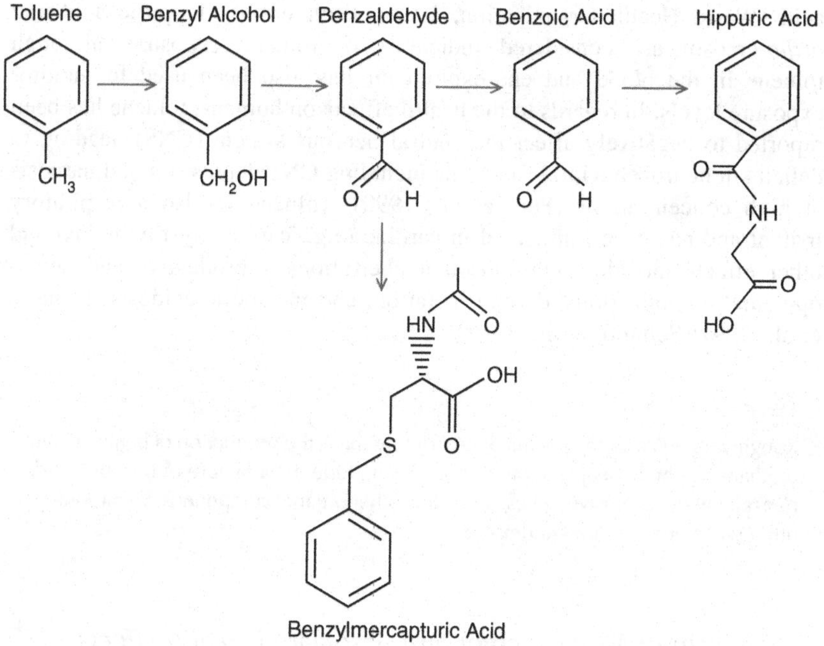

Toluene Benzyl Alcohol Benzaldehyde Benzoic Acid Hippuric Acid

CH_3 CH_2OH

Benzylmercapturic Acid

FIGURE 9.6 Metabolic pathway of toluene initially driven by the hydroxylation of toluene to benzyl alcohol by five associates of the cytochrome P450 (CYP) family.

metabolized and excreted via the common pathways discussed above (Wallen, et al., 1985). To understand the health effects, a true comprehension of its metabolism is first necessary. Review this process carefully as it is depicted in Figure 9.6. The major pathway is initially driven by the hydroxylation of toluene to benzyl alcohol by five associates of the cytochrome P450 (CYP) family. There is some debate as to how benzyl alcohol (the first intermediate) is metabolized to benzaldehyde, but the majority attributes this to alcohol dehydrogenase. Others believe this is mediated by cytochrome P450 directly. Benzaldehyde is in turn metabolized to benzoic acid, primarily by aldehyde dehydrogenase-2 (ALDH-2), with a small portion metabolized by cystolic ALDH-1. Benzoic acid that, upon conjugation with glycine, results in hippuric acid (also referred to as benzoylaminoacetic acid), the major urinary metabolite.

Hippuric acid hydroxylation to cresols (*ortho-* and *para-*, but not *meta-*cresol) is considered a minor pathway in the metabolism of toluene. In addition, it has been shown that benzylmercapturic acid is a minor metabolite of toluene. Several substances have been shown to interfere with the toluene biotransformation scheme depicted, including the presence of appreciable levels of benzene, which results in reduced excretion of hippuric acid (World Health Organization, 1985). Both urinary hippuric acid and *ortho*-creosol are considered indicators of toluene exposure, although toluene in the blood and end-expired air has also been used to monitor exposure levels. In regards to the health effects on humans, toluene has been reported to negatively affect the central nervous system (CNS), leading to deficits in neurobehavioral functions including CNS depression and narcosis at high concentrations (Foo, et al., 1990). Toluene is also a respiratory irritant and has been implicated in cardiac (e.g., cardiac dysrhythmias) and other effects including chromosomal aberrations, reproductive and developmental complications, dermal irritation, and metabolic acidosis (Donald, et al., 1991; Schmid, et al., 1985).

TEXTBOX 9.4

Roughly 90–95% of the chemical use for toluene is the production of benzene and urethane via hydrodealkylation. Other uses include manufacture of benzoic acid, benzaldehyde, explosives, dyes, and many other organic compounds. (Data source: http://water.epa.gov/drink/index.cfm.)

Metabolism of Methoxychlor and Associated Health Effects

As introduced in Chapter 8, methoxychlor is a member of the organochlorine class of pesticides. Recent findings suggesting that exposure to methoxychlor is implicated in endocrine disruptor activity has led to increased examination of the methoxychlor metabolic pathway (Bulger, et al., 1985; Hu and Kupfer, 2002). An **endocrine disruptor** is a synthetic chemical that when absorbed into the body has the propensity to either mimic or block hormones, thus disrupting the body's normal functions. In fact, a key investigation found that both methoxychlor and its metabolites have estrogen-like activity with several metabolites having proestrogen activity (Kupfer and Bulger, 1987). Members of this same investigative group have proposed a metabolic pathway of methoxychlor (Figure 9.7). The major activation of methoxychlor is through cytochrome P450-catalyzed demethylation. More specifically, cytochrome P450 monooxygenase isozymes (CYP1A1, CYP1A2, and CYP3A4, among others) are mainly responsible for the degradation of methoxychlor in the human liver.

Following the proposed pathway highlighted in Figure 9.7, methoxychlor *O*-demethylation to 1,1,1-trichloro-2-(4-hydroxyphenyl)-2-(4-methoxyphenyl) ethane (mono-OH-M) occurs. Subsequently, mono-OH-M is either *ortho*-hydroxylated to form catechol-M or is further *O*-demethylated to yield 1,1,1-trichloro-2, 2-bis(4-hydroxyphenyl)ethane (bis-OH-M). The final product, 1,1,1-trichloro-2-(4-hydroxyphenyl)-2-(3, 4-dihydroxyphenyl)ethane (tris-OH-M), is subsequently formed from bis-OH-M via ortho-hydroxylation and from catechol-M by *O*-demethylation. Except for bis-OH-M, these metabolites are

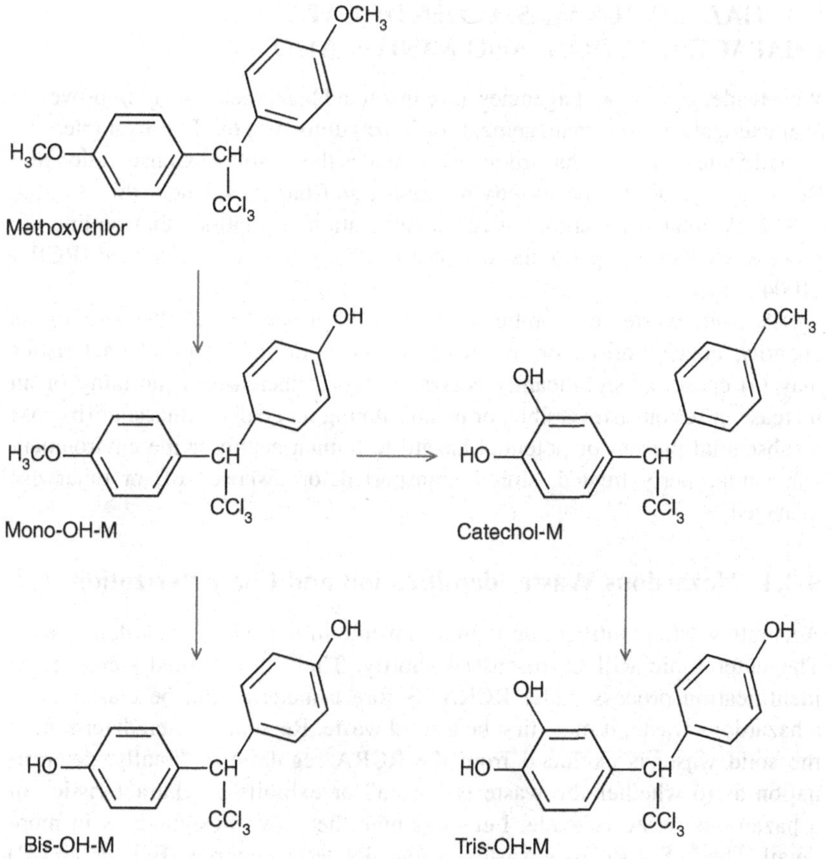

FIGURE 9.7 Proposed metabolic pathway of methoxychlor by human P450s. *Adapted from Hu and Kupfer (2002) with permission from the American Society for Pharmacology & Experimental Therapeutics.*

chiral compounds consisting of two enantiomeres. Many of today's biologically active compounds, for example, agro-chemicals and pharmaceuticals, are chiral. Understand that one of the two **enantiomers** of a chiral compound may well be more toxic than the other. Consequently, to predict the exact toxicities of compounds under study, accurate determination of the concentrations of both enantiomers is paramount.

TEXTBOX 9.5

A **chiral** molecule is one that is not superimposable on its mirror image. **Enantiomers** are isomeric molecules that are non-superimposable mirror images. Therefore, a molecule that is superimposable on its mirror image is achiral.

9.3. HAZARDOUS WASTE IDENTIFICATION, CHARACTERIZATION, AND MINIMIZATION

Worldwide, government agencies have instituted legal measures to improve the characterization and management of **hazardous waste**. For example, the United States regulates hazardous waste under the Resource Conservation and Recovery Act (RCRA) as briefly mentioned in Chapter 4. Under this Act, the U.S. EPA created hazardous waste identification regulations that outline the process whether any particular material is a hazardous waste, defined (RCRA §1004(5)) as:

"A solid waste, or combination of solid waste, which because of its quantity, concentration, or physical, chemical, or infectious characteristics may (a) cause, or significantly contribute to, an increase in mortality or an increase in serious irreversible, or incapacitating reversible, illness; or (b) pose a substantial present or potential hazard to human health or the environment when improperly treated, stored, transported, or disposed of, or otherwise managed."

9.3.1. Hazardous Waste Identification and Characterization

Accurate waste identification is indispensible in managing hazardous waste. The latter topic will be discussed shortly. There are defined steps in the identification process under RCRA. Before a material can be classified as a hazardous waste, it must first be a solid waste. Regulators then determine if the solid waste is excluded from the RCRA regulations. Finally, determination as to whether the waste is "listed" or exhibits a "characteristic" of a hazardous waste is made. Let's examine these two designations in more detail. The U.S. EPA lists wastes in four defined categories (F, K, P, and U) according to industrial wastestreams. The F list assigns particular solid wastes from conventional industrial or manufacturing processes as hazardous. The K list designates particular solid wastes from specified industries as hazardous. Both the P and U list pure or commercial grade formulations of specified unused chemicals as hazardous. Each waste on the F, K, P, and U lists is assigned a waste code and hazard code. Table 9.3 provides information that brings such designations into context. For brevity purposes, example wastes from each list with relevant hazard codes are listed. Note that wastes that have not been expressly listed by the U.S. EPA may still be deemed hazardous if they exhibit one of four characteristics (as defined under RCRA):

1. **Ignitability (D001)**: Ignitable wastes can create fires under certain conditions, are spontaneously combustible, or have a flash point less than 60°C. Examples of D001 wastes include ethylene dichloride, isopropanol, kerosene, petroleum solvents, and naphtha.

TABLE 9.3 Representative Hazardous Waste Listings with Appropriate Hazard Codes

U.S. EPA Listing	Hazardous Waste/Waste Type/Waste Stream	Hazard Code or Other Designation*
F001	Spent halogenated solvents	T
F005	Spent non-halogenated solvents	I, T
F007	Spent cyanide plating bath solutions	R, T
F011	Spent cyanide solutions from salt bath pot cleaning	R, T
K009	Distillation bottoms from the production of acetaldehyde from ethylene	T
K006	Wastewater treatment sludge from the production of chrome oxide green pigments	T
K107	Column bottoms from product separation from the production of 1,1-dimethylhydrazine (UDMH) from carboxylic acid hydrazines	C, T
K176	Baghouse filters from the production of antimony oxide, including filters from the production of intermediates (e.g., antimony metal)	E
P011	Arsenic oxide (As_2O_5)	A_1
P050	Endosulfan	A_1
P071	Methyl parathion	A_1
P105	Sodium azide	A_1
U008	Acrylonitrile	A_2
U023	Benzotrichloride	C, R, T
U210	Ethene, tetrachloro	A_2
U161	Methyl isobutyl ketone	I

** A_1 = absence of letter indicating that the compound is only listed for acute toxicity; A_2 = absence of letter indicating that the compound is only listed for toxicity; C = corrosive waste; E = toxicity characteristic waste; H = acute hazardous waste; I = ignitable waste; R = reactive waste; T = toxic waste.*

2. **Corrosivity (D002)**: Corrosive wastes are acids or bases (pH ≤ 2 or ≥ 12.5) that are capable of corroding metal containers. Examples of D002 wastes include anhydrous ammonia, phosphoric acid, hydrochloric acid, hypochlorous acid, potassium hydroxide, and acetic acid.
3. **Reactivity (D003)**: Reactive wastes are unstable under "ordinary" conditions. They are known to initiate explosions, toxic gases, or vapors when heated, compressed, or mixed with water. Example D003 wastes include chromic acid, sulfides, organic peroxides, and perchlorates.
4. **Toxicity (D004-D043)**: Toxic wastes are injurious or fatal when ingested or absorbed. Example D004-D043 wastes include arsenic, total cresols, methoxychlor, toxaphene, trichloroethylene, and vinyl chloride.

Consider cyanide-bearing wastes, which when classified under characteristics, fall under the D003 category for reactivity. A majority of the cyanide used in industrial processes is in the form of sodium cyanide (NaCN). Sodium cyanide hydrolyzes to form hydrogen cyanide (HCN) and sodium hydroxide (NaOH):

$$NaCN + H_2O \rightarrow HCN + NaOH.$$

Hydrogen cyanide is a characteristic weak acid with a dissociation constant of 4.8×10^{-10} (p$Ka = 9.32$) at 25°C. At a pH < 7, most of the cyanide is present as dissolved hydrogen cyanide acid gas. It evaporates quickly from the solution as a toxic gas. Now consider a wastestream containing two DOO2 type chemicals: ammonia and hypochlorous acid. In water, they will react to form chloroamines:

$$HOCl + NH_3 \rightleftharpoons NH_2Cl + H_2O$$

$$HOCl + NH_2Cl \rightleftharpoons NHCl_2 + H_2O$$

$$HOCl + NHCl_2 \rightleftharpoons NCl_3 + H_2O.$$

Both pH and the Cl_2/NH_3 ratio affect the relative concentrations of chloroamine species. For instance, the formation of trichloroamine (a known eye and nose irritant) above is favored by a low pH and a large ratio of Cl_2 to NH_3.

Example Problem 9.3

Consider the dissociation of sodium cyanide as presented in the text. Discuss hydrogen cyanide's (HCN) dissociation potential (and thus toxic effects) as the pH is increased.

Answer: At higher pH values, HCN dissociates into ions (with complete dissociation at pH 11). At a pH of 9.32, 50% of the HCN is in the form of free cyanide (CN^-). In wastewater, cyanide typically exists in the form of CN^- and HCN, or in the form of a complex ion.

Recall from Chapter 5 that chlorination is an effective means of waste conversion/treatment to less toxic substances. As discussed in Example Problem 9.3, CN^- ions are formed in standard wastewater processes. Consider the alkaline dechlorination of CN^-, which first results in cyanogen chloride (CNCl):

$$CN^- + Cl_2 \rightleftharpoons CNCl + Cl^-.$$

Under a pH of 8, CNCl is oxidized to cyanate:

$$CNCl + 2OH^- \rightleftharpoons CNO^- + Cl^- + H_2O.$$

Lastly, conversion to CO_2 and N_2 occurs:

$$2CNO^- + 4OH^- + 3Cl_2 \rightleftharpoons 2CO_2 + N_2 + 6Cl^- + 2H_2O.$$

How do waste characterization efforts in the United States compare with other countries and multi-governmental bodies? The European Union, for example, has provided itself with similar legal framework (Council Directive 91/689/EEC) aimed at characterizing hazardous waste based on 14 established chemical and physical properties. In order to simplify Union legislation, the provisions of Directive 91/689/EEC have been integrated into the Waste Framework Directive 2008/98/EC. Like the U.S. EPA, the Union adopted a common encoding system for waste characteristics for a broad variety of purposes including classification of hazardous wastes, transport of waste, installation permits, and decisions about recyclability of the waste or as a basis for waste statistics. Both the U.S. EPA and the Union have created a system for sharing information with **waste generators** and **waste transporters** on best practice regarding waste prevention and to develop guidelines in order to assist waste prevention programs.

9.3.2. Hazardous Waste Minimization

Waste minimization supports efforts that encourage a more sustainable civilization, reduce the amounts of waste generated, and lower the toxicity of wastes that are generated by numerous sources. The cornerstone of waste prevention is the five-stage waste management hierarchy:

1. Reduce pollution at the source whenever feasible;
2. Recover and reuse onsite whenever feasible;
3. Recycle offsite in an environmentally-safe manner whenever feasible;
4. Treat pollution in an environmentally-safe manner whenever feasible;
5. Disposal or release into the environment as a last resort.

Figure 9.8 displays these strategies in a waste prevention pyramid, with desirability decreasing as it moves toward the pinnacle. Waste prevention includes equipment or technology modifications, process or procedure modifications, product reformulation or redesign, and raw material substitutions. (Chapter 10 will provide expanded coverage of green chemistry concepts and sustainable chemical processes.)

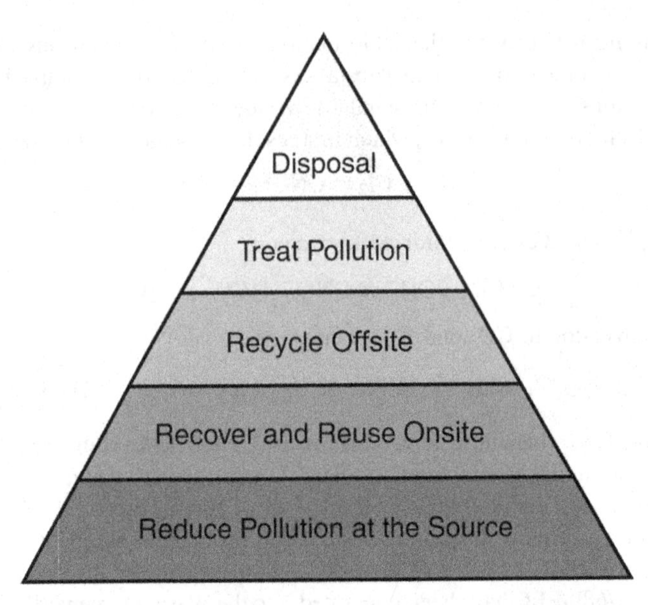

FIGURE 9.8 Defined strategiesin a waste prevention pyramid, with desirability decreasing as it moves toward the pinnacle.

Waste minimization, one of the first initiatives of pollution prevention, is a national policy that was first presented in the 1984 U.S. Hazardous and Solid Waste Amendments (HSWA) to RCRA. Waste minimization involves the reduction of any solid or hazardous waste that is generated or subsequently treated, stored, or disposed of. This policy also includes the concepts of recycling, general volume reduction, as well as reduction in the quantity of toxic components. Chemical and thermal destruction methods, among others, can be used to achieve toxicity reduction. Conventional thermal processes destroy chemical agents either by **incineration** or by **pyrolysis**—a process whereby chemical agents are heated in the absence of oxygen. For example, pyrolysis transforms hazardous organics (semi-volatile organic compounds and fuels) into gaseous components, liquid, and a solid residue (coke) containing fixed carbon and ash. Subsequent production of combustible gases, including carbon monoxide, hydrogen, and methane, results. This process is especially important when dealing with organic radioactive waste generated in nuclear facilities. Such waste must be treated to attain a volume reduction and concurrently produce a stable form suitable for temporary storage or disposal. In terms of data dissemination, the U.S. EPA, in partnership with the states, biennially collects information regarding the generation, management, and final disposition of hazardous wastes regulated under RCRA, as amended. The purpose of such a report is to communicate the findings of EPA's hazardous waste reporting data collection efforts to the public, government agencies, and the regulated community. Included in this

TABLE 9.4 Quantity of RCRA Hazardous Waste Managed (Based on 2005 Data)

Management Method	Tons Managed	Percentage of Quantity
Deepwell or underground injection	21,846,692	49.7
Other treatment	4,221,555	9.6
Other disposal	3,433,990	7.8
Aqueous organic treatment	3,356,122	7.6
Landfill/surface impoundment	2,037,543	4.6
Energy recovery	1,719,390	3.9
Aqueous inorganic treatment	1,705,585	3.9
Incineration	1,437,996	3.3
Metals recovery	1,420,320	3.2
Fuel blending	1,174,625	2.7
Sludge treatment	516,002	1.2
Stabilization	425,931	1.0
Other recovery	328,180	0.7
Solvent recovery	296,681	0.7
Land treatment/application/farming	3,248	0.0
Total	**43,923,861**	**100.0**

Data source: National Analysis: The National Biennial RCRA Hazardous Waste Report, 2005.

report are data on the quantity of hazardous waste managed (Table 9.4) from storage, treatment, disposal, and energy capture.

9.4. RISK ASSESSMENT

When briefly comparing the U.S. EPA and European Union regulation framework, it should be obvious that waste management programs seek to protect human health and the environment from the probable risks of hazardous chemicals and radionuclides. Having discussed the criteria that make wastes hazardous, it is the methods by which health risks associated with exposure are determined that are of the greatest public concern. The use of environmental **risk assessment** before establishing regulations is a common and necessary practice.

In the context of evaluating health and environmental impacts, we can define risk to be the chance of harmful effects to human health or to ecological systems resulting from exposure to an environmental stressor. Risk assessment can thus be used to characterize the nature and enormity of health risks to humans and ecological receptors from chemical contaminants and other stressors introduced into the environment (Bascietto, et al., 1990). Risk managers then take this information to develop guidelines and safety measures to protect sensitive entities from stressors or contaminants. There are four defined steps of the risk assessment process derived from concepts previously discussed:

1. Hazard identification;
2. Dose-response assessment;
3. Exposure assessment;
4. Risk characterization.

The goal of risk assessment is to come up with uniform guidelines that would provide standard procedures for estimating health effects on individuals or populations from hazardous chemicals and drugs. In hazard identification we assess what health problems are caused by the hazardous chemical(s) of interest. Methods used to screen chemical inventories for hazardous substances that may pose risks to humans and the environment are routinely utilized. Hazardous chemicals are identified by four separate criteria: persistence (P), bioaccumulation (B), toxic properties (T), and quantity (Q) information (Arnot and Mackay, 2008), as depicted in Figure 9.9. The first three are

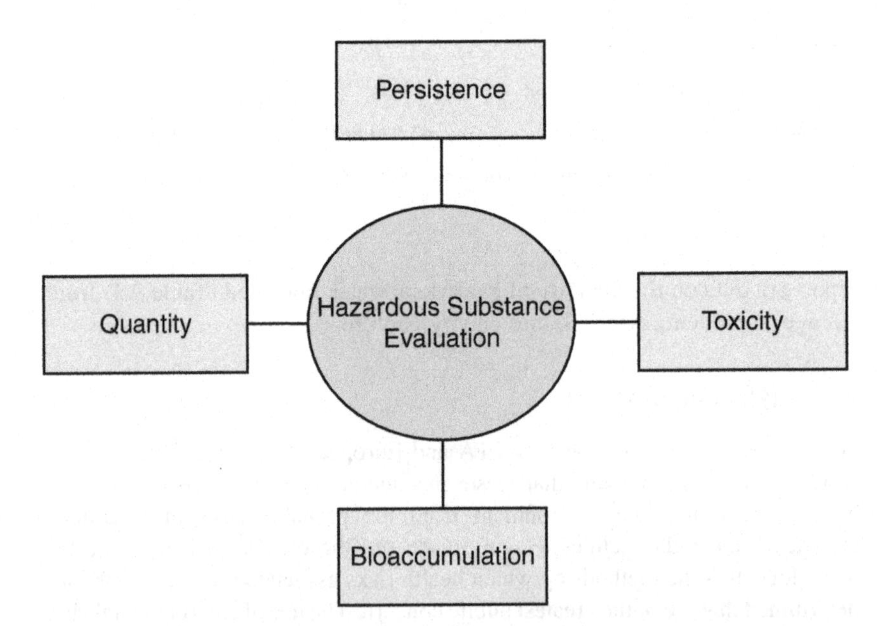

FIGURE 9.9 Hazard substances screening from a holistic mass balance point of view.

used to assess the property of the chemical in question. The latter, Q, is an extensive property used to assess risk based on the quantity of chemical(s) released. Although varying definitions and criteria are demonstrated, multiple countries or governmental bodies implement this scheme in some form or another.

It is important to document the dose-response relationship(s) over the range of observed doses (i.e., the doses that are reported in the experimental data collected). Extrapolation to approximate the risk (probability of adverse effect) outside the lower range of available experimental data is needed in order to make inferences about the critical region where the dose level begins to cause the adverse effect in the individual or population. Exposure assessment involves determining how much of the substance individuals or populations (and their numbers) were exposed to over a specified range of time. Finally, the risk characterization process estimates the incidence and severity of harm to human health and the environment that may occur as a result of exposure to contaminants (Santos, 1987). It also describes the uncertainties and limitations of the overall process. The importance of statistical analysis/validation and data interpretation as part of the health risk assessment process needs to be better understood by environmental health professionals so they can be effectively communicated to the general public. This involves proper sampling and sample treatment, established and calibrated analytical methods, certified reference materials, adequate and statistically-worked data collection, and laboratory quality control. Consider the example statistical approach to calculating estimated risk associated with exposure in section 9.4.1.

Mathematical models are often used in addition to or in lieu of monitoring data to estimate environmental concentrations and to support regulatory standards. Multivariate statistical techniques that account for variability in species and toxicity tests can be valuable tools in assessment. For example, many of the multivariate and graphical methods (e.g., principal component analysis and one-way ANOVA) highlighted in Chapter 2 can be used to assess, for example, the relationship between particulate matter (PM) concentrations and cell toxicity. Recall from Chapter 7 that particulate sources, composition, and geographical location play roles in PM toxicity to humans. However, little is known about how these components specifically participate (individually or in groups) in the toxicity and specific outcomes at the cellular level. In total, over 40 exposure assessment models are currently supported and used by the U.S. EPA for regulatory and research purposes (Williams, et al., 2010). U.S. EPA scientists and regulators have used these models in studying the source-to-outcome continuum (Figure 9.10), which helps assess the movement and transformation of xenobiotics in the environment and outcomes that address which receptors have the potential for exposure. Finally, greater elucidation of toxicity pathways will support the risk assessment paradigm. This will allow health professionals to link

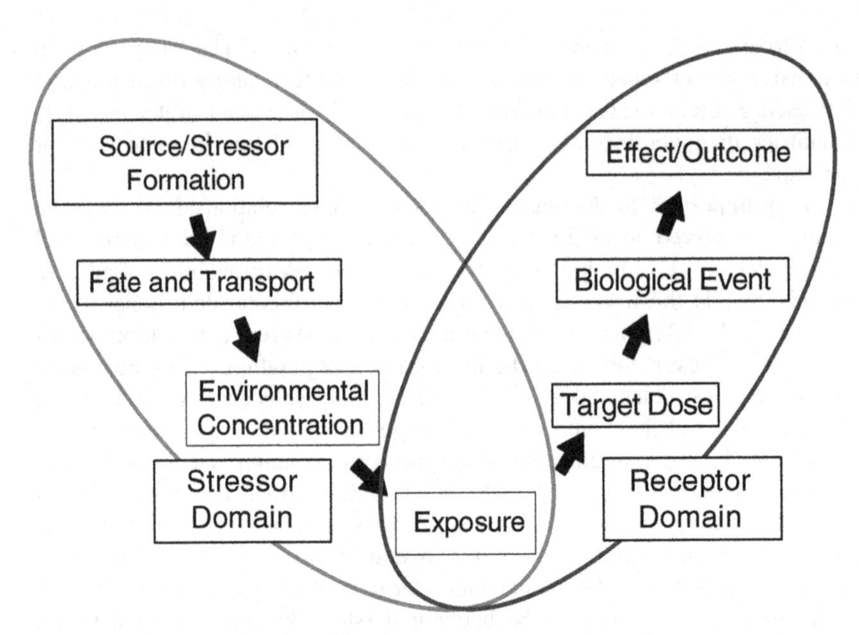

FIGURE 9.10 Source-to-outcome continuum used in the exposure assessment process. This continuum involves a series of steps across the sciences that link sources to health and associated environmental impacts. *Modified with permission from ILM Publications (Williams, et al., 2010).*

toxicological responses across levels of biological organization, including systems, molecular, and cellular biology.

9.4.1. Calculating Statistically-Estimated Risk Associated with Exposure

In the environmental health and medical fields, **risk ratios** (often referred to as relative risk) are used to estimate the probable risk that a member of an exposed group will develop a disease relative to the probability that a member of an unexposed group will develop that same disease (King and Zeng, 2002). Risk

TABLE 9.5 Independent Groups' Notation

	Disease Present	Disease Absent	Total
Exposure to factor	A	B	A + B
Non-exposure to factor	C	D	C + D
Total	A + B	B + D	n = A + B + C + D

ratios are simply ratios of proportions and are calculated (using the independent groups' notation in Table 9.5) as follows (Bewick, et al., 2004):

$$R_{ratio} = \frac{A/(A+B)}{C/(C+D)}. \tag{9.2}$$

In the above equation, the numerator represents the estimated risk in the exposed individuals and the denominator the risk in the non-exposed individuals. A risk ratio of >1 or <1 typically means that being exposed to a certain substance or factor either increases or decreases the risk of disease. Calculation of the 95% confidence interval is necessary given that estimates of R_{ratio} will not follow a normal distribution (Bewick, et al., 2004). This calculation will be covered shortly.

The **odds ratio** is another commonly used metric to estimate risk to exposed individuals. It is the ratio of the odds of the disease in a group exposed to the factor, to the odds of the disease in the non-exposed group, estimated by (Bewick, et al., 2004)

$$O_{ratio} = \frac{A/B}{C/D}. \tag{9.3}$$

As with the R_{ratio}, an approximate 95% confidence interval for the true population O_{ratio} can be calculated by first considering the standard error (SE) of natural logarithm (ln) O_{ratio}:

$$SE_{(\ln O_{ratio})} = \sqrt{\frac{1}{A}+\frac{1}{B}+\frac{1}{C}+\frac{1}{D}}. \tag{9.4}$$

The 95% confidence interval for the population ln O_{ratio} can then be calculated:

$$\ln O_{ratio} \pm (1.96)SE_{(\ln O_{ratio})}. \tag{9.5}$$

Note: The antilog (e^x) of the lower and upper limits is needed to obtain the 95% confidence interval. The SE (ln R_{ratio}) is calculated similarly, with a slight modification of the denominator to correct for the differences in ratio calculations. Generally, when the risk of the disease in the non-exposed is low, the O_{ratio} approximates to the R_{ratio} (Bewick, et al., 2004).

These methods have been used in a number of studies examining the epidemiological risk associated with exposure to various hazardous substances. For example, the odds ratio was instituted in quantitatively assessing the association between benzene exposure and leukemia from historical hygiene data for industrial workers in Ohio (Rinsky, 1989). The study evaluated the rate of mortality experienced by a cohort of 10 exposure groups occupationally exposed to benzene. It was found that a strongly positive exposure-response relationship existed between benzene and leukemia, and that this relationship extended to exposure levels of <1 ppm.

Research Application XI: Pesticide Exposure Alters Sperm Chromatin Structure

As showcased in Chapter 8, organophosphorus pesticides are esters of phosphoric or thiophosphoric acids and are highly toxic to mammals. Although there is extensive knowledge on the neurotoxic effects, less is known about their impact on male reproduction, which could have consequences to those in high-risk groups (e.g., agricultural workers). To help answer this, a 2004 study by Mexican scientists evaluated whether exposure to a mixture of organophosphorus pesticides altered sperm chromatin structure and semen quality in agricultural workers in Villa Juarez, State of Durango, Mexico (Sánchez-Peña, et al., 2004). Study participants were 18–50 years of age and had no history of chemotherapy, radiotherapy, or chronic illnesses. Participants living in the region but not occupationally exposed were also tested.

In this study, chromatin susceptibility was assessed by a sperm chromatin structure assay (SCSA) and urinary levels of diakylphosphates (DAP) (organophosphorus pesticide metabolites) used to assess exposure. In fact, six metabolites in total were measured. As expected, there were significant differences in urinary DAP concentrations among participants. The investigators reported a significant association between DAP urinary concentrations and the number of spermatozoa demonstrating greater susceptibility to acid denaturation of DNA in samples from agricultural workers exposed to organophosphorus pesticides. This was in spite of the fact that rapid metabolism and excretion of pesticides mired full exposure assessment.

In the end, sperm chromatin was in fact found to be a sensitive target to pesticide exposure, even after only short time periods. This is speculated to contribute to adverse reproductive outcomes as increased susceptibility of sperm to acid-induced denaturation has been correlated with semen quality and infertility in mammals.

9.5. END OF CHAPTER PROBLEMS

9.1 Discuss why it is imperative to establish direct exposure routes when assessing individuals exposed to a toxic substance.

9.2 Explain why the liver is considered a target organ when assessing the toxic effects of xenobiotics.

9.3 What is the most critical factor in determining if a substance will be an acute or chronic toxicant?

9.4 Refer to Table 9.2. Explain why there are noticeable differences in LD_{50} values among rats and mice for a given chemical substance.

9.5 Given the following toxicity data for a typical response (% mortality) over dose (concentration that results in 50% of the measured effect) of a given xenobiotic, plot the data in Microsoft Excel and discuss the relevance of the distribution of data. Is the response surprising? Expected? Provide reasoning.

Dose	0.25	0.5	1.0	1.5	2.0	2.5	3.0	3.5	4.0
Mortality (%)	0.0	1.0	3.5	11.0	39.0	46.0	48.5	50	50

9.6 From the graph generated in problem 9.5, roughly estimate the LD_{50}.

9.7 Acute toxicity testing requires the use of statistical analysis for reporting purposes. From knowledge gathered from Chapter 2, discuss why such analyses (give specific examples) are particularly important when computing the LD_{50}.

9.8 Consider the hydrolysis of acetyl chloride (CH_3COCl). Write the net reaction in H_2O and determine which RCRA-defined characteristic for non-listed wastes (D001, D002, D003, or D004-D043) it will fall under.

9.9 Anhydrous liquid ammonia (LNH_3) is frequently transported in bulk on inland and/or coastal waters. Show the dissociation of NH_3 in water and discuss the resulting species in terms of RCRA characteristics for non-listed wastes.

9.10 As defined under RCRA, list the four characteristics of wastes that deem them hazardous even if they are not classified.

9.11 In section 9.3.1, we discussed the formation of chloroamines in a wastewater stream housing two example RCRA listed chemicals. Generate a balanced reaction for the oxidation of one of those species, $NHCl_2$ (in water).

9.12 What is the significance of the free chlorine generated in problem 9.11?

9.13 Review Equation 9.4, which relates to the standard error (*SE*) of natural logarithm (ln) O_{ratio} when estimating the ratio of the odds of the disease in a group exposed to a given factor, to the odds of the disease in the non-exposed group. Rearrange this equation to produce the *SE* of natural logarithm (ln) R_{ratio}.

9.14 Methylene diphenylisocyanate (MDA) is extensively used in industry to manufacture polyurethane foam products. To assess the odds risk of industrial workers chronically exposed to this compound versus those non-exposed developing liver disease, the following observed frequency table was generated as part of an example occupational study. Calculate the odds risk ratio O_{ratio} for this study. Discuss what this value means in relation to potential risk.

	Disease Present	Disease Absent	Total
Exposure to factor	13	199	212
Non-exposure to factor	2	175	177
Total	15	374	389

9.15 Calculate the 95% confidence interval for the data presented in problem 9.14.

9.16 Using the same data in problem 9.14, estimate the risk ratio R_{ratio} and discuss the significance of the calculated value.

9.17 Using the equation generated in problem 9.13, calculate the $SE_{(\ln R_{ratio})}$ from the values estimated in problem 9.14.

9.18 In regards to Research Application XI, search your institution's research journal databases for similar articles of interest. Do other investigations corroborate these results? Do organochlorine pesticides produce similar effects?

9.6 REFERENCES

Alexander, R. E. (1988). Health effects from radiation. *Environmental Science & Technology, 22*, 144–145.

Arnot, J. A., & Mackay, D. (2008). Policies for chemical hazard and risk priority setting: Can persistence, bioaccumulation, toxicity, and quantity information be combined? *Environmental Science & Technology, 42*, 4648–4653.

Bascietto, J., Hinckley, D., Plafkin, J., & Slimak, M. (1990). Ecotoxicity and ecological risk assessment. *Environmental Science & Technology, 24*, 10–15.

Bewick, V., Cheek, L., & Ball, J. (2004). Statistics review II: Assessing risk. *Critical Care, 8*, 287–291.

Bulger, W. H., Feil, V. J., & Kupfer, D. (1985). Role of hepatic monooxygenases in generating estrogenic metabolites from methoxychlor and from its identified contaminants. *Molecular Pharmacology, 27*, 115–124.

Cuddihy, R. G., McClellan, R. O., & Griffith, W. C. (1979). Variability in target organ deposition among individuals exposed to toxic substances. *Toxicology and Applied Pharmacology, 49*, 179–187.

Donald, J. M., Hooper, K., & Hopenhayn-Rich, C. (1991). Reproductive and developmental toxicity of toluene: A review. *Environmental Health Perspectives, 94*, 237–244.

Foo, S. C., Jeyaratnam, J., & Koh, D. (1990). Chronic neurobehavioral effects of toluene. *British Journal of Industrial Medicine, 47*, 480–484.

Fourches, D., Muratov, E., & Tropsha, A. (2010). Trust, but verify: On the importance of chemical structure curation in cheminformatics and QSAR modeling research. *Journal of Chemical Information and Modeling, 50*, 1189–1204.

Guengerich, F. P. (2006). Cytochrome P450s and other enzymes in drug metabolism and toxicity. *AAPS Journal, 8*, 101–111.

Hu, Y., & Kupfer, D. (2002). Metabolism of the endocrine disrupter pesticide methoxychlor by human P450s: Pathways involving a novel catechol metabolite. *Drug Metabolism and Disposition, 30*, 1035–1042.

Jakoby, W. B. (1978). The glutathione S-transferases: A group of multifunctional detoxification proteins. *Advanced Enzymology, 46*, 383–414.

King, G., & Zeng, L. (2002). Estimating risk and rate levels, ratios and differences in case-control studies. *Statistics in Medicine, 21*, 1409–1427.

Kupfer, D., & Bulger, W. H. (1987). Metabolic activation of pesticides with proestrogenic activity. *Federal Proceedings, 46*, 1864–1869.

Pelkonen, P., Lang, M. A., Negishi, M., Wild, C. P., & Juvonen, R. O. (1997). Interaction of aflatoxin B_1 with cytochrome P450 2A5 and its mutants: Correlation with metabolic activation and toxicity. *Chemical Research in Toxicology, 10*, 85–90.

Rand, G. (1995). *Fundamental of Aquatic Toxicology: Effects, Environmental Fate, and Risk Assessment*. Boca Raton, FL: CRC Press.

Rinsky, R. A. (1989). Benzene and leukemia: An epidemiologic risk assessment. *Environmental Health Perspectives, 82*, 189–191.

Sánchez-Peña, L. C., Reyes, B. E., López-Carrillo, L., Recio, R., Morán-Martínez, J., Cebrián, M. E., & Quintanilla-Vega, B. (2004). Organophosphorus pesticide exposure alters sperm chromatin structure in Mexican agricultural workers. *Toxicology and Applied Pharmacology, 196*, 108–113.

Santos, S. L. (1987). Risks: How to get more science in assessments. *Environmental Science & Technology, 21*, 239–240.

Schmid, E., Bauchinger, M., & Hauf, R. (1985). Chromosome changes with time in lymphocytes after occupation exposure to toluene. *Mutation Research, 142*, 37–39.

Wallen, M., Holm, S., & Nordqvist, M. B. (1985). Coexposure to toluene and *p*-xylene in man: Update and elimination. *British Journal of Industrial Medicine, 42*, 111–116.

Williams, P. R. D., Hubbell, B. J., Weber, E., Fehrenbacher, C., Hrdy, D., & Zartarian, V. (2010). An overview of exposure assessment models used by the U.S. Environmental Protection Agency. In G. Hanrahan (Ed.), *Modeling of Pollutants in Complex Environmental Systems*. St. Albans, UK: ILM Publications.

World Health Organization (WHO). (1985). *Toluene*. Environmental Health Criteria, No. 52.

Zbinden, G., & Flury-Roversi, M. (1981). Significance of the LD50 test for the toxicological evaluation of chemical substances. *Archives of Toxicology, 47*, 77–99.

Chapter 10

Abstracts and Keywords

Abstract

This chapter presents material related to the burgeoning area of green chemistry. It starts with coverage of the 12 Principles of Green Chemistry and their relation to chemical and industrial processes. The next section deals with life cycle assessment (LCA) and its four complementary phases. Next, information on example sustainable chemical and related processes is presented, including emerging environmental technologies and sustainability metrics. Subsequently, pollution prevention and waste minimization practices are detailed. Finally, example sustainable materials and application areas are covered.

Keywords

Green chemistry, 12 Principles of Green Chemistry, Life cycle assessment (LCA), Sustainable technology, Renewable energy, Sustainability metrics

Chapter 10

Green Chemistry and Sustainable Chemical Processes

The three Rs—reduce, reuse, and recycle—all help to cut down on the amount of personal waste we throw away. Such wastes include hazardous household products that contain corrosive, toxic, ignitable, or reactive ingredients that can pollute the environment and pose a threat to human health. In addition, nearly every industrial process, from manufacturing consumer goods to generating energy, produces varying types and quantities of usable and end-product waste materials. Similar to municipal solid wastes, such as cardboard, newspapers, and beverage containers, these industrial materials are also valuable commodities that can be reduced, reused, and recycled. For example, 80% of all industrial lubricants used in production operations are being disposed of prematurely, and often improperly. Fortunately, chemical engineers are actively involved in developing ways to extend the useful life of lubricants through fundamental chemical concepts to minimize the aging process. Experts are also actively engaged in changing the way that industries think about waste to see the value of a spent material as a product or commodity, not as a waste. Positive economic rewards for such activities and favorable environmental results are moving facilities toward more waste reduction and materials management.

FIGURE 10.1 The ubiquitous three Rs symbol representing the concepts of reduce, reuse, and recycle.

10.1. PRINCIPLES OF GREEN CHEMISTRY

Green chemistry refers to the application of minimization principles to eliminate negative environmental impacts of chemicals and chemical processes. Such principles are based on redesigning current industrial and laboratory-based methods to adhere to defined (and well-established) principles of green chemistry. The **12 Principles of Green Chemistry**, originally published by Anastas and Warner (1998), and now incorporated into the U.S. Environmental Protection Agency's (EPA) Green Chemistry Program, are outlined in Figure 10.2 and further defined below.

The 12 Principles of Green Chemistry

1. **Prevent waste**: Design chemical syntheses to prevent waste;
2. **Maximize atom economy**: Design syntheses so that the final product contains the maximum proportion of the starting materials (few wasted atoms!);
3. **Design less hazardous chemical syntheses**: Design syntheses to use and generate substances with minimal environmental toxicity;
4. **Design safer chemicals and products**: Design chemical products that have minimal environmental toxicity;

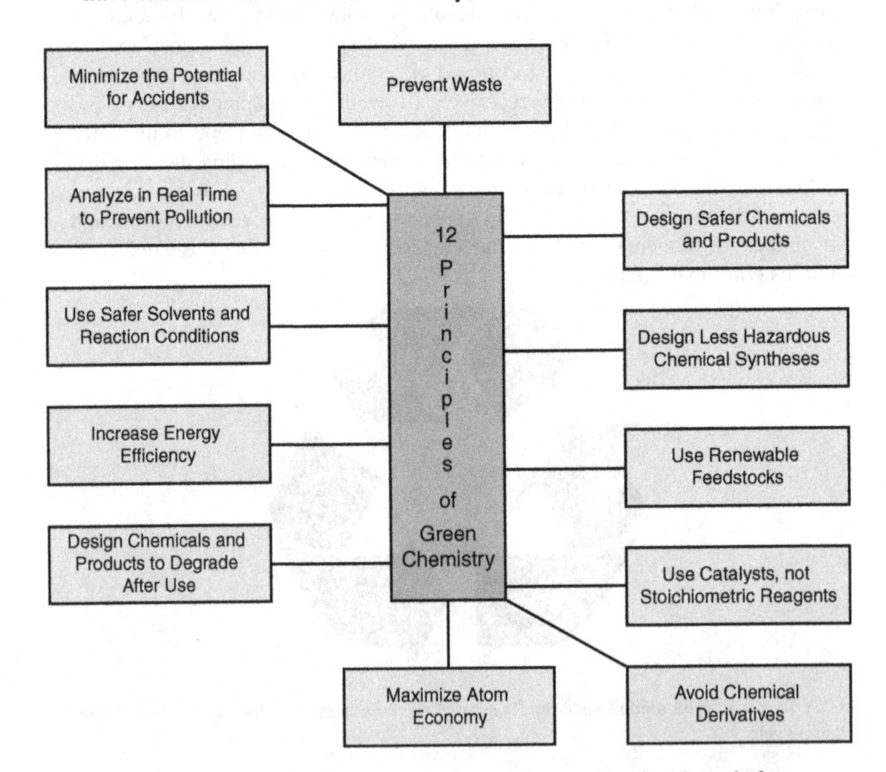

FIGURE 10.2 The 12 Principles of Green Chemistry outlined in schematic form.

5. **Use safer solvents/reaction conditions**: The goal of avoiding solvents, separation agents, or other auxiliary chemicals at all costs. If unavoidable, use innocuous chemicals;
6. **Increase energy efficiency**: Attempt to run reactions at ambient temperature and pressure whenever possible;
7. **Use renewable feedstocks**: Attempt to use raw materials and feedstocks (starting materials) that are renewable rather than depleting. For example, renewable feedstocks are often made from agricultural products or are the wastes of other processes;
8. **Avoid chemical derivatives**: Attempt to avoid using blocking or protecting groups or any temporary modifications whenever possible;
9. **Use catalysts, not stoichiometric reagents**: Try to minimize waste by using catalytic reactions that can carry out a single reaction multiple times;
10. **Design chemicals and products that degrade after use**: The design of chemical products that break down to innocuous substances upon use by the consumer or industrial process;
11. **Analyze in real time to prevent pollution**: Use and encourage the use of in-process real-time monitoring and control during syntheses. This minimizes or eliminates byproducts;
12. **Minimize the potential for accidents**: Design chemicals to minimize the potential for chemical accidents including explosions, fires, and environmental releases.

Let's consider a few examples that apply fundamental concepts of the 12 Principles of Green Chemistry. There is growing urgency in the field of organic chemistry to not only develop cost-effective reactions that can achieve high yields, but also to center on the "greenness" of the processes involved. In such a capacity, experimental yields can be calculated by the following, general expression:

$$\text{Yield} = \frac{X_{actual}}{X_{theoretical}} \times 100\%, \tag{10.1}$$

where X can represent, for example, mass. At the basic level, simply developing and using alternative solvents can offer immediate economical and environmentally-friendly advantages over traditionally-used solvents. For instance, 2-methyltetrahydrofuran (2-Me THF; Figure 10.3), made from renewable resources, is a green alternative to tetrahydrofuran for organometallic reactions (e.g., Grignard and metal-catalyzed coupling) and dichloromethane for biphasic reactions (e.g., alkylation and nucleophilic substitution). As a green alternative, 2-Me THF demonstrates higher boiling points and a low heat of vaporization, which leads to reduced reaction times and energy savings during distillation and recovery, respectively (Avcock, 2007). In Chapter 1 we discussed the properties of water that contribute to its role as one of nature's greatest

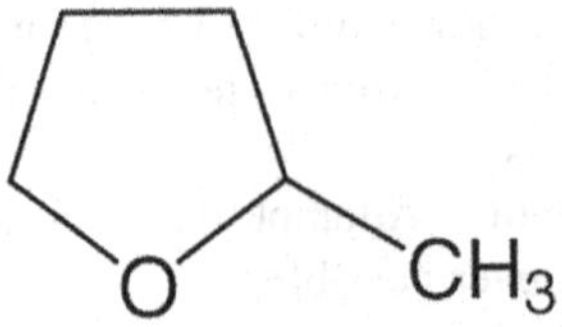

FIGURE 10.3

solvents. However, the use of water as a medium for organic reactions is only a recently developed practice. Those realizing this potential have found improved selectivities and yield (up to 85–90%) in modern organic reactions. For example, Diels-Alder reactions (see example in Figure 10.4) have been shown to be greatly accelerated by using water as a solvent. It was found that, because of hydrophobic effects, using water as a solvent not only accelerated reaction rates but also enhanced reaction selectivities, even when the reactants were sparingly soluble or insoluble in the medium (Rideout and Breslow, 1980).

Metathesis has very important applications in industrial chemical processing as well as fine chemical production. The term metathesis means to transpose or to change places, with a metathesis reaction being defined as "the interchange of atoms or groups of atoms between two molecules." In fact, metathesis has been compared to a dance in which the couples change partners. In olefin metathesis (note that the term olefin is another name for alkene, a carbon chain that exhibits double bonding), the double-bonding atom groups will change places with one another (Figure 10.5a). This occurs with the assistance of special catalyst molecules that form metal carbine structures with the reactants, thus making the overall reaction favorable.

Since the discovery of the metathesis reaction in 1971 by Yves Chauvin, the process has become more and more synthetically useful with many different types of metathesis reactions developed, illustrating the power of available catalysts, including those developed by Drs. R. Schrock and R. H. Grubbs. The development of a catalyst with the metal ruthenium by Grubbs is of particular interest given its stability in air and high selectivity (Trinka and Grubbs, 2001). The 1st generation Grubbs catalyst is pictured in Figure 10.5b. It has the unique properties of being able to initiate metathesis in the presence of alcohols, water, and carboxyl acids. The developments of 2nd generation Grubbs catalysts have since been developed and are commercially available

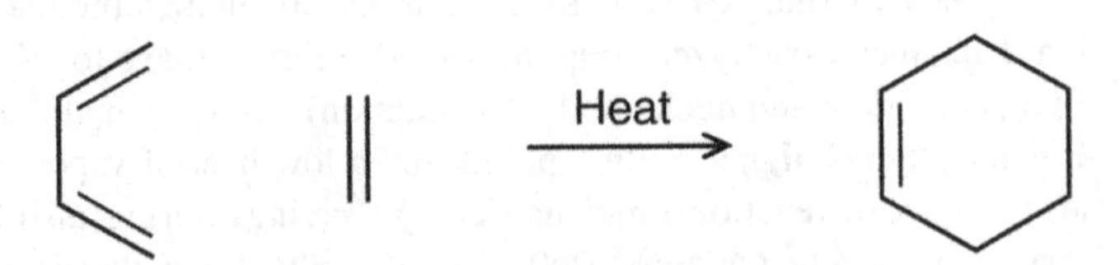

FIGURE 10.4 A simple Diels-Alder reaction of 1,3-butadiene with ethene to form cyclohexene.

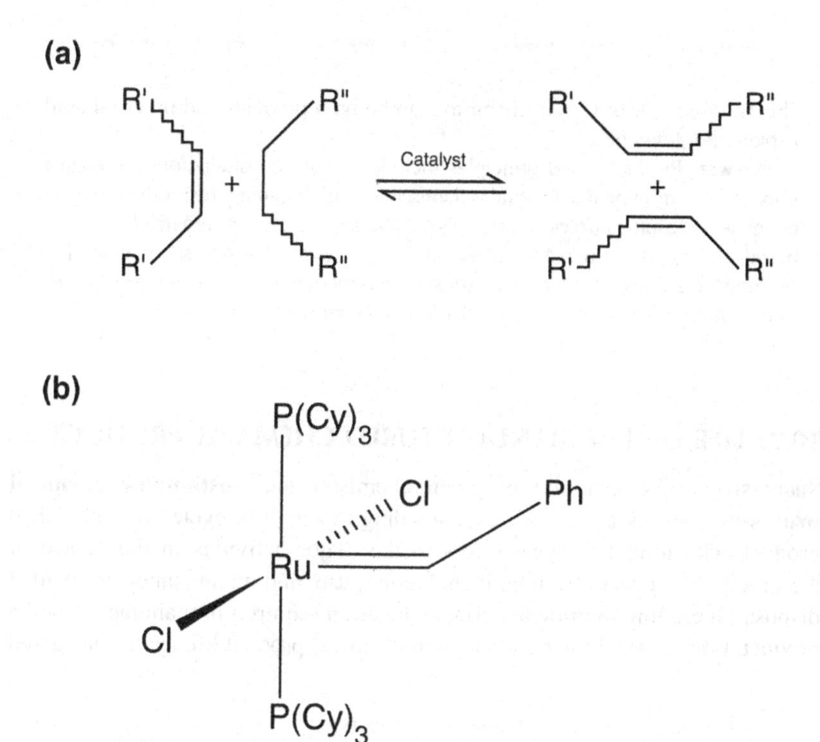

FIGURE 10.5 (a) A generalized olefin metathesis reaction that allows the exchange of substituents between different olefins. (b) Grubbs 1st generation catalyst (benzylidene bis(tricyclohexylphosphine)dichlororuthenium) used in organic synthesis to achieve olefin cross-metathesis and other ring-opening and ring-closing metatheses.

today. As a result, metathesis technology has been labeled as a great step forward for green chemistry by reducing potentially hazardous waste through smarter production. It lends itself well to many applications including making compounds during the drug discovery process, the development of agro- and petrochemicals, and in the manufacturing of flavor and fragrance chemicals. For example, ring-closing metathesis (RCM) is now routinely used in one of the synthesis steps of mevinolin, a drug used to lower cholesterol rates in the blood. If we again consider our discussion of greener agricultural practices, metathesis can be used to improve the synthesis of insect pheromones, which are a more environmentally-conscience alternative to pesticides.

As shown above, the 12 Principles of Green Chemistry can provide a roadmap for scientists and technologists in their quest to reduce or eliminate the use and generation of hazardous-type chemicals and to design appropriate technology to increase economic, environmental, and societal improvements. Basic components of these principles will be incorporated into subsequent sections of this chapter to further stress the importance of green chemistry in modern industrial efforts and to facilitate student learning.

Example Problem 10.1

The 12 Principles of Green Chemistry can be both product- and process-based. Explain the differences.

Answer: Product-based principles include toxicity considerations, ecological impacts, determining if a material is flammable, and designing the product so it can either degrade into harmless substances or be easily recycled. Alternatively, process-based principles include the development of chemical processes that limit or exclude hazardous substances, use substitutions for overly reactive reagents, reduce or eliminate solvent use, and consider limiting energy usage and costs.

10.2. LIFE OF THE MANUFACTURED CHEMICAL PRODUCT

Successful implementation of green chemistry and **sustainable chemical processes** requires a detailed understanding of the "life cycle" of a chemical product. The term life cycle refers to the major activities in the course of the product's life span from its manufacture, use, and maintenance, to its final disposal, including the raw material acquisition required in manufacturing the product. Figure 10.6 illustrates a typical chemical product life cycle considered

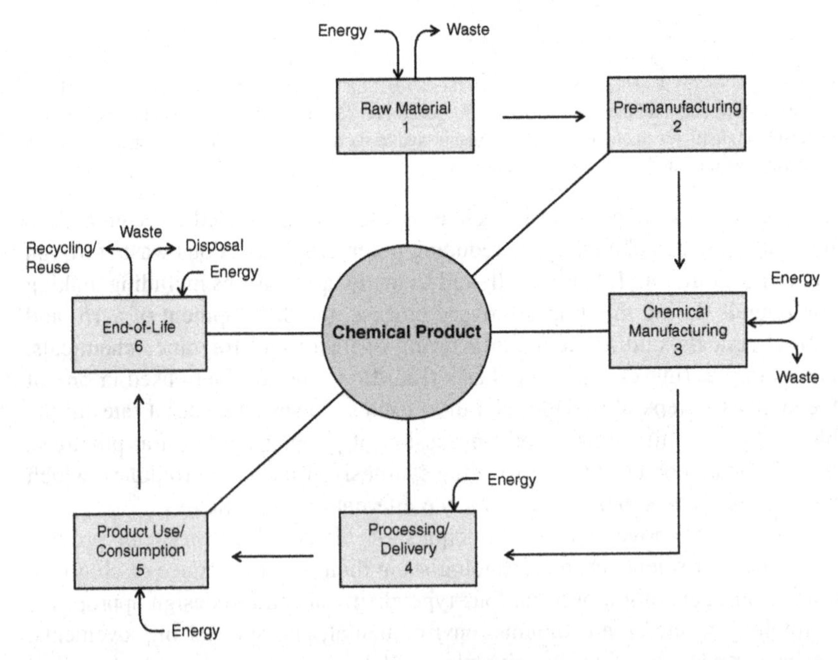

FIGURE 10.6 Chemical product life cycle considered in a life cycle assessment (LCA). Shown is a systematic approach for investigating the inputs/outputs of materials and energy and the associated environmental impacts directly correlated to the product through its full life cycle.

in a **life cycle assessment (LCA)**. The LCA consists of four complementary phases: i) goal and scope, ii) life cycle inventory, iii) life cycle impact assessment, and iv) interpretation.

As evident in the schematic, LCA requires energy materials and balances for all stages of the product life cycle. Full LCA is thus considered a "**cradle-to-grave**" approach for assessing industrial systems. As shown in Figure 10.6, "cradle-to-grave" begins with the gathering of raw materials to create the product and ends at the point when all materials are recycling/reused or properly disposed of. By including the impacts throughout the product life cycle, LCA provides a comprehensive view of the environmental aspects of the product or process and a more accurate picture of the true environmental impacts (e.g., energy input and waste generation) in product and process selection. Minimizing or eliminating impacts is the ultimate goal of sustainable design ventures.

TEXTBOX 10.1

Sustainability, in the broadest sense, is an umbrella term describing the interrelationship between the human social system (including economic considerations) and the physical environment. This chapter puts sustainable development in a chemical and process engineering context. Topics such as sustainability, resource use, and energy are common threads that run throughout chapter topics.

Let's consider LCA for the example process of aluminum smelting: the reduction of refined alumina to metallic aluminum by the electrolytic separation of aluminum from its oxide. What is its environmental impact? Are the starting products recovered during the process? To answer this, we first have to examine the steps in the smelting process, as outlined below in a series of electrolytic cells carrying direct current:

1. Alumina is dissolved in a molten batch of cryolite (an electrolyte) and aluminum fluoride at 1,000°C;
2. The cryolite is placed in an iron vat lined with graphite (Note: The vat serves as the cathode);
3. Carbon anodes (immersed in cryolite) carry the current to the solution, and on to the next cell. Electric current is then passed through the molten material;
4. At the cathode, electrolysis reduces aluminum ions to aluminum metal. At the anode, carbon is oxidized to form carbon dioxide gas. The overall reaction is $2Al_2O_3 + 3C \rightarrow 4Al + 3CO_2$;
5. Molten metal sinks to the bottom of the cell and is drained off as product.
6. In this process, the cryolite and aluminum fluoride are generally recovered with little or no loss. And, as shown in the above reaction, the carbon anodes are consumed during the process. The principal products of the reaction are CO_2, which is released as a gas, and elemental aluminum, which as mentioned, settles and is collected for end use.

A brief cradle-to-grave assessment of the production of liquid steel reveals a more substantial process involving the following stages:

1. Mining, extraction, and processing of raw materials (including iron ore, coal, and limestone);
2. Reduction of iron (III) oxide (main component of iron ore) to produce iron with the help of coal and coke in blast furnaces:
$Fe_2O_3 + CO \rightarrow 2\ Fe(I) + 3CO_2$;
3. Oxygen is used to lower the hot metal's carbon content before it becomes steel;
4. Contaminants are removed through the use of lime.

The production process above generates large quantities of heat, gas, and slag, which are utilized to a large extent. Apart from carbon dioxide, steel production generates emissions of nitrogen dioxide, sulfur oxides, and dust. Significant sources of gases include the heating ovens for slags and stoves for the blast furnaces. In addition, copious amounts of water are needed for cooling furnaces, coke, and steel products. This can in turn result in wastewater that will need some form of end-use treatment.

By implementing a variety of simple engineering changes or product modifications, industrial entities can affect multiple stages of the life cycle of a given chemical product or process, whether it be the reduction in energy use, curbing greenhouse gas emissions, or end-use waste minimization. Moreover, innovative use of alternative feedstocks (e.g., bio-based, municipal waste) and energy sources offering reduced environmental impact are being routinely considered. For instance, **transesterification** is the process of converting plant-derived oils into biodiesel fuel. Because it is produced from renewable sources, biodiesel is an attractive alternative to standard diesel and a viable means of reducing petroleum dependence and greenhouse gas emissions (Adi, et al., 2009). Conveniently, biodiesel is friendly to conventional diesel engine systems and has been shown to deliver similar performance and engine durability to standard diesel fuel. In the transesterification process, one mole of triglyceride (main constituent of plant oils) reacts with three moles of alcohol (typically methanol) in the presence of a strong catalyst (e.g., acid, base, or enzymatic), producing a mixture of fatty acid alkyl esters (biodiesel) and glycerol (Figure 10.7) (Ma and Hanna, 1999). In many processes, alkaline catalysis (use of NaOH) has been shown to achieve transesterification of triglycerides with high speed and efficiency. It is routinely reported to be more effective than acid catalysis in this role. Other researchers have experimented with the use of alkaline-earth oxide compounds, such as CaO and SrO. Such compounds are slightly soluble in methanol and demonstrate good catalytic activity and a long catalyst lifetime. More recently, calcium ethoxide produced by the following reaction

$$Ca + 2CH_3CH_2OH \rightleftharpoons Ca(OCH_2CH_3)_2 + H_2\uparrow$$

$$
\begin{array}{l}
CH_2-O-\overset{\overset{\displaystyle O}{\|}}{C}-R' \\
| \\
CH_2-O-\overset{\overset{\displaystyle O}{\|}}{C}-R'' \\
| \\
CH_2-O-\overset{\overset{\displaystyle O}{\|}}{C}-R'''
\end{array}
+ 3CH_3OH \longrightarrow
\begin{array}{l}
CH_3-O-\overset{\overset{\displaystyle O}{\|}}{C}-R' \\
\\
CH_3-O-\overset{\overset{\displaystyle O}{\|}}{C}-R'' \\
\\
CH_3-O-\overset{\overset{\displaystyle O}{\|}}{C}-R'''
\end{array}
+
\begin{array}{l}
CH_2OH \\
| \\
CH-OH \\
| \\
CH_2-OH
\end{array}
$$

FIGURE 10.7 Transesterification of triglyceride with short-chain alcohol (methanol) and a catalyst.

has been shown to be an effective solid base catalyst to aid the transesterification process.

Dow Corning and Genencor, two leaders in the chemical and biotechnology sectors, have formed an alliance to make bio-based organosilicones. Here, molecular biologists are taking advantage of marine organisms that produce large relative masses of biogenic silica to isolate genes and proteins controlling silica biosynthesis and nanofabrication (Morse, 1999). Silicon biotechnology (e.g., Figure 10.8) is expected to yield unique biologically-mediated organosilicone-based materials for applications in diagnostics, biosensors, controlled delivery of active ingredients, and personal care products (Jenck, et al., 2004).

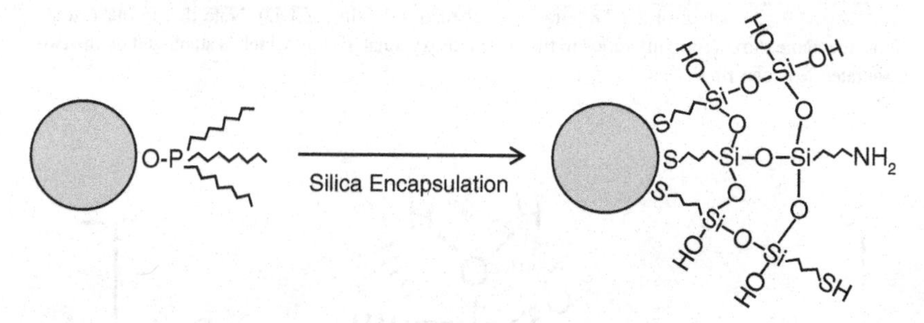

FIGURE 10.8 Functional organosilicone molecules containing $-NH_2$ or $-SH$ incorporated into the shell of a quantum dot to provide surface functionalities for biomedical applications. Quantum dots are fluorescent semi-conductor nanocrystals used for biological imaging.

Research Application XII: Green Oxidation Chemistry

Oxidative bleaching agents such as chlorine (Cl_2), chlorine dioxide (ClO_2), and sodium hypochlorite (NaOCl) have traditionally been used in household cleaners and in the pulp and paper industries. While extremely effective in their application, chlorine-containing agents can lead to the formation of hazardous byproducts, including dioxins (e.g., 2,3,7,8-tetrachlorodibenzo-*p*-dioxin (Figure 10.9)).

This byproduct is also known as TCDD, or simply, dioxin. Dioxin is extremely toxic and can lead to chloracne, cirrhosis of the liver, heart and kidney damage, and can act as a teratogen resulting in possible birth defects.

Alternatives such as hydrogen peroxide (H_2O_2) have been shown to oxidize compounds without the use of chlorine atoms. Initial attempts to use H_2O_2 were problematic given two important limitations: i) H_2O_2 requires higher temperatures and pressures resulting in increased energy usage and costs, and ii) peroxide oxidation can be unselective due to the presence of free radicals. To remedy these limitations, Dr. Terry Collin and his colleagues at Carnegie Mellon University developed tetraamido macrocyclic ligands (TAML® (Figure 10.10)) that, when coordinated with a transition metal (e.g., Fe), effectively catalyze hydrogen peroxide oxidation (Collins, 2002). Dr. Collins reported that the iron catalysts developed can be composed exclusively of biochemically-common elements. They are efficient and selective peroxide activators for many areas of technology, are water-soluble, and are effective in minute quantities (nanomolar to low micromolar) over a broad pH range. For his environmentally-friendly developments, Dr. Collins received the 1998 Presidential Green Chemistry Challenge Award.

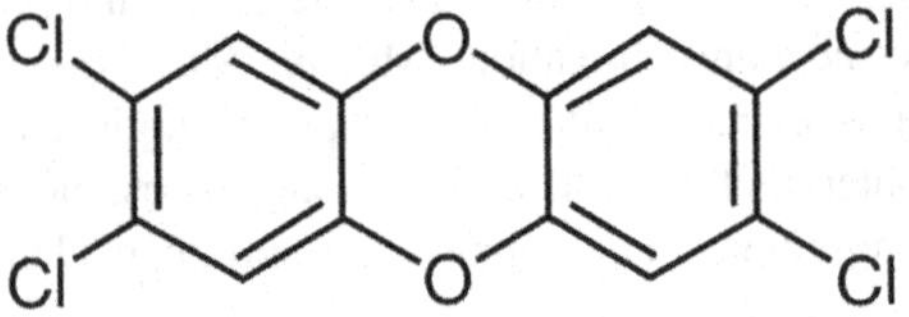

Cl O Cl
Cl O Cl

FIGURE 10.9 Structure of 2,3,7,8-tetrachlorodibenzo-*p*-dioxin (TCDD). Note that in our discussions, the name dioxin formally refers to the central dioxygenated ring, which is stabilized by the two associated benzene rings.

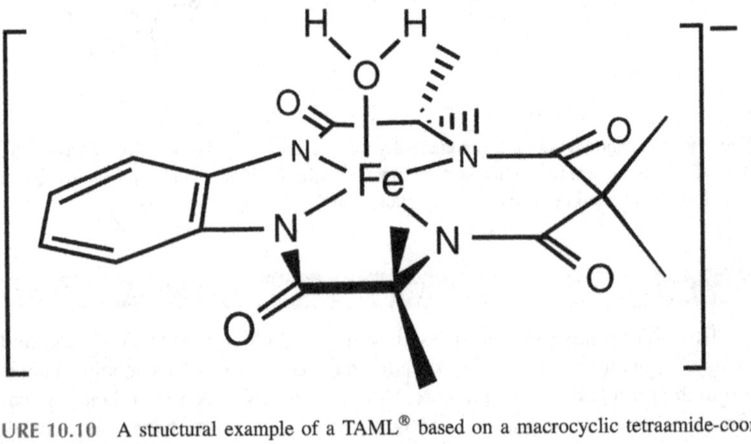

H H
O
O N Fe N O
N N O
O
−

FIGURE 10.10 A structural example of a TAML® based on a macrocyclic tetraamide-coordinated ligand and an iron center.

10.3. SUSTAINABLE CHEMICAL AND RELATED PROCESSES

This section is devoted to process or product innovations where modern developments have played a key role in decision-making and project justification, including new innovations offering enhanced "sustainability." What are the challenges for industry? How do we measure the environmental impacts of industrial processes? How do we know if industry is following the 12 Principles of Green Chemistry? We will answer these important and timely questions in subsequent sections.

10.3.1. Design of Sustainable Technology

The call for chemical processes to be safe, compact, flexible, energy efficient, environmentally benign, and conducive to the rapid commercialization of new products poses new challenges for chemical engineers and industrial managers intent on following the 12 Principles of Green Chemistry. In order to achieve such goals the design and development of sustainable technologies that use less energy, fewer limited resources, and can be reused or recycled are needed. For example, feedstock recycling of plastic waste into chemicals useful as fuels or raw materials is now routinely demonstrated in the industrial setting. The use of biomass as feedstock is also of expanding interest, and of great environmental significance. For instance, there is movement toward the development of alternative fuels, including ethanol produced from cellulosic materials like grasses and wood chips. Structurally, cellulose is an extensive polymer of glucose and a great potential biofeedstock. In theory, two moles of ethanol can be produced from one mole of glucose:

$$C_6H_6H_{12} \rightarrow 2\,C_2H_5OH + 2CO_2.$$

However, obtaining this product is a complicated process involving either an enzymatic approach or biomass hydrolysis. The latter involves the depolymerization of the biomass polysaccharides to fermentable sugars. Other examples of sustainable technology in use include industrial process routes with a reduced **carbon (CO_2) footprint**, higher carbon efficiency, and/or reduced use or generation of hazardous components.

TEXTBOX 10.2

A **carbon footprint** is a measure of the impact our activities have on the environment, and in particular, climate change. More specifically, it is a measurement of all greenhouse gases individually produced with units of kg (or often tons) of CO_2 equivalent.

According to an International Energy Agency (2008) report, energy demand is projected to rise by over 60% until 2030. By then, more than 80% of the energy sources are expected to be fossil fuel–based. The negative environmental effects of continued use of fossil fuels can only be offset by the development of new, sustainable technology solutions. Limiting CO_2 emissions from energy generation is key to reducing emissions of greenhouse gases, and hence the carbon capture and storage concepts discussed earlier in this textbook could play an essential role in the development of a more sustainable energy system. However, as discussed, technology presently available is costly, entails high operating costs, or is not proven to work in large-scale commercial installations.

The creation of eco-industrial parks is also of great consideration. These are settings where industries are grouped such that they can exchange raw materials, waste materials, and energy amongst each other, thereby reducing the net inputs and outputs of the park. Such parks are based on the concepts of **Industrial Ecology**—a sustainable relationship between industrial and ecological systems guided by conservation of non-renewable resources, pollution prevention, and intersocietal equity. Again referring to Figure 10.6, the study of the inputs and outputs of the industrial "ecosystem" (material flow) can reveal the efficiency of the processes involved and help in optimizing the LCA process. A prime example is the TiagaNova Eco-Industrial Park located in Fort McMurray, Alberta, Canada. This modern industrial complex is actively involved in energy efficiency, byproduct synergy, and routinely considers strategies to reduce resource use, reduce waste generation, and increase land use efficiency to enhance the quality of life for nearby residents.

TEXTBOX 10.3

According to the National Academy of Sciences, liquid fuels from **biomass** have the potential to reduce the use of petroleum fuels and CO_2 emissions from the U.S. transportation sector over the next 25 years (Ma and Hanna, 1999). According to the report, biofuels will likely be produced from dedicated grass crops, agricultural and forestry residues, and municipal solid wastes rather than food crops such as corn, grain, or soybeans. It is estimated that the U.S. could produce 550 million dry tons of biomass annually by 2020.

10.3.2. Emerging Environmental Technologies

The need for cost-effective, innovative technologies that minimize adverse environmental effects are paramount. Recognizing technology's key role, modern research and development efforts that nurture the acceptance and use of equipment, processes, and systems critical to improved environmental performance are underway. Table 10.1 lists a variety of emerging or recently developed technologies with their potential or proven environmental benefits

TABLE 10.1 Representative Emerging Environmental Technologies

Technology	Example Applications	Potential or Proven Environmental Benefits
Biofuels	Transportation sector and energy storage.	Reduction in greenhouse gas emissions, less fossil fuel consumption per BTU produced, and non-toxic and biodegradable. A renewable energy source.
Micro- and nano-technologies	Solvent-free mixing, *in situ* reagent generation, wireless communication, nanosensing of pollutants, and photo catalysts for solar applications.	Cuts energy and material inputs, less reagent consumption, and provides for potentially safer experiments.
Membrane bioreactors (MBR)	Aerobic treatment of landfill leachate and industrial waste streams.	Compact and versatile environment for the biological degradation of wastewater components. Low energy requirements and cost effective.
Hydrogen economy	Energy storage potential.	Potential for less fossil fuel consumption. Hydrogen also has a natural compatibility with fuel cells.
Geothermal, solar, tidal, hydro, wave, and wind power	Energy storage.	Less fossil fuel consumption and pollutant release. Renewable energy sources.
Bio- and synthetic catalysts	Enable the use of alternative reagents and feedstocks.	Offer natural and synthetic alternative pathways that are consistent with the principles of green chemistry.
Real-time, field-based, or *in situ* monitors	Field-based environmental monitoring of pollutants.	High temporal and spatial environmental data acquisition. Decreased reagent/sample consumption, decreased waste, potentially safer experiments and chemical reactions, portability, and low energy consumption.

FIGURE 10.11 Wind turbines winding through the San Gorgonio Pass near Palm Springs, California. This area houses nearly 4,500 turbines collectively harnessing 650 megawatts of pure, renewable energy.

and applications. Although not conclusive, this list is representative of areas that have been well received by chemists, engineers, environmental scientists, and industrial managers alike.

A large portion of the popular emerging technologies are in the area of **renewable energy**—energy generated from natural resources whose use does not result in detrimental greenhouse gas or pollution emission. According to a 2009 report (Renewable Energy Policy Network for the 21st Century, 2009), nearly 18% of 2008's global final energy came from renewable sources and was broken down as follows: traditional biomass (9.6%), hydro (2.3%), modern biomass (1.9%), solar (1.8%), wind (1.3%), geothermal (0.45%), and tidal (0.09%). In the four years from end-2004 to end-2008, solar photovoltaic (PV) capacity increased sixfold to more than 16 gigawatts (GW), wind power capacity (Figure 10.11) increased 250% to 121 GW, and total power capacity from new renewables increased 75% to 280 GW, including significant gains in hydro, geothermal, and biomass power generation. During the same period, solar heating capacity doubled to 145 gigawatts-thermal (GWth), while bio-diesel production increased sixfold to 12 billion liters per year and ethanol production doubled to 67 billion liters per year. Among new renewables (excluding large hydropower), wind power was the largest addition to world-wide renewable energy capacity. Existing wind power capacity grew by 29% in 2008 to reach 121 GW, more than double the 48 GW that existed in 2004. The 2008 increase was led by high growth in the strongest markets of the United States (8.4 GW), China (6.3 GW), India (1.8 GW), and Germany (1.7 GW) (National Academy of Sciences, 2009).

Example Problem 10.2

A non-renewable energy resource:

a. Can be used over and over again
b. Will ultimately be depleted
c. Will not work for modern electrical devices
d. Is often rechargeable

Answer: b. It will ultimately be depleted.

Green nanotechnology initiatives are also currently in full swing with applications near commercialization, including nanosensors and nanoscale coatings to replace thicker, more wasteful polymer coatings that prevent corrosion, nanosensors for detection of aquatic toxins and pollutants, nanoscale biopolymers for improved decontamination and recycling of heavy metals, smart particles for environmental monitoring and purification, nanoparticles as a novel photocatalyst for solar applications, and other nano-based environmental catalysts. We are already beginning to see the environmental benefits of green nanotechnology in the areas of fuel cell research and microelectronics. However, as nanomaterials are developed, the environmental and human health impacts of both the products themselves and the processes to make them must be considered to ensure their long-term economic viability.

10.3.3. Sustainability Metrics

The ability to systematically identify and evaluate the economic, environmental, and societal impacts of chemical processes relies on well-developed **sustainability metrics**. Note that there is no standard method of measuring successes and therefore, it is extremely difficult to compare one chemical industry working toward sustainability with another. Sustainability indicators (e.g., water consumption and discharge quality, NO_x emissions, and energy consumption) are useful in such evaluations. The conversion of absolute measures into relative measures, such as ratios, which screen out statistical "noise" such as differences in size or output, and focus on relationships, is encouraged.

Evaluation methods can also include qualitative aspects of examination and the development of defined standards. For example, the International Organization for Standardization (ISO) has developed LCA standards to consolidate procedures and methods for stakeholder and community review. Core ISO standards for LCA and a brief description of each are listed in Table 10.2. These standards lay out a framework for the four phases of LCA defined previously. They are instrumental in evaluating the magnitude and significance of potential impacts of given chemical product/manufacturing systems.

The American Institute of Chemical Engineers (AIChE) has also instituted practical LCA benchmarks and well-defined metrics through the development of the AIChE Sustainability Index (SI). The AIChE SI uses publicly available

TABLE 10.2 Core ISO Standards Used in Conducting the LCA Process

ISO Standard	Description
ISO 14040	Outlines the framework for LCA from which the remaining standards are derived.
ISO 14041	Establishes the goals, purpose, audience, scope, and stakeholders that will be impacted by the process.
ISO 14042	Outlines the life cycle impact assessment (LCIA) phase to evaluate the magnitude and environmental impacts of a given process.
ISO 14043	The interpretation phase of the LCA.
ISO 14047	Provides technical report assistance on how to apply the LCIA.
ISO 14048	Provides LCA data documentation format.
ISO 14049	Provides examples of technical reports to apply goal and scope definitions.

data to assess the sustainability performance of representative companies in the chemical industry with respect to strategic commitment, sustainability innovation, environmental performance, safety performance, product stewardship, social responsibility, and value chain management (Calvin, et al., 2009). AIChE SI metrics are based on seven assessment areas (Figure 10.12) that allow companies to measure their performance and compare with the overall sector.

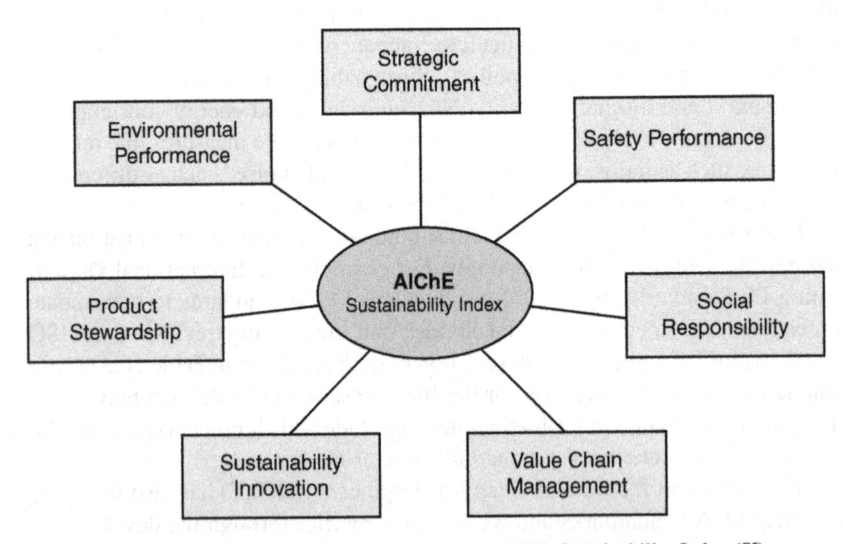

FIGURE 10.12 Seven assessment areas of the AIChE Sustainability Index (SI).

By and large, sustainability metrics provide methodology to measure parameters in a specific chemical process that identify operational difficulties, process sensitivities, and/or unused resources, and, through their analysis, identify opportunities for process optimization and improvements. Metrics can also be employed to measure the inherent safety of a given process that bodes well with attempting to meet and maintain the 12th Principle of Green Chemistry stated above.

10.4. POLLUTION PREVENTION AND WASTE MINIMIZATION

A number of the 12 Green Chemistry Principles detail some form of **pollution prevention** and/or **waste minimization**. Pollution prevention is a mission based on the five strategies (in order of desirability) below (Habicht, 1992):

1. Reduce pollution at the source whenever feasible;
2. Recover and reuse on-site whenever feasible;
3. Recycle off-site in an environmentally safe manner whenever feasible;
4. Treat pollution in an environmentally safe manner whenever feasible;
5. Disposal or release into the environment as a last resort.

Figure 10.13 displays these strategies in a pollution prevention pyramid, with desirability decreasing as it moves toward the pinnacle. Pollution prevention includes equipment or technology modifications, process or procedure modifications, product reformulation or redesign, and raw material substitutions.

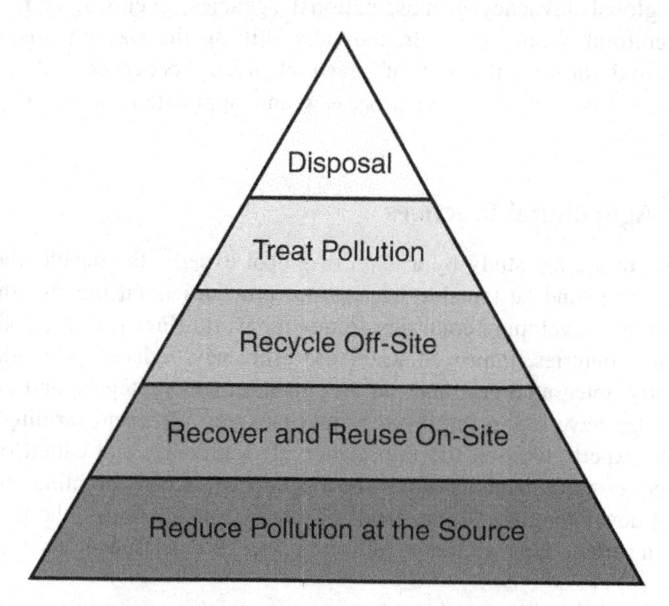

FIGURE 10.13 Five strategies of pollution prevention with desirability decreasing as it moves toward the pinnacle of the pyramid.

Waste minimization, one of the first initiatives of pollution prevention, is a national policy that was first presented in the 1984 U.S. Hazardous and Solid Waste Amendments (HSWA) to RCRA (recall our discussion in Chapter 9). Waste minimization involves the reduction of any solid or hazardous waste that is generated or subsequently treated, stored, or disposed of. This policy also includes the concepts of recycling, general volume reduction, as well as reduction in the quantity of toxic components. As also discussed in Chapter 9, chemical and thermal destruction methods, among others, can be used to achieve toxicity reduction. Recall that conventional thermal processes destroy chemical agents either by incineration or by pyrolysis—a process whereby chemical agents are heated in the absence of oxygen.

10.5. SUSTAINABLE MATERIALS AND APPLICATION AREAS

Since World War II, the production of chemical substances for industry and scientific research has increased 400-fold and is currently estimated at over 500 million tons (McCoy, 2006). Of the total worldwide use, 80% are consumed by the developed world. The remaining 20% are consumed by developing countries as they search for greater economic growth through industrial sectors, agriculture, and services. At the European Community (EC) level, approximately 2,000 substances are on the List of Dangerous Substances. In the United States, 2,000–3,000 substances are notified each year under the Toxic Substances Control Act (TOSCA) (Chandler, 2005). An array of global advocacy groups, national agencies, chemical companies, and agricultural alliances are dramatically shifting the way chemicals are manufactured through the use of green chemistry concepts and greener marketing initiatives. Example processes and application areas are highlighted below.

10.5.1. Agricultural Practices

According to a 2006 study by a team of global experts, the development of safer pesticides and sustainable agricultural practices is on the rise in both developed and developing countries around the world (Pretty, et al., 2006). In developing countries, improved water use efficiency, reduced pesticide use, agroforestry, integrated pest and nutrient management strategies, and conservation tillage have led to improved farm yields and “greener” farming practices. The experts stressed the importance of community and called out the carbon sequestration benefits of the above strategies as well, pointing out that, if 25% of developing world farmlands adopted such techniques, the potential increase in carbon sequestration would amount to 100 (±4) megatons of carbon annually.

Community-level action provides a sense of great opportunity and hope for building communities in which economic and social development and environmental protection are reinforcing components of sustainable development. These actions are derived, likely on an unconscious level, from rethinking and retooling fundamental concepts of science, economics, and social structure. Consider the work of the Sustainable Rural Enterprise (SRE), a multi-awarded non-governmental organization (NGO) based in Aklan, Philippines, that focuses on productive, or income-generating, uses of renewable energy. This NGO, under the leadership of Perla Manapol, dually operates and manages a coconut-based rural enterprise that processes waste coconut byproducts into high-value commodities: clean burning charcoal and activated carbon from the shelves, erosion control mats, mulch, media filter, and fiberboard from the husks. These labor-intensive processes employ more than 500 families of low-income coconut farmers and former guerrilla insurgents, plus an additional 350 families of former Muslim insurgents working at an SRE satellite operation in Mindanao, Philippines. The majority of the workers are women and the traditionally unemployable such as the aged and physically disabled (Figure 10.14).

FIGURE 10.14 Philippine women twining ropes from cocofiber. One twine-making unit can provide jobs for five, and the twines themselves are woven into nets by another set of workers. These nets cover the foothills surrounding the communities with the sole purpose of providing shelter from devastating typhoons and landslides, two occurrences becoming more prevalent in the Philippines. Photo courtesy of Perla Manapol.

Another high-value, environmentally-friendly product of this community effort is cocopeat. Although a waste product of the coconut-producing industry, cocopeat is a highly prized water filtration media. Workers dry and layer the cocopeat in a lined box with an effluent distribution-piping network on top. Effluent is pumped from an aeration chamber through a piping network and distributed amongst the cocopeat. As the effluent passes through the cocopeat media, the downward flow draws atmospheric oxygen into the pore spaces of the media. This naturally-aerated media filtration system provides a highly treated and polished effluent suitable for reuse in the community. In Europe, its use in horticulture—in lieu of peat moss—is highly encouraged. In addition, cocopeat is pressed into binder-free fiberboard, material that is being increasingly used in "green construction."

In the more developed countries, there is a push toward organic farming, integrated pest management (IPM) activities, and the development and production of reduced-risk synthetic pesticides and biopesticides. IPM relies on the life cycle of pests and crops to economically control unwanted pests by withholding the use of pesticides until potential damage reaches a set threshold level. Under various international initiatives (e.g., the U.S. Food Quality Protection Act of 1996), environmentally-persistent pesticides are being phased out for greener, reduced-risk options. For example, products such as simazine (Figure 10.15a), a triazine-based herbicide that has been associated with a number of human health issues, is being replaced with glyphosate (Figure 10.15b), an aminophosphate analog of the amino acid glycine that is

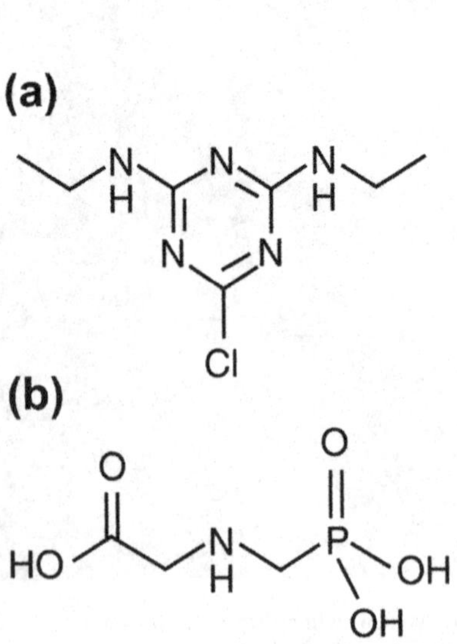

FIGURE 10.15 (a) Chemical structure of the triazine-based herbicide simazine. (b) Chemical structure of glyphosate, an aminophosphate analog of the amino acid glycine that is less persistent in the environment.

less persistent in the environment and generally considered less toxic than aspirin, caffeine, and table salt. In more quantitative terms, U.S. EPA maximum contaminant levels for drinking water are reported to be 4 ppb and 700 ppb for simazine and glyphosate, respectively. Biopesticides such as the fungicide Regalia, derived from plants, bacteria, fungi, and insects, currently account for half of the new pesticides being registered by the U.S. EPA. They induce plants to produce cell strengtheners, antioxidants, phenolics, and specialized proteins that are known to fight pathogens that affect plants.

10.6. END OF CHAPTER PROBLEMS

10.1 Define the term "green chemistry" and discuss how the 12 Principles of Green Chemistry are incorporated.

10.2 Choose one of the 12 Principles of Green Chemistry, briefly define it, and provide evidence for it being either product- or process-based.

10.3 Define the term life "cycle assessment" (LCA) and list its four complementary phases.

10.4 Of the items listed below, which one may be utilized as a renewable feedstock source for organic chemicals?
a. Coal
b. Petroleum
c. Natural gas
d. Biomass

10.5 Provide reasoning why catalysts are preferable over stoichiometric reagents in chemical reactions. Consider the 9th Principle of Green Chemistry when answering.

10.6 In lieu of our discussion of emerging environmental technologies, renewable energy resources currently generate:
a. More energy than fossil fuels
b. Less energy than fossil fuels
c. The same amount of energy as fossil fuels

10.7 Recall the transesterification process outlined in the text:
The reaction between the triglyceride and the alcohol is a reversible reaction. Explain why alcohol must be added in excess in this reaction.

10.8 In Research Application XII we learned that hydrogen peroxide (H_2O_2) has been shown to effectively oxidize compounds of interest. List and describe other oxygen-containing compounds that have been shown to exhibit oxidative properties. Note: This may require consultation from external sources of information.

10.9 Describe how nanotechnology has contributed to the 12 Principles of Green Chemistry. What are possible drawbacks of nanotechnology?

10.10 In your own words, define the term "sustainability."

10.11 Describe both quantitative and qualitative aspects of evaluating the economic, environmental, and social impacts of chemical processes.

10.12 List the five strategies of pollution prevention. What role does waste minimization play in pollution prevention?
10.13 Describe example steps being taken by developing countries toward sustainable agricultural practices. How do these steps differ from those being instituted by more developed countries?
10.14 Briefly describe how biopesticides are effective in fighting plant pathogens.
10.15 Briefly define the term "olefin metathesis." Describe how this process is important in modern-day green chemistry applications.

10.7 REFERENCES

Adi, G., Hall, C., Snyder, D., Bunce, M., Satkoski, C., Kumar, S., Garimella, P., Stanton, D., & Shaver, G. (2009). Soy-biodiesel impact on NO_x emissions and fuel economy for diffusion-dominated combustion in a turbo-diesel engine incorporating exhaust gas recirculation and common rail fuel injection. *Energy Fuels, 23*, 5821–5829.

Anastas, P., & Warner, J. (1998). *Green chemistry: Theory and practice*. Oxford, UK: Oxford University Press.

Avcock, D. (2007). Solvent applications of 2-methyltetrahydrofuran in organometallic and biphasic reactions. *Organic Process Research & Development, 11*, 156–159.

Calvin, C., Schuster, D., Beloff, B., & Dicksen, T. (2009). The AIChE sustainability index: The factors in detail. *Chemical Engineering Progress*, January, 60–63.

Chandler, A. D. (2005). *Shaping the industrial century: The remarkable story of the evolution of the modern chemical and pharmaceutical industries*. Cambridge, MA: Harvard University Press.

Collins, T. J. (2002). TAML oxidant activators: A new approach to the activation of hydrogen peroxide for environmentally significant problems. *Accounts of Chemical Research, 35*, 782–790.

Habicht, H. F. (1992). *Memorandum: EPA definition of pollution prevention*. May 28. U.S. Environmental Protection Agency.

International Energy Agency (IEA). (2008). *Key world energy statistics*. Paris: International Energy Agency Report.

Jenck, J. F., Agterberg, F., & Droescher, M. J. (2004). Products and processes for a sustainable chemical industry: A review of achievements and prospects. *Green Chemistry, 6*, 544–556.

Ma, F., & Hanna, M. A. (1999). Biodiesel production: A review. *Bioresource Technology, 70*, 1–15.

McCoy, M. (2006). Facts and figures of the chemical industry. *Chemical & Engineering News, 84*, 35–72.

Morse, D. E. (1999). Silicon biotechnology: Harnessing biological silica production to construct new materials. *Trends in Biotechnology, 17*, 230–232.

National Academy of Sciences. (2009). *Liquid transportation fuels from coal and biomass: Technological status, costs, and environmental impacts*. Washington, D.C: National Academies Report.

Pretty, J. N., Noble, A. D., Bossio, D., Dixon, J., Hine, R. E., Penning de Vries, F. W. T., & Morison, J. I. L. (2006). Resource-conserving agriculture increases yields in developing countries. *Environmental Science & Technology, 40*, 1114–1119.

Renewable Energy Policy Network for the 21st century. (2009). *Renewable global status report, 2009 Update.*

Rideout, D. C., & Breslow, R. (1980). Hydrophobic acceleration of Diels-Alder reactions. *Journal of the American Chemical Society, 102*, 7816–7817.

Trinka, T. M., & Grubbs, R. H. (2001). The development of $L_2X_2Ru{=}CHR$ olefin metathesis catalysts: An organometallic success story. *Accounts of Chemical Research, 34*, 18–29.

Appendix I

Common Excel Shortcuts and Key Combinations

Shortcuts and combination keys	Function
F1	Excel help is opened
F2	Activates a given cell
F3	Activates the **Paste** command
F4	Repeats most recent worksheet action
F5	Opens the **Go To** dialog box
F6	Moves to the next pane
F7	Spell check is activated
F8	Extends a given selection
F9	Calculates all worksheets in all open workbooks
F10	Selects the **Menu** bar
F11	Creates a chart of the current data
F12	Opens the **Save As** dialog box
Shift+F2	Edits a cell comment
Shift+F3	Opens the **Function** dialog box
Shift+F5	Activates the **Find** command
Shift+F6	Switches to the previous pane in a worksheet
Shift+F8	Enables the addition of non-adjacent cells to another selection of cells
Shift+F9	Calculates the entire worksheet
Shift+F10	Displays the **Shortcut** menu for the selected item
Shift+F12	Saves the active worksheet
Shift+Ins	Pastes the data from the clipboard
Shift+Home	Selects the cells from the active cell to the left edge of the current row
Shift+PgUp	Selects the cells from the active cell to the top of the current column
Ctrl+F1	Closes and reopens the current task pane
Ctrl+F3	Opens the **Define Name** dialog box
Ctrl+F4	Closes the workbook window
Alt+D	Displays the **Data** menu
Alt+E	Displays the **Edit** menu
Alt+F	Displays the **File** menu

(*Continued*)

Shortcuts and combination keys	Function
Alt+H	Displays the **Help** menu
Alt+I	Displays the **Insert** menu
Alt+O	Displays the **Format** menu
Alt+T	Displays the **Tools** menu
Alt+V	Displays the **View** menu
Alt+W	Displays the **Windows** menu

Appendix II

Values of Student's *t* at the Various Confidence Levels

Degrees of Freedom	90% (0.10)	95% (0.05)	98% (0.02)	99% (0.01)	99.5% (0.005)	99.9% (0.001)
	Confidence level and *p*-values (in parentheses)[a]					
1	6.31	12.71	31.82	63.66	127.3	636.6
2	2.92	4.30	6.96	9.92	14.08	31.60
3	2.35	3.18	4.54	5.84	7.45	12.92
4	2.13	2.78	3.74	4.60	5.59	8.61
5	2.02	2.57	3.36	4.03	4.77	6.87
6	1.94	2.44	3.14	3.71	4.31	5.96
7	1.89	2.36	2.99	3.50	4.02	5.41
8	1.86	2.31	2.90	3.36	3.83	5.04
9	1.83	2.26	2.82	3.25	3.69	4.78
10	1.81	2.22	2.76	3.17	3.58	4.59
25	1.70	2.06	2.48	2.78	3.08	3.45
50	1.68	2.01	2.40	2.68	2.94	3.26

[a]*Note: The critical values of $|t|$ listed are for a two-tailed test. For a one-tailed test, the value is taken from the column twice the desired p-value.*

Appendix III

Critical Values of F for a One-Tailed Test with $p = 0.05$

	1	2	3	4	5	6	7	8	9	10	12	15	20
1	161.4	199.5	215.7	224.6	230.2	234.0	236.8	238.9	240.5	241.9	243.9	245.9	248.0
2	18.51	19.00	19.16	19.25	19.30	19.33	19.35	19.37	19.38	19.40	19.41	19.43	19.45
3	10.13	9.552	9.277	9.117	9.013	8.941	8.887	8.845	8.812	8.786	8.745	8.703	8.660
4	7.709	6.944	6.591	6.388	6.256	6.163	6.094	6.041	5.999	5.964	5.912	5.858	5.803
5	6.608	5.786	5.409	5.192	5.050	4.950	4.876	4.818	4.772	4.735	4.678	4.619	4.558
6	5.987	5.143	4.757	4.534	4.387	4.284	4.207	4.147	4.099	4.060	4.000	3.938	3.874
7	5.591	4.737	4.347	4.120	3.972	3.866	3.787	3.726	3.677	3.637	3.575	3.511	3.445
8	5.318	4.459	4.066	3.838	3.687	3.581	3.500	3.438	3.388	3.347	3.284	3.218	3.150
9	5.117	4.256	3.863	3.633	3.482	3.374	3.293	3.230	3.179	3.137	3.073	3.006	2.936
10	4.965	4.103	3.708	3.478	3.326	3.217	3.135	3.072	3.020	2.978	2.913	2.845	2.774
11	4.844	3.982	3.587	3.363	3.204	3.095	3.012	2.948	2.896	2.854	2.788	2.719	2.646
12	4.747	3.885	3.490	3.259	3.106	2.996	2.913	2.849	2.796	2.753	2.687	2.617	2.544
13	4.667	3.806	3.411	3.179	3.025	2.915	2.832	2.767	2.714	2.671	2.604	2.533	2.459
14	4.600	3.739	3.344	3.112	2.958	2.848	2.764	2.699	2.646	2.602	2.534	2.463	2.388
15	4.543	3.682	3.287	3.056	2.901	2.790	2.707	2.641	2.588	2.544	2.475	2.403	2.328
16	4.494	3.634	3.239	3.007	2.852	2.741	2.657	2.591	2.5438	2.494	2.425	2.352	2.276
17	4.451	3.592	3.197	2.965	2.810	2.699	2.614	2.548	2.494	2.450	2.381	2.308	2.230
18	4.414	3.555	3.160	2.928	2.773	2.661	2.577	2.510	2.456	2.412	2.342	2.269	2.191
19	4.381	3.522	3.127	2.895	2.740	2.628	2.544	2.477	2.423	2.378	2.308	2.234	2.155
20	4.351	3.493	3.098	2.866	2.711	2.599	2.514	2.447	2.393	2.348	2.278	2.203	2.124
30	4.171	3.316	2.922	2.689	2.534	2.421	2.334	2.266	2.211	2.165	2.092	2.015	1.932

Appendix IV

Critical Values of F for a Two-Tailed Test with $p = 0.05$

	1	2	3	4	5	6	7	8	9	10	12	15	20
1	647.8	799.5	864.2	899.6	921.8	937.1	948.2	956.7	963.3	968.6	976.7	984.9	993.1
2	38.51	39.00	39.17	39.25	39.30	39.33	39.36	39.37	39.39	39.40	39.41	39.43	39.45
3	17.44	16.04	15.44	15.10	14.88	14.73	14.62	14.54	14.47	14.42	14.34	14.25	14.17
4	12.22	10.65	9.997	9.605	9.364	9.197	9.074	8.980	8.905	8.844	8.751	8.657	8.560
5	10.01	8.434	7.764	7.388	7.146	6.978	6.853	6.757	6.681	6.619	6.525	6.428	6.329
6	8.813	7.260	6.599	6.227	5.988	5.820	5.695	5.600	5.523	5.461	5.366	5.269	5.168
7	8.073	6.542	5.890	5.523	5.285	5.119	4.995	4.899	4.823	4.761	4.666	4.568	4.467
8	7.571	6.059	5.416	5.053	4.817	4.652	4.529	4.433	4.357	4.295	4.200	4.101	3.999
9	7.209	5.715	5.078	4.718	4.484	4.320	4.197	4.102	4.026	3.964	3.868	3.769	3.667
10	6.937	5.456	4.828	4.468	4.236	4.072	3.950	3.855	3.779	3.717	3.621	3.522	3.419
11	6.724	5.256	4.630	4.275	4.044	3.881	3.759	3.664	3.588	3.526	3.430	3.330	3.226
12	6.554	5.096	4.474	4.121	3.891	3.728	3.607	3.512	3.436	3.374	3.277	3.177	3.073
13	6.414	4.965	4.347	3.996	3.767	3.604	3.483	3.388	3.312	3.250	3.153	3.053	2.948
14	6.298	4.857	4.242	3.892	3.663	3.501	3.380	3.285	3.209	3.147	3.050	2.949	2.844
15	6.200	4.765	4.153	3.804	3.576	3.415	3.293	3.199	3.123	3.060	2.963	2.862	2.756
16	6.115	4.687	4.077	3.729	3.502	3.341	3.219	3.125	3.049	2.986	2.889	2.788	2.681
17	6.042	4.619	4.011	3.665	3.438	3.277	3.156	3.061	2.985	2.922	2.825	2.723	2.616
18	5.978	4.560	3.954	3.608	3.382	3.221	3.100	3.005	2.929	2.866	2.769	2.667	2.559
19	5.922	4.508	3.903	3.559	3.333	3.172	3.051	2.956	2.880	2.817	2.720	2.617	2.509
20	5.871	4.461	3.859	3.515	3.289	3.128	3.007	2.913	2.837	2.774	2.676	2.573	2.464
30	5.567	4.182	3.589	3.245	3.027	2.867	2.746	2.651	2.575	2.511	2.412	2.307	2.195

Appendix V

Common Matrix Notation and Calculations

A matrix is a table of numbers consisting of n rows and m columns (i.e., a $n \times m$ matrix):

$$\mathbf{A} = \begin{pmatrix} a_{11} & a_{12} & a_{1m} \\ a_{21} & a_{22} & a_{2m} \\ \vdots & & \vdots \\ a_{n1} & a_{n2} & a_{nm} \end{pmatrix} \text{ or with real numbers: } \mathbf{A} = \begin{pmatrix} 1 & 4 & 6 \\ 2 & 9 & 7 \\ 3 & 5 & 8 \end{pmatrix},$$

where an individual element of $\mathbf{A} = a_{ij}$. The first subscript in a matrix refers to the row and the second to the column. A square matrix consists of the same number of rows and columns (i.e., $n \times n$ matrix). Note that matrix $\mathbf{A}$ above is square but matrix $\mathbf{B}$ below is not:

$$\mathbf{B} = \begin{pmatrix} 1 & 4 & 6 \\ 2 & 9 & 7 \end{pmatrix}.$$

A vector is a type of matrix that has only one row (i.e., a row vector) or one column (i.e., a column vector). Below, $\mathbf{a}$ is a column vector, while $\mathbf{b}$ is a row vector:

$$\mathbf{a} = \begin{pmatrix} 6 \\ 2 \end{pmatrix} \quad \mathbf{b} = \begin{pmatrix} 3 & 1 & 6 \end{pmatrix}.$$

A symmetric matrix is a square matrix in which $a_{ij} = a_{ji}$, for all i and j, and presented as follows:

$$\mathbf{A} = \begin{pmatrix} 3 & 4 & 1 \\ 4 & 8 & 5 \\ 1 & 5 & 0 \end{pmatrix}.$$

Don't confuse this with a diagonal matrix, a symmetric matrix where all the off diagonal elements are 0:

$$\mathbf{D} = \begin{pmatrix} 3 & 0 & 0 \\ 0 & 1 & 0 \\ 0 & 0 & 5 \end{pmatrix}.$$

An identity matrix is a diagonal matrix with 1s on the diagonal:

$$\mathbf{I} = \begin{pmatrix} 1 & 0 & 0 \\ 0 & 1 & 0 \\ 0 & 0 & 1 \end{pmatrix}.$$

In a transposed matrix A', the rows and columns are interchanged as follows. Compare that to the first matrix presented above:

$$\mathbf{A} = \begin{pmatrix} a_{11} & a_{21} & a_{n1} \\ a_{12} & a_{22} & a_{n2} \\ \vdots & & \vdots \\ a_{1m} & a_{2m} & a_{nm} \end{pmatrix} \text{ or with real numbers: } \mathbf{A} = \begin{pmatrix} 1 & 2 & 3 \\ 4 & 9 & 5 \\ 6 & 7 & 8 \end{pmatrix}.$$

For matrix addition, each element of the first matrix is added to the corresponding element of the second to produce a result. Note that the two matrices must have the same number of rows and columns:

$$\begin{pmatrix} 0 & 2 & -1 \\ 3 & 1 & 3 \\ 1 & 6 & 0 \end{pmatrix} + \begin{pmatrix} 10 & -1 & 3 \\ 2 & 0 & 2 \\ 3 & 1 & 1 \end{pmatrix} = \begin{pmatrix} 10 & 1 & 2 \\ 5 & 1 & 5 \\ 4 & 7 & 1 \end{pmatrix}.$$

Note that matrix subtraction works in the same way, except that elements are subtracted instead of added.

Multiplication of an $n \times n$ matrix **A** and an $n \times n$ matrix **B** given a result of $n \times n$ matrix **C**:

$$\begin{pmatrix} 2 & 1 & 3 \\ -2 & 2 & 1 \end{pmatrix} \begin{pmatrix} 2 & 1 \\ 3 & 2 \\ -2 & 2 \end{pmatrix} = \begin{pmatrix} 1 & 10 \\ 0 & 4 \end{pmatrix}.$$

Note that in multiplication, $\mathbf{A} \times \mathbf{B}$ does not generally equal $\mathbf{B} \times \mathbf{A}$. In other words, matrix multiplication is not commutative. Considering associative and distributive laws: $(\mathbf{A} \times \mathbf{B}) \times \mathbf{C} = \mathbf{A} \times (\mathbf{B} \times \mathbf{C})$; $\mathbf{A} \times (\mathbf{B} + \mathbf{C}) = \mathbf{A} \times \mathbf{B} + \mathbf{A} \times \mathbf{C}$ and $(\mathbf{B} + \mathbf{C}) \times \mathbf{A} = \mathbf{B} \times \mathbf{A} + \mathbf{C} \times \mathbf{A}$.

Typically, one takes the multiplication by an inverse matrix as the equivalent of matrix division. The inverse of a matrix is that matrix which, when multiplied by the original matrix, gives an identity (**I**) matrix with the inverse of denoted by a superscripted −1:

$\mathbf{A}^{-1}\mathbf{A} = \mathbf{A}\mathbf{A}^{-1} = \mathbf{I}$.

Note that to have an inverse, a matrix must be square. Consider the following matrix:

$$\mathbf{A} = \begin{pmatrix} a_{11} & a_{12} \\ a_{21} & a_{22} \end{pmatrix}.$$

The inverse of this matrix exists if $a_{11}a_{22} - a_{12}a_{21} \neq 0$. If the inverse exists, it is then given by

$$\mathbf{A} = \frac{1}{a_{11}a_{22} - a_{12}a_{21}} \begin{pmatrix} a_{22} & -a_{12} \\ -a_{21} & a_{11} \end{pmatrix}.$$

Note for covariance and correlation matrices, an inverse will always exist, provided that there are more subjects than there are variables and that every variable has a variance greater than 0. The existence of the inverse is dependent upon the determinant, a scalar-values function of the matrix. For example:

$$\det \mathbf{A} = \begin{pmatrix} a_{11} & a_{12} \\ a_{21} & a_{22} \end{pmatrix} = a_{11}a_{22} - a_{12}a_{21}.$$

For covariance and correlation matrices, the determinant is a number that often expresses the generalized variance of the matrix. Here, covariance matrices with small determinants denote variables that are highly correlated (see discussions on factor analysis or regression analysis).

Defining the determinant can help in formalizing the general form of the inverse matrix:

$$\mathbf{A}^{-1} = \frac{1}{\det \mathbf{A}} \text{adj } \mathbf{A},$$

where adj $\mathbf{A}$ = the adjugate of $\mathbf{A}$. There are a couple of ways to compute an inverse matrix, with the easiest typically being in the form of an augmented matrix $(\mathbf{A}|\mathbf{I})$ from $\mathbf{A}$ and $\mathbf{I}_n$, then utilizing Gaussian elimination to transform the left half into **I**. Once completed, the right half of the augmented matrix will be $\mathbf{A}^{-1}$. In addition, one can compute the i,j th element of the inverse by using the general formula

$$\mathbf{A}_{ji} = \frac{\mathrm{C}_{ij}\,\mathbf{A}}{\det \mathbf{A}},$$

where $\mathrm{C}_{ij} = i,j$, the cofactor expansion of matrix **A**.

An orthogonal matrix has the general form $\mathbf{A}\mathbf{A}^{t} = \mathbf{I}$. Thus, the inverse of an orthogonal matrix is simply the transpose of that matrix. Orthogonal matrices are very important in factor analysis. Matrices of eigenvectors are orthogonal matrices. Note that only square matrices can be orthogonal matrices. As discussed, the eigenvalues and eigenvectors of a matrix play an important part in multivariate analysis. General concepts regarding eigenvalues and eigenvectors include:

1. Eigenvectors are scaled so that **A** is an orthogonal matrix;
2. An eigenvector of a linear transformation is a non-zero vector that is either left unaffected or simply multiplied by a scale factor after transformation;
3. The eigenvalue of a non-zero eigenvector is the scale factor by which it has been multiplied;
4. An eigenvalue reveals the proportion of total variability in a matrix associated with its corresponding eigenvector;
5. For a covariance matrix, the sum of the diagonal elements of the covariance matrix equals the sum of the eigenvalues;
6. For a correlation matrix, all the eigenvalues sum to n, the number of variables;
7. The decomposition of a matrix into relevant eigenvalues and eigenvectors rearranges the dimensions in an n dimensional space so that all axes are perpendicular.

Index

Page numbers followed by f indicate figures, t indicate tables and b indicate text box.

B

C

D

F

G

H

J

K

L

N

O

P

Q

R

S

T

U

V

X

Y

Z

Printed in the United States
By Bookmasters